Molecular Science of Fluctuations Toward Biological Functions

Masahide Terazima • Mikio Kataoka •
Ryuichi Ueoka • Yuko Okamoto
Editors

Molecular Science of Fluctuations Toward Biological Functions

Springer

Editors
Masahide Terazima
Kyoto University
Kyoto
Japan

Mikio Kataoka
Nara Institute of Science and Technology
Ikoma
Japan

Ryuichi Ueoka
Sojo University
Kumamoto
Japan

Yuko Okamoto
Nagoya University
Nagoya
Japan

ISBN 978-4-431-55838-5 ISBN 978-4-431-55840-8 (eBook)
DOI 10.1007/978-4-431-55840-8

Library of Congress Control Number: 2016930393

Springer Tokyo Heidelberg New York Dordrecht London

Printed on acid-free paper

Springer Japan KK is part of Springer Science+Business Media (www.springer.com)

Foreword

Biologists are entranced by the stunning representations of biomolecules derived from X-ray crystallography. Yet these images give a false impression of constancy and rigidity. In solution or in a living cell, biological macromolecules are not static, but rather they are in constant motion because of thermal energy. These fluctuations have a wide range of distance, from short-range oscillations to large-scale domain movements of proteins. They also cover a wide range of times, from picoseconds to milliseconds to minutes and beyond.

If these fluctuations had no functional consequence, they would still be of interest to physicists and physical chemists, but not so much to biologists. But some of these fluctuations have profound functional consequences. A key example occurs in enzymatic catalysis, where the binding of substrates and then the molecular motions required to attain the transition state would be impossible without fluctuations.

Thus, this volume on the *Molecular Science of Fluctuations Toward Biological Functions* serves an important purpose. It discusses many biological systems in which fluctuations are important. It describes many methods, ranging from NMR to computer simulations, that are used to measure fluctuations. And finally it is intensely interdisciplinary, drawing from many disciplines and appealing to biophysicists, physicists, physical chemists, and biomedical scientists.

My own experience with the fluctuations community in Japan is still fresh in my mind. On January 7 and 8, 2012, I participated in the 5th International Symposium on Molecular Science of Fluctuations toward Biological Fluctuations, which was held in the Todaiji Culture Center in Nara. There, in the shadow of the Big Buddha, an international group of scientists and students shared the latest information about how to measure fluctuations in biomolecules and how to understand their functional consequences. It was particularly exciting that the group was so international, because those of us from the West get too few opportunities to interact with our Japanese colleagues. We need to "fluctuate" more often.

Thomas R. Cech
Distinguished Professor of Chemistry
and Biochemistry, BioFrontiers Institute,
University of Colorado Boulder,
Boulder, CO, USA

Preface

Understanding biological reactions at the molecular level is one of major targets in many scientific fields, such as physical chemistry, biochemistry, biophysics, and medical science. Structural information about biomolecules is certainly important. However, due to the thermal motion of solvent molecules, biological molecules in solution at room temperature should be always fluctuating even during their functions, and this motion cannot be or should not be ignored. Static structures are not sufficient, and, without knowledge of this motion of "fluctuation", we cannot truly understand the reactions or their functions. Hence, it has been recently recognized that the nature of "fluctuation" and the dynamics should be an essential factor. Why can such biological molecules perform selective and efficient reactions for their functions of life? Does the fluctuation really play an important role for the reactions? Can we use the fluctuation to control the reactions? For revealing the importance of fluctuation, scientists in a variety of fields (physics, chemistry, pharmacology, medicine, etc.) have to collaborate and elucidate the relationship between fluctuations and the reactions that are related with various biological functions. In this book, we review recent developments in the field of "fluctuation and molecular science of biological molecules toward the biological function".

To establish the field of "fluctuation", there are several elements we should develop and/or clarify: how to detect the fluctuation, how to regulate the fluctuation, what is connection between the fluctuations and the biological functions, and how to use this fluctuation for applications.

In this book, first, some detection methods of the fluctuation are described. One of important ways of detection is thermodynamics, which is closely related with the fluctuation of the system in principle. Meanwhile, time-resolved methods are similarly important to reveal the connection between the fluctuation and the reaction. Combining the thermodynamic measurement with a spectroscopic method, time-resolved thermodynamics recently has become possible. The principle and applications are described in Chap. 1. Among various thermodynamic properties, pressure serves as an important tool to explore volume fluctuation. The principle of the relatively recently developed pressure perturbation calorimetry is reviewed (Chap. 2). How pressure affects the structure, phase behavior, and dynamics of lipid membranes demonstrates the promising role of this tool to study high-energy conformational and functional substates in proteins. It also shows how to use this method for understanding protein aggregation and fibrillation phenomena.

For monitoring structural information in a time domain, a unique and powerful time-resolved X-ray scattering method has been developed. The method, picosecond time-resolved Laue crystallography, was applied to reveal a detailed mechanism of reaction of a blue light photosensor protein, PYP (Chap. 3). Nuclear magnetic resonance (NMR) spectroscopy is another useful approach to characterize the conformational dynamics of biomolecules over a broad range of time scales. Several NMR strategies to deal with dynamic conformational equilibria and ensembles are described by using model molecules of monomeric and dimeric ubiquitins and the oligosaccharide moieties of gangliosides (Chap. 4).

To analyze the structure–function relationship of proteins, replacing amino acids (mutation techniques) is very useful. However, the replacement of amino-acid residues is usually restricted within the 20 types of natural amino acids. A unique method to introduce nonnatural amino acids into proteins in a site-specific manner is described in Chap. 5. The powerfulness of this technique is shown by the site-specific incorporation of fluorescent amino acids into proteins.

The recent discovery of intrinsically disordered proteins (IDPs) (also known as natively unfolded proteins) is probably a good example to demonstrate the importance of the fluctuation. The unfolded or highly disordered structures under physiological conditions can still specifically recognize their partner molecules and fold into the defined tertiary structures upon binding to their target molecules. A description of IDPs and the place of these exceptionally interesting “creatures” within the protein kingdom are introduced in Chap. 6. Combined with the mutation technique, the reaction mechanism of ligand-induced folding and the role of unfolded structure are described (Chap. 7).

Not only these experimental studies, but also theoretical approaches are essential for understanding the role of fluctuation at the molecular level. In this book, two approaches are presented, based on the statistical mechanical theory and molecular dynamics simulation. The recent 3D-RISM/RISM theory was successfully applied to the molecular recognition. It showed that the structural fluctuation plays a crucial role in the process where a protein expresses its function (Chap. 8). Although molecular dynamic simulation is powerful, conventional simulations in the canonical ensemble encounter great difficulty, because they tend to get trapped in states of local minima. A simulation in a generalized ensemble performs a random walk in potential energy, volume, and/or other physical quantities and can overcome this difficulty (Chap. 9).

The fluctuations of the protein energy landscape inside cells will be interesting for practical applications. The spatio-temporal fluctuations in protein folding rates, protein stability, protein structure, protein function, and protein interactions are the subject of a recent generation of experiments in in-cell modeling. Some specific examples of how the cell modulates energy landscapes and leads to fluctuations are described in Chap. 10. Applications to medical science are also very important targets. This book contains two approaches. It was found that hybrid liposomes (HL) (a mixture of vesicular and micellar molecules) have inhibitory effects on the growth of various tumor cells in vitro. In clinical applications, prolonged survival and remarkable reduction of neoplasm were attained in patients with lymphoma after treatment with HL without any side effects (Chap. 11). It is quite interesting to know that this prominent property is related with the fluctuation of the cancer cell. A complex formed by human α-lactalbumin and oleic acid (HAMLET) has a unique apoptotic activity like HL for the selective killing of tumor cells. HAMLET provides a unique example, in which a single globular protein has two independent biological functions. The historical background and recent progress of the studies on HAMLET and related protein–fatty acid complexes are reviewed (Chap. 12).

We thank the authors for their outstanding contributions to this book. The cooperation of each contributor has made this book a special issue. We are also grateful for the financial support from a Grant-in-aid for Scientific Research on Innovative Areas (Research in a proposed research area) (Number of Area: 2006) from the Ministry of Education, Culture, Sports, Science and Technology in Japan.

We hope that this book will be useful for future development of the field "fluctuation of biomolecules toward biological function".

Kyoto, Japan — Masahide Terazima
Ikoma, Japan — Mikio Kataoka
Kumamoto, Japan — Ryuichi Ueoka
Nagoya, Japan — Yuko Okamoto
August 2015

Contents

Chapter 1
Time-Resolved Detection of Protein Fluctuations During Reactions

Masahide Terazima

Abstract Biological molecules fluctuate at room temperature in solution because of thermal motion not only at the ground state but also during their functions. Why can biological molecules perform selective and efficient reactions in life? A field of "fluctuation and molecular science of biological molecules" can be established to link the relationship between fluctuations and biological reactions, and for that purpose, studies that examine fluctuations and dynamics of biological molecules are essential. To detect protein fluctuations during reactions, we have developed time-resolved methods to measure thermodynamic properties, which are closely related with fluctuations. In this chapter, examples are presented that show enhanced fluctuations of proteins during reactions.

Keywords Protein reaction • Time-resolved thermodynamics • Fluctuation

1.1 Introduction

1.1.1 Fluctuations of Biomolecules

The activity of life essentially consists of a series of various chemical reactions and a number of biological molecules cooperating to carry out these reactions. To understand the biological function at the molecular level, the reactions have to be elucidated step by step. For relatively small molecules, modern advanced techniques and molecular theories may allow us to understand the molecular mechanism of the reactions in detail. Intermediate species, even very short-lived species with lifetimes of picoseconds, can be identified during the reaction. However, as biomacromolecules are often flexible, not only elucidating the reaction scheme but also determining the structure can be very difficult. A variety of techniques have been developed for obtaining structural information, for example, crystallographic X-ray scattering, NMR, CD, IR, and Raman scattering. Because of these efforts, the

M. Terazima (✉)
Department of Chemistry, Graduate School of Science, Kyoto University, Kyoto 606-8502, Japan
e-mail: mterazima@kuchem.kyoto-u.ac.jp

M. Terazima et al. (eds.), *Molecular Science of Fluctuations Toward Biological Functions*, DOI 10.1007/978-4-431-55840-8_1

three-dimensional structure of biomolecules has increased dramatically over the last few decades, and structural databases continue to expand.

Although structures of biological molecules provide important information for understanding the mechanisms of the reactions, this information essentially represents a starting point. Static structures alone are not sufficient. This is because reactions are always dynamic in nature, and the detection of dynamics is essential. In particular, we focus our research attention on one key factor, "fluctuation." Because most biological molecules function in water at ambient temperature, they undergo thermal fluctuation. Biological molecules are always fluctuating because of thermal energy even during their function. Naively, random movement and fluctuations appear to not relate with the reaction along the reaction coordinate, such that dynamics could represent undesirable or even a cumbersome feature that prohibits efficient reactions.

For example, Fischer proposed the "lock-and-key model" to explain enzymatic activities [1]. In this model, the conformation of the rigid substrate complements the conformation of the rigid active site of the enzyme. The simple lock-and-key model implies a tightly structured rigid enzyme. (This model has been sometimes modified, e.g., the shape of the active site changes to adjust to the shape of the transition state of the reaction. This change is a dynamical one, but along the reaction coordinate.) This model has successfully explained the selectivity of an enzyme. If this model is literally correct, fluctuations should interfere or hamper enzyme activity. However, it has been recognized that this model cannot explain a significant body of existing data that describe a variety of reactions. Thus, fluctuations may explain a variety of reactions of enzymes.

Fluctuations involve random movements without changing the average structure and energy. The native state of a protein should be a global minimum of the Gibbs energy surface. However, there must exist many local (shallow) minima around the global minimum (Fig. 1.1), and there must be many jumps among the minima due to thermal energy. The importance of fluctuations has been sometimes taken into account for explaining enzyme activity. For example, one of the models is an induced fit model. In this model, the enzyme possesses an active site conformation, which is different from the shape of the substrate, and then undergoes a change in its shape to interact with the substrate. Another typical model is the conformation selection model. In this model, the protein adopts a variety of conformational substates, and one of the conformations among the ensemble of the conformations interacts with the substrate. Furthermore, even though fluctuations are a random movement not along the reaction coordinate, these dynamics may enhance the efficiency of the reaction bypassing a transition state along the reaction coordinate (Fig. 1.1). By considering protein fluctuations, we can potentially understand many phenomena. Moreover, there may be additional questions to answer, including whether fluctuations really play an important role during reactions.

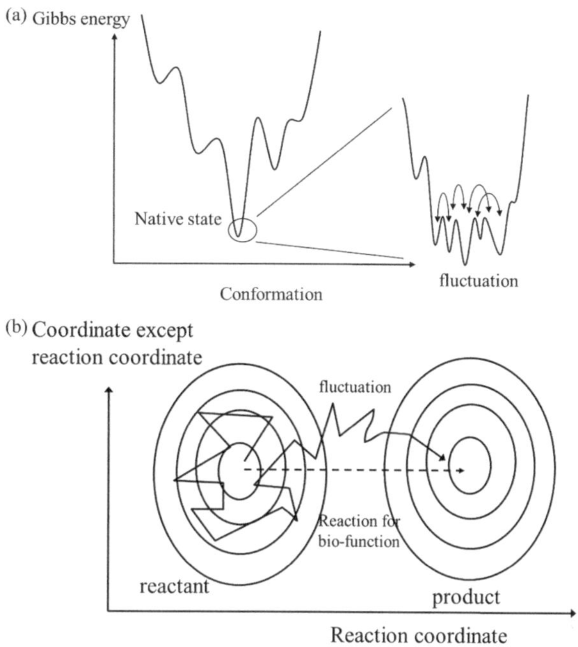

Fig. 1.1 (**a**) The native state of a protein is a global minimum of the Gibbs energy surface. Around the global minimum, there should be many local minima. Because proteins undergo dynamic fluctuations, transitions among the minima are observed. (**b**) The concept of fluctuations of a protein reaction. Although the fluctuation is a random movement that is not along the reaction coordinate, these dynamics may enhance the efficiency of the reaction

1.1.2 Time-Resolved Detection of Fluctuations

Although it is difficult to experimentally detect fluctuations even for stable species, some experimental detection approaches have been developed to study conformational fluctuations. For example, single molecule detection, or NMR techniques using hydrogen-deuterium (H-D) exchange, relaxation dispersion, and high-pressure methods, may be applicable for the detection of fluctuations in biomolecules. However, it is quite difficult to detect fluctuations during a reaction.

Thermodynamics may represent an additional approach to detect fluctuations. For example, thermal expansion represents a structural fluctuation [2, 3]. This value is proportional to the cross-correlation of the volume (V) and entropy (S) fluctuations:

$$< SV- < S >< V >> \quad = k_{\mathrm{B}}\mathrm{TV}\alpha_{\mathrm{th}} \tag{1.1}$$

where $< >$ indicates the ensemble average and k_{B} is the Boltzmann constant. Similarly, the heat capacity reflects the fluctuation of the energy. The partial molar isothermal compressibility $\left(\overline{K}_T = -\left(\partial\overline{V}/\partial P\right)_T\right)\right)$ is directly linked to the mean-square fluctuation of the protein partial molar volume by $\left\langle\left(\overline{V} - \left\langle\overline{V}\right\rangle\right)^2\right\rangle \equiv \delta\overline{V}^2 = k_{\mathrm{B}}T\overline{K}_T$. Therefore, by using these quantities, one can obtain information on the fluctuation. A more detailed description is given in a later chapter.

However, traditional thermodynamic measurement techniques require reactions to be under equilibrium, so that it is almost impossible to apply these techniques to irreversible reactions or reactions involving short-lived species. On this point, our group has demonstrated the usefulness of the pulsed laser-induced transient grating (TG) and transient lens (TrL) methods for quantitative measurements of the enthalpy change (ΔH) and the partial molar volume change (ΔV) during photochemical reactions in time domains at one temperature without assumptions [4, 5]. Using the advantages of this technique, one can determine other thermodynamic properties, such as the heat capacity change (ΔC_p) and changes in the thermal expansion coefficient ($\Delta\alpha_{\mathrm{th}}$) in the time domain. Hence, these parameters can now be evaluated for short-lived intermediate species using the TG and TrL methods. Here, we review thermodynamic properties of short-lived species during chemical reactions of particular proteins.

1.2 Methods

The thermodynamic properties in different time regions can be determined by the TG and TrL methods. These techniques detect the refractive index changes following photoexcitation [4–9]. Because the thermal energy released from the photoexcited molecules and the partial molar volume changes induce a change in the refractive index, one can determine these quantities in a time-resolved manner.

1.2.1 The Transient Grating (TG) Method

In the TG method, two pulsed laser beams are crossed at an angle θ within the coherence time so that an interference (grating) pattern is created with a wave number q (Fig. 1.2):

$$q = 2\pi/\Lambda = 4\pi\sin(\theta/2)/\lambda_{\mathrm{ex}}, \tag{1.2}$$

where Λ is the fringe length and λ_{ex} is the wavelength of the excitation laser. The wave number q can be varied by varying θ. Photosensor proteins are photoexcited by this grating light and the chemical reaction is initiated. When a probe beam is introduced to the photoexcited region, a part of the light is diffracted as the TG signal. Under a weak diffraction condition, the TG signal intensity ($\mathrm{I_{TG}}(t)$) is given by:

$$\mathrm{I_{TG}}(t) = \alpha\delta n(t)^2 + \beta\delta k(t)^2 \tag{1.3}$$

where α and β are constants representing the sensitivity of the detection system, and $\delta n(t)$ and $\delta k(t)$ are the refractive index change and the absorption change at the

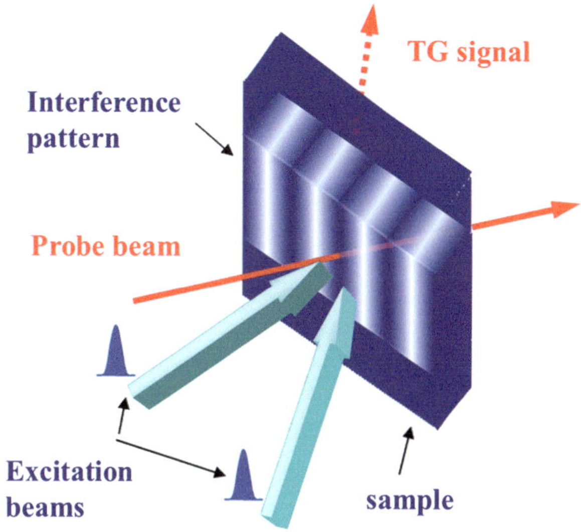

Fig. 1.2 Principles of the transient grating (TG) method. Two pulsed laser light is crossed inside the sample solution. Molecules are photoexcited by the interference pattern of the light. The time development of the spatial modulation of the refractive index is monitored by the diffraction of the probe light

probe wavelength, respectively. When the absorption change at the probe wavelength is negligible, the TG signal intensity is proportional to the square of the refractive index change $\delta n(t)$:

$$\mathrm{I_{TG}(t)} = \alpha\delta n(t)^2 \tag{1.4}$$

Two main factors contribute to the refractive index change: the thermal effect (thermal grating (ThG), $\delta n_{\mathrm{th}}(t)$) and a change in chemical species by the reaction (species grating, $\delta n_{\mathrm{spe}}(t)$):

$$\mathrm{I_{TG}}(t) = \alpha\left\{\delta n_{\mathrm{th}}(t) + \delta n_{\mathrm{spe}}(t)\right\}^2 \tag{1.5}$$

The temporal profile of $\delta n_{\mathrm{th}}(t)$ is determined by the convolution between the thermal diffusion decay and intrinsic temporal evolution of the thermal energy (Q (t)):

$$\delta n_{\mathrm{th}}(t) = (\mathrm{d}n/\mathrm{d}T)\left(\mathrm{W}\Delta N/\rho C_{\mathrm{p}}\right)(\partial Q(t)/\partial t)^{*}\exp\left(-D_{\mathrm{th}}q^2 t\right), \tag{1.6}$$

where * represents the convolution integral, dn/dT is the refractive index change due to variations of the temperature variation of the solution, W is the molecular weight (g/mol), ρ is the density (g/cm^3) of the solvent, ΔN is the molar density of the excited molecule (mol/cm^3), and D_{th} is the thermal diffusivity. Because the value of $D_{\mathrm{th}}q^2$ under particular experimental conditions is known from the ThG signal of the calorimetric reference (CR) sample, which converts all absorbed photon energy into the thermal energy within the time response of the detection system, the ThG component of a sample can be extracted from the observed TG signal by curve fitting with the function of $\exp(-D_{th}q^2t)$. The amplitude of ThG (δn_{th}) can be accurately determined from the pre-exponential factor. Once Q is determined from the ThG intensity, the enthalpy change ΔH is calculated by:

$$Q = h\nu - \Phi_f h\nu_f - \Delta H\Phi, \tag{1.7}$$

where $h\nu$ is the photon energy for excitation, Φ_f is the fluorescence quantum yield, $h\nu_f$ is the averaged photon energy of fluorescence, and Φ is the quantum yield of the reaction. Here, ΔH is defined by the enthalpy difference relative to the ground state.

The decay lifetime of the ThG signal ($(D_{th}q^2)^{-1}$) can be varied from 100 ns to 100 μs depending on the q^2 value. If the formation time of a transient species is longer than this, the ThG signal associated with this formation becomes very weak, and Q cannot be accurately determined from the signal. For measuring ΔH for relatively slowly created species, the TrL method is used.

1.2.2 *The Transient Lens (TrL) Method*

In the TrL experiment, a spatially Gaussian-shaped laser beam is used for excitation (Fig. 1.3) [10, 11]. The photo-induced refractive index distribution with the Gaussian shape is monitored by the light intensity change at the central part of another probe beam. The signal intensity ($I_{TrL}(t)$) is proportional to the photo-induced refractive index change, $\delta n(t)$:

$$I_{TrL}(t) = \alpha' \delta n(t), \tag{1.8}$$

where α' is a constant representing the sensitivity of the system. The origins of the refractive index change are the same as for the TG case, the thermal effect (thermal lens, ThL), and a change in chemical species by the reaction (species lens), which comprises the volume lens and the population lens contributions [10].

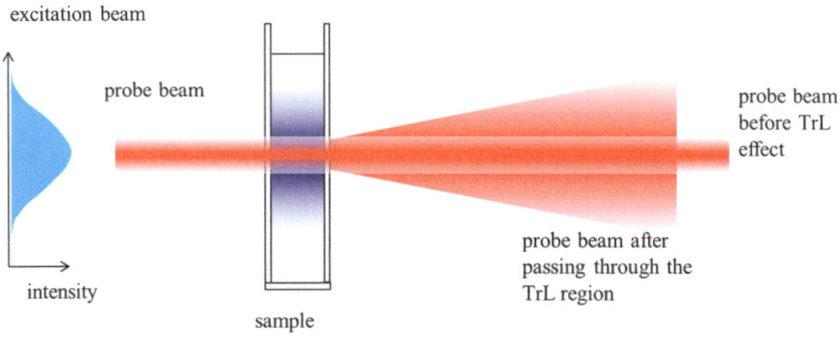

Fig. 1.3 Illustrations of the principles of the transient lens (*TrL*) method. The sample solution is irradiated by a Gaussian spatial-shaped excitation light. The spatial profile of the refractive index is monitored by focusing (or defocusing) the probe beam

$$I_{\mathrm{TrL}}(t) = \alpha' \{\delta n_{\mathrm{th}}(t) + \delta n_{\mathrm{spe}}(t)\} \tag{1.9}$$

Similar to the situation for the TG case, the temporal profile of the ThL signal is determined by the heat releasing and thermal diffusion processes. Expressing the impulsive response of the ThL signal as $f(t)$, one may find that the temporal profile of the ThL signal is given by:

$$I_{\mathrm{ThL}}(t) = \alpha'(\mathrm{d}n/\mathrm{d}T)(W/\rho C_p)(\partial Q(t)/\partial t)^* f(t), \tag{1.10}$$

where $f(t)$ was determined from the ThL signal of the CR sample. As for the ThG method, the released thermal energy, and hence the ΔH, is calculated from the amplitude of the ThL signal by comparison with that of the CR sample. The thermal diffusion process is much slower than that for the TG method, because the excitation beam is larger than the fringe length of the transient grating. This TrL method is used for quantitative measurement of ΔH for a species that is created in less than 100 ms.

1.2.3 Thermodynamical Properties

The heat capacity change is measured from the temperature dependence of ΔH using the following definition of the heat capacity:

$$\partial \Delta H/\partial T = \Delta C_p \tag{1.11}$$

Assuming that ΔC_p does not depend on the temperature over a small temperature range, one may obtain:

$$\Delta H(T) = \Delta H(T_0) - (T_0 - T)\Delta C_p, \tag{1.12}$$

where T_0 is any reference temperature. In this study, ΔC_p is calculated from the slope of ΔH plotted against T.

The refractive index change due to the volume change (volume grating signal, δn_{v}) is given by:

$$\delta n_{\mathrm{v}} = V\frac{\mathrm{d}n}{\mathrm{d}V}\Delta V \Delta N \tag{1.13}$$

where $V\mathrm{d}n/\mathrm{d}V$ is the refractive index change due to the molecular volume change. Taking a ratio of δn_{v} to δn_{th} of the CR sample with the known solvent property ($V\mathrm{d}n/\mathrm{d}V$), ΔV is determined.

According to the definition of the thermal expansion coefficient, the change in the thermal expansion coefficient ($\Delta\alpha_{\mathrm{th}}$) can be measured from the temperature dependence of the volume change:

$$(1/V)(\partial \Delta V/\partial T) = \Delta\alpha_{\mathrm{th}} \tag{1.14}$$

Hence, the thermal expansion volume $V\Delta\alpha_{\mathrm{th}}$ was determined from the slope of the $\Delta V - T$ plot.

1.3 Fluctuations During Reactions

1.3.1 Ligand Dissociation of Carbonmonoxymyoglobin (MbCO)

The relationship between the structural properties of proteins and their reactions is essential for understanding the biological functions of proteins. Myoglobin (Mb) has been used as a model system for experimental and theoretical studies of such structure-kinetics relationship [12–23]. A heme is embedded within the protein, and a small ligand (e.g., O_2, CO, NO) is reversibly bound to the sixth coordination site, on the distal side, of the heme. By photoexcitation of the heme, the ligand-metal bond is photodissociated (Fig. 1.4). Because there is no route to pass the ligand through the protein to the outer solvent, it is postulated that the protein structure has to change or fluctuate to release the ligand from the protein.

Extensive research has been conducted to understand the dissociation kinetics and subsequent protein deformation of Mb. The transient absorption signal following photoexcitation of MbCO rises within a few picoseconds and shows a weak (c.a. 4 %) decay component with a lifetime of 180 ns, which represents the geminate recombination kinetics (Fig. 1.4a). The signal remains almost constant during the microsecond timescale and decays to the baseline with a lifetime of c.a. 3 ms by a bimolecular recombination. Although absorption spectroscopic studies revealed structural changes around the heme chromophore, structural changes in the other parts of Mb and the dynamics of the protein have remained unclear. These difficulties may be solved using a time-resolved spectroscopy that does not use the optical transition of the heme.

The time profile of the TG signal shows more complex features [24–27]. Figure 1.4b depicts the TG signal of sperm whale MbCO at 25 °C [26]. The TG signal rises within the excitation pulse width (10 ns), followed by a decay component. The signal then shows a growth-decay-growth curve within a few tens of microseconds. On a longer timescale, the signal again rises during several hundred microseconds and finally decays to the baseline with a lifetime of several milliseconds. By comparing the TG signal after the photoexcitation of a calorimetric reference (CR) sample and also the transient absorption signal after photoexcitation of MbCO, the signal components were identified as described below.

The time profile during 10^{-7}–10^{-4} s range can be reproduced very well with Eq. (1.15):

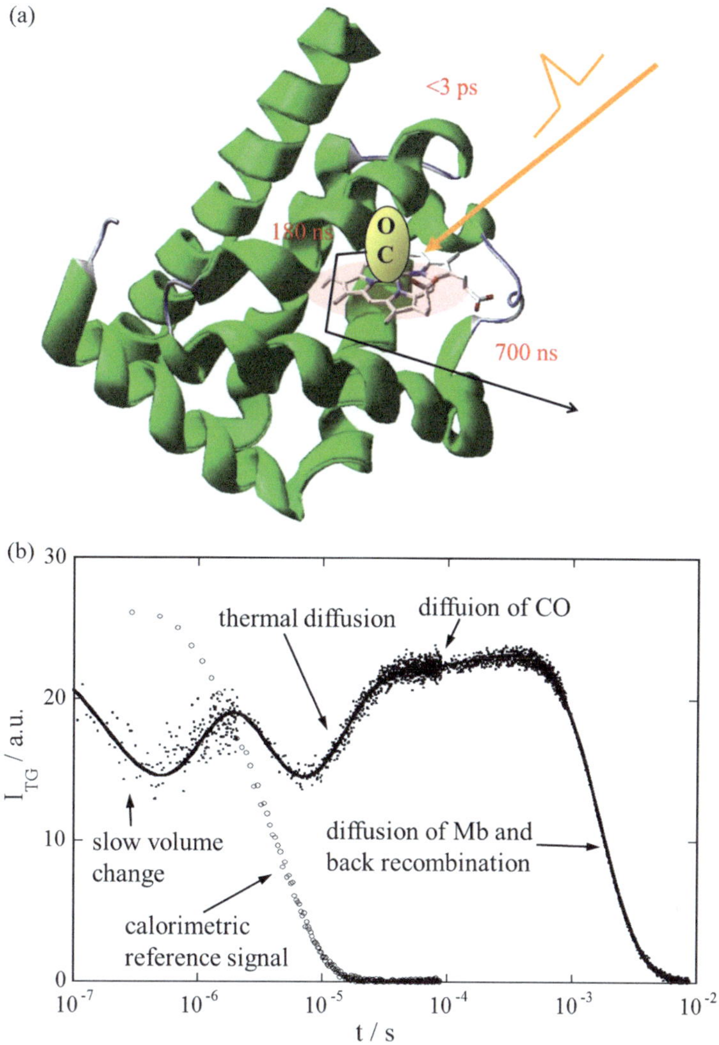

Fig. 1.4 (**a**) Photodissociation of CO from MbCO after light irradiation. Approximate time scales of these processes are also shown in the figure. (**b**) Typical TG signal observed after photoexcitation of MbCO. Assignments of the dynamical components are shown in the figure

$$
\begin{aligned}
I_{\mathrm{TG}}(t) = {} & \\
& \alpha\Big[\delta n_{\mathrm{s}}\exp(-k_{\mathrm{s}}t) + \delta n_{\mathrm{th}}\exp(-D_{\mathrm{th}}q^2t) + \delta n_{\mathrm{CO}}\exp(-(\mathrm{D}_{\mathrm{CO}}q^2 + k_{\mathrm{b}})t) \\
& +\delta n_{\mathrm{Mb}}\exp(-(\mathrm{D}_{\mathrm{Mb}}q^2 + k_{\mathrm{b}})t)\Big]^2 \\
& +\beta\big(\delta k_{\mathrm{g}}\exp(-k_{\mathrm{g}}t) + \delta k_{\mathrm{Mb}}\exp(-(\mathrm{D}_{\mathrm{Mb}}q^2 + k_{\mathrm{b}})t)\big)^2
\end{aligned}
\tag{1.15}
$$

where k_{s} denotes a rate constant ($(700\ \mathrm{ns})^{-1}$) of the first decaying component. The second term (δn_{th}) of the right-hand side represents the thermal grating component.

The first rise-decay curve represents the interference between δn_s and δn_{th}. The first term in the amplitude grating (β) term represents the absorption change, which includes the geminate recombination process with the rate constant of k_g and the ligand recombination rate (Mb + CO → MbCO) with a rate constant of k_b. Using least squares fitting, the signal can be separated into δn and δk contributions, and the decay rate constants were determined [24–27]. Because the diffusion constant of CO should be much larger than those of Mb or MbCO ($D_{CO} >> D_{Mb}$), the rising component (~submillisecond) is attributed to the diffusion of CO and the slower one (~ms) to Mb and MbCO ($D_{MbCO} \sim D_{Mb}$). $k_b^{-1} = 3$ ms is reasonably close to that measured by the laser flash photolysis experiment probed at 633 nm under the same conditions (2.8 ms). The diffusion constants of Mb and CO were determined to be $D_{CO} = 2.6 \times 10^{-9}$ m^{2}s^{-1} and $D_{Mb} = 1.2 \times 10^{-10}$ m^{2}s^{-1}.

The kinetics with the lifetime of k_s^{-1} (700 ns) observed in the TG signal has not been detected by the transient absorption technique. Hence, this observed 700 ns kinetics in the TG signal is not associated with an absorption change but is attributed to the kinetics of a molecular volume change and/or the thermal grating component. These spectrally silent dynamics in the TG signal were assigned to the escaping process of the CO from the second heme pocket to the solvent (Fig. 1.4a).

From the magnitude of the absolute thermal grating signal intensity, the enthalpy change of the reaction for MbCO → Mb + CO was determined. The absolute magnitude of the signal intensity was calibrated by the thermal grating signal intensity of the CR sample under the same conditions. The thermal energy released after the 700 ns kinetics was $Q = 159$ kJ/mol. Because the quantum yield of the photodissociation reaction of MbCO is $\phi = 0.96$, the enthalpy difference between the initial MbCO and the photodissociated species (Mb + CO) was determined to be $\Delta H_s = (h\nu - Q)/\phi = 70$ kJ/mol.

There are two different grating contributions to δn_s: the volume and thermal gratings. To determine the volume change of this reaction, the TG signal using a TG setup with a high wave number was measured [26]. Analyzing the signal, the thermal energy release by the ligand escape process was determined to be $Q_s = -30$ kJ/mol (endothermic process). ΔV_s was obtained from the magnitude of δn_s at the temperature of $dn/dT = 0$ (i.e., $\delta n_{th} = 0$), and the molecular volume expansion was found to be 14.5 cm^3/mol during the process of Mb(CO) → Mb + CO.

To examine the temperature dependence of the volume change more clearly, the TG signal of MbCO was measured as a function of temperature over the range of 0–25 °C. Some of the signals in the range of 0–10 μs are depicted in Fig. 1.5. As the temperature was decreased, the intensity of the bump in the microsecond time range decreases, and it eventually becomes a shallow dip. With further decreases in temperature, the dip becomes shallower, and finally around 0 °C, this component disappears. This temperature-dependent feature is consistently explained by the temperature dependence of $|dn/dT|$ of water. Because the intensity of the thermal grating component is proportional to $|dn/dT|$ (Eq. 1.6) and the absolute magnitude of $|dn/dT|$ is smaller at lower temperatures ($dn/dT = 0$ at ~0 °C), the interference effect between the thermal grating and the $\delta n_s + \delta n_{spe}$ component decreases with decreasing temperature. This explains the temperature-dependent feature described above.

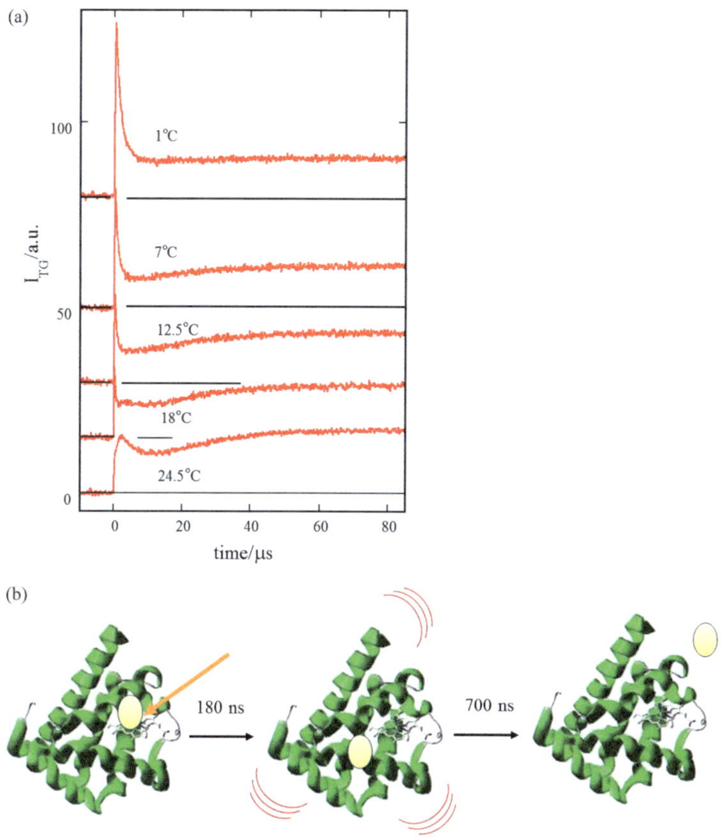

Fig. 1.5 (**a**) Temperature dependence of the TG signal observed after photoexcitation of MbCO. (**b**) Schematic illustration of the ligand escape process from Mb. The intermediate species has a larger fluctuation, which should aid release of the ligand

Furthermore, from the temperature dependence of the signal, the amplitude of the volume grating $\delta n_{\Delta V\mathrm{s}}$ was obtained at various temperatures, and the results showed that the volume change was dependent on the temperature. The volume change with the rate constant of k_s (ΔV_s) was 14.7 cm^3/mol at room temperature, and it increased to 16.8 cm^3/mol at 0 °C. The slope of the plot, ΔV vs. T, corresponds to $\Delta\alpha_{\mathrm{th}}$ ($\Delta\alpha_{\mathrm{th}} = (\partial\Delta V/\partial T)_{\mathrm{p}}$). Hence, the temperature dependence of ΔV indicates that $\Delta\alpha_{\mathrm{th}}$ is not zero, i.e., the thermal expansion coefficient depends on the state. The important point derived from this set of experiments is that this value increased ($\Delta\alpha_{\mathrm{th}} = 0.1$ cm^3/mol K > 0) for the reaction intermediate.

This is an interesting observation, because α_{th} is proportional to the cross-correlation of the volume (V) and entropy (S) fluctuations [1]. Previous studies on the thermodynamic properties of proteins have already clearly showed that the partial molar volume of the unfolded state increases more significantly with temperature than that of the folded state. The difference in α_{th} between the folded and unfolded metmyoglobin was reported to be $\Delta\alpha_{th} = +2.4$ cm^3/mol K [28]. The larger α_{th} for unfolded protein is intuitively understandable if one considers that the unfolded protein has a loose protein structure, which is expected to be more sensitive to the temperature. Because the observed temperature dependence of ΔV of this MbCO photodissociation reaction is not so large (0.1 cm^3/mol K) compared with this value, loosening of the protein part should not be so significant following ligand dissociation from the heme. However, one may also consider that the estimated increase is large, because the fluctuation change should not be spread over the whole protein, but rather localized in a small area that is related with the ligand escape route. As stated above, the ligand bound site is surrounded by amino acid residues, and there is no route for the ligand to pass through to the outer solvent. Thus, conformational fluctuations should be essential to facilitate dissociation of the ligand from the protein (Fig. 1.5b).

1.3.2 Fluctuation During the Photoreaction of PYP

Enhanced fluctuations during an enzymatic reaction were also observed for particular photosensor proteins. For example, PYP is a blue-light sensor protein isolated from the purple sulfur bacterium *Ectothiorhodospira halophila* [29, 30]. It is a relatively small (14 kDa) water-soluble protein, and the protein structure has been determined by X-ray crystallography [31]. The chromophore of PYP is 4-hydroxycinnamic acid, which is covalently bound to the side chain of Cys 69 via a thioester linkage [32, 33]. After photoexcitation, PYP shows a complete photocycle triggered by the photoisomerization of this chromophore [34, 35]. Upon flash excitation of the chromophore, the ground state (pG, $\lambda_{max} = 446$ nm) is converted into a red-shifted intermediate (pR, $\lambda_{max} = 465$ nm) in less than 2 ns. Subsequently, pR decays in the submillisecond timescale range into a blue-shifted intermediate (pB, $\lambda_{max} = 355$ nm), which returns to pG in a sub-s timescale [36–38]. The pR state possesses the *cis* conformation of the chromophore, and the process of *trans* to *cis* isomerization was accompanied by the specific formation of new hydrogen bonds.

The reaction scheme was revealed by the TG method [39–41]. Figure 1.6 depicts the TG signal of the PYP solution [39]. The signal rises quickly following excitation and then slowly with about a 1 μs time constant decays with a lifetime of several microseconds, which depends on the grating wave number. After the decay, the signal shows a growth-decay feature twice during a few milliseconds timescale. From measurement at different q values, it was found that the decay of the thermal grating and the kinetics of the second growth-decay signal depend on q. Because there is no optical absorption from the original species (pG) nor from any intermediates of PYP at the probe wavelength (633 nm), the observed TG signal must come from the refractive index change after photoexcitation. It was found that the TG

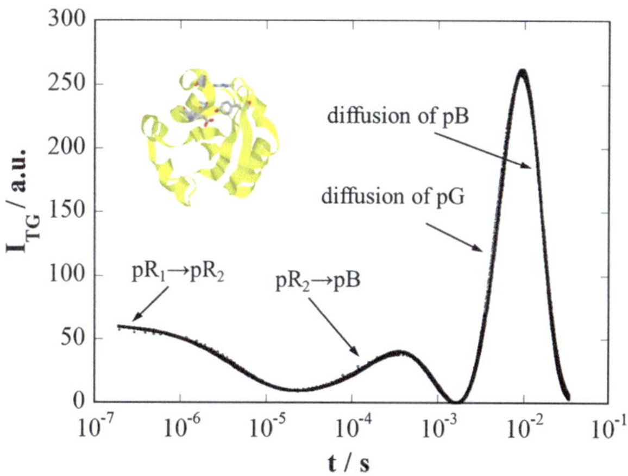

Fig. 1.6 Typical TG signal observed after photoexcitation of PYP. Assignments of the dynamical components are shown in the figure. Inlet: structure of PYP

signal in the entire region can be well reproduced by the square of the summation of six exponential functions [39]:

$$I_{\mathrm{TG}} = \alpha\{\delta n_{\mathrm{s}}\exp(-k_{\mathrm{s}}t) + \delta n_{\mathrm{th}}\exp(-q^2 D_{\mathrm{th}}t) + \delta n_1\exp(-k_1 t) + \delta n_2\exp(-k_2 t) + \delta n_{pG}\exp(-q^2 D_{pG}t) + \delta n_{pB}\exp(-q^2 D_{pB}t)\}^2 \quad (1.16)$$

where k_{s} is the rate constant of the slow rising component, and D_{pG} and D_{pB} are the diffusion coefficients of pG and pB, respectively. The kinetics, which does not depend on q (k_1 and k_2), should represent the reaction dynamics of PYP. Considering the timescale, these dynamics should be the pR → pB process. The temporal profile associated with the pR → pB transformation can be fitted with lifetimes of 170 μs and 1.0 ms.

The slow rise with the rate constant of k_{s} in the TG signal represents an intermediate species between the pR and pB states. The absorption spectrum of this species is the same as that of pR, and this new species was called pR_2. If this rising δn_{s} component comes entirely a change in volume, then the volume change (ΔV) in these dynamics can be calculated from the intensity of this component using Eq. (1.13), $\Delta V = +5$ cm^3/mol. Further measurements for some site-directed mutants of PYP showed that this spectrally silent dynamics is sensitive to any amino acid change located far from the chromophore. The result of the mutant dependence also supports the conclusion that these dynamics reflect global movement of residues.

The second term of Eq. (1.16) represents the ThG signal. The intensity of this component should be proportional to the thermal energy released by the first step pG^* → pR. The amount of thermal energy is determined by single exponential fitting. Comparing the thermal grating signal intensity with that of a calorimetric reference sample, the stored energy in pR_2, i.e., the enthalpy change for pG → pR_2

(ΔH_1), was obtained as $\Phi\Delta H_1 = 57$ kJ/mol. Using the quantum yield of the reaction $\Phi = 0.35$ [42], ΔH_1 was determined to be 160 kJ/mol.

The enthalpy of the long-lived species, the pB state (ΔH_2), was determined by the TrL method. The decay of the TrL signal is in the order of tens of milliseconds, and the thermal energy released by the pR → pB process can be detected as a rise in the signal. By comparing the ThL signal intensity with that of the CR sample, the enthalpy difference between pG and pB (ΔH_2) was determined to be $\Delta H_2 = 60$ kJ/mol.

The result of ΔH ($\Delta H_1 = 160$ kJ/mol, $\Delta H_2 = 60$ kJ/mol) clearly indicates that the energy of the pR state is astonishingly high and that large energetic stabilization takes place during the pR → pB process. In a time-resolved X-ray study [43], it was shown that the *trans* to *cis* isomerization is completed in less than 1 ns. From an energetic viewpoint, these observations suggest that the protein part has not yet relaxed in the pR_2 state to adopt the new conformation of the chromophore, that is, the protein structure is strained in the pR state. This strain may cause the large enthalpy of the pR species. It is reasonable to speculate that the following step (pR → pB) is driven by this energy stored in pR. In pB, the whole protein structure is relaxed to adopt the *cis* form of the chromophore, and this relaxation causes the lower enthalpy of pB.

When the solution temperature was decreased, the ThG signal intensity decreased. This decrease is mostly due to the temperature dependence of the dn/dT term (Eq. 1.6). Comparing the signal intensity of the PYP sample with that of the CR sample at the same temperature, ΔH_{pR} at various temperatures was determined and plotted in Fig. 1.7a. This value did not depend on the temperature within experimental accuracy. This result means that the heat capacity of pR_2 is the same as that of pG (within ± 0.5 kJ/mol K).

Figure 1.7b shows the temperature dependence of ΔH_{pB}. The ΔH_{pB} value decreases with decreasing temperature. This temperature dependence of ΔH_{pB} indicates a nonzero ΔC_p value of the pB species. From linear fitting over the observed temperature range, $\Delta C_p = 2.7$ kJ/mol K was obtained. The determined ΔH and ΔC_p, including the activation parameters along the PYP photoreaction coordinate, is schematically shown in Fig. 1.8.

Using the time-resolved TG method, a rather large temperature-dependent ΔV was found [39]. The slope of the plot, ΔV vs. T, corresponds to $\Delta\alpha_{\mathrm{th}}$, which is the change in the coefficient of thermal expansion for the initial (pG) and the intermediate (pR_2) forms. This indicates that the thermal expansivity of pR_2 is larger than that of pG. As stated above, a larger α_{th} is an indication of larger structural and/or entropy fluctuation. Therefore, even in this initial intermediate species pR_2, which is created 1 μs following photoisomerization of the chromophore, the protein structure is loosened. Apparently, the structural change is not restricted around the chromophore. As stated above, the difference of α_{th} between the folded and unfolded metmyoglobin was $\Delta\alpha_{\mathrm{th}} = +2.4$ mL/mol·K, whereas $\Delta\alpha_{\mathrm{th}}$ for pG and pR_2 of PYP was + 0.6 mL/mol·K. Hence, this observation, temperature-dependent ΔV, provides a resemblance of the pR_2 state to the unfolded state of protein.

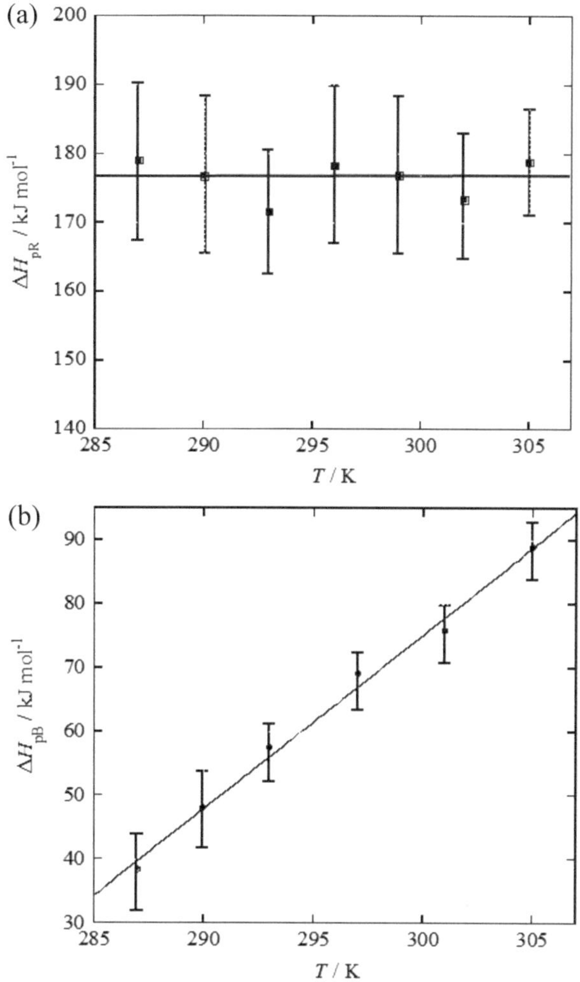

Fig. 1.7 (**a**) ΔH of pR_2 (*squares*) against temperature calculated from the ThG signal intensities. (**b**) ΔH of pB (*circles*) against temperature calculated from the ThL signal intensities

A time-resolved FT-IR experiment did not show a large backbone amide group movement in pR [44]. Furthermore, ΔV described in the previous section from the TG measurement is much smaller than expected from the unfolding process. These results (TG and FT-IR) suggest that the conformation of pR is not altered significantly from pG. Conversely, the thermal expansivity significantly changed. Therefore, the larger thermal expansivity in pR_2 indicates that the rigidity of the structure is weakened with a small conformational change. Apparently, this weakening is caused by the strain induced by the photoisomerization of the chromophore, as suggested from the large enthalpy of this species. This larger conformational flexibility in pR_2 may cause the next larger structural change to the protein in $pR_2 \rightarrow pB$.

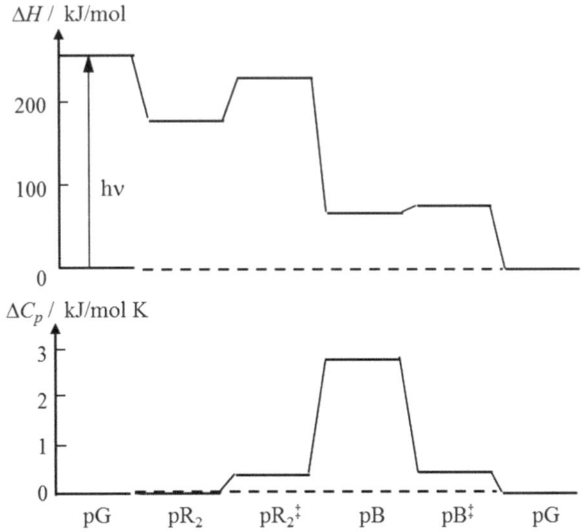

Fig. 1.8 ΔH and ΔCp along the reaction coordinate of PYP

An NMR structure of a PYP lacking 25 N-terminal residues (Δ25) was reported [45] and showed that the central β-sheet and a small part of one α-helix in the helical connector region were still intact but that the remainder of the protein was highly flexible in the pB state. Conformational change was also detected by monitoring the diffusion coefficient (D) of the intermediate species, pB. The smaller D_{pB} was attributed to a significant conformational change that increased the friction for translational diffusion [39, 40]. A later study on certain N-terminal truncated PYPs concluded that the conformational change (unfolding process) within the N-terminal region was the main cause for the change in D and that the solvation structure around this group was significantly altered [46, 47]. An analysis of the signal based on the time-dependent diffusion coefficient clearly showed that the protein diffusion changed with a time constant of 170 μs, corresponding to the $pR_2 \rightarrow pB'$ transition. Hence, it is reasonable to consider that the loosened structure suggested by the thermal expansivity is the main driving force to induce the unfolding of the N-terminal α-helices.

1.3.3 *Fluctuation During the Photoreaction of the Phototropin LOV Domain*

Another example of enhanced fluctuation during a reaction is observed for a blue-light sensor protein, phototropins (phots). Phots regulate phototropism [48, 49], chloroplast relocation [50, 51], and stomatal opening [52]. *Arabidopsis* has two types of phot, phot1 and phot2, both of which have two photosensitive domains, LOV (light, oxygen, voltage)1 and LOV2, in their N-terminal half [53, 54] and a serine/threonine kinase domain in the C-terminal half [55]. Both LOV domains non-covalently contain a blue-light-absorbing chromophore, flavin mononucleotide (FMN).

The photochemical reaction proceeds with several intermediate species. In the dark state, the LOV domain has an absorption spectrum with a peak at ~447 nm (D_{447}). Upon illumination with blue light, the photoexcited triplet state (L_{660}) was created within 30 ns, then within a few microseconds forms an intermediate with a blue-shifted absorption spectrum (S_{390}) [56]. The LOV domain on the S_{390} state forms a covalent bond between the C(4a) of the isoalloxazine ring of FMN and a cysteine of the protein moiety [57–60]. After the absorption spectrum change from L_{660} to S_{390}, spectrally silent kinetics have been identified in the time domain for the phot2LOV2 domain with the linker by monitoring changes in the diffusion coefficient (D) [61, 62]. These dynamics were attributed to an unfolding process of the linker region. The final state adopts a different protein conformation but has the same absorption spectrum as S_{390} and is referred to as the T_{390} state. The photo-adduct state has a lifetime of seconds or minutes before the original D_{447} state is recovered [63]. An NMR study on oat phot1LOV2 revealed the presence of an α-helix, known as the J-helix, in a part of the linker region [64]. The NMR (NOE) signal under light illumination showed that the interaction between the linker and the LOV2 core disappeared upon photoreaction. These observations are consistent with the findings of the above mentioned time-resolved studies.

1.3.3.1 LOV2 Domain

For clarifying the role of the fluctuation and the conformational dynamics of the linker region, the reaction scheme and the fluctuation of the phot2LOV2 domain (without the linker) were first described [65]. The TG signal after photoexcitation of the LOV2 sample (without the linker) (Fig. 1.9a) rises quickly within the time response of ~20 ns. The signal then decayed within a few microseconds, rose, and decayed again. After the decay within the tens of microseconds time range, the signal showed a weak rise and decay to the baseline in the sub-second time range. This temporal profile over a broad time range was reproduced well using the sum of four exponential functions:

$$I_{TG}(t) = \alpha\left\{\delta n_1 \exp(-k_1 t) + \delta n_{th} \exp\left(-D_{th} q^2 t\right) - \delta n_3 \exp(-k_3 t) + \delta n_4 \exp(-k_4 t)\right\}^2, \quad (1.17)$$

where δn_i ($i = 1$, 3, and 4) are pre-exponential factors representing the refractive index changes. By measuring the TG signal at various q values, it was found that the largest decay rate constant ($k_1 = (0.9\ \mu s)^{-1}$) did not depend on q^2, whereas the other rate constants did. The q^2 independence of k_1 indicated that these dynamics represented a chemical reaction rate (not diffusion), and these dynamics were attributed to the adduct formation of the chromophore. The last weak rise-decay

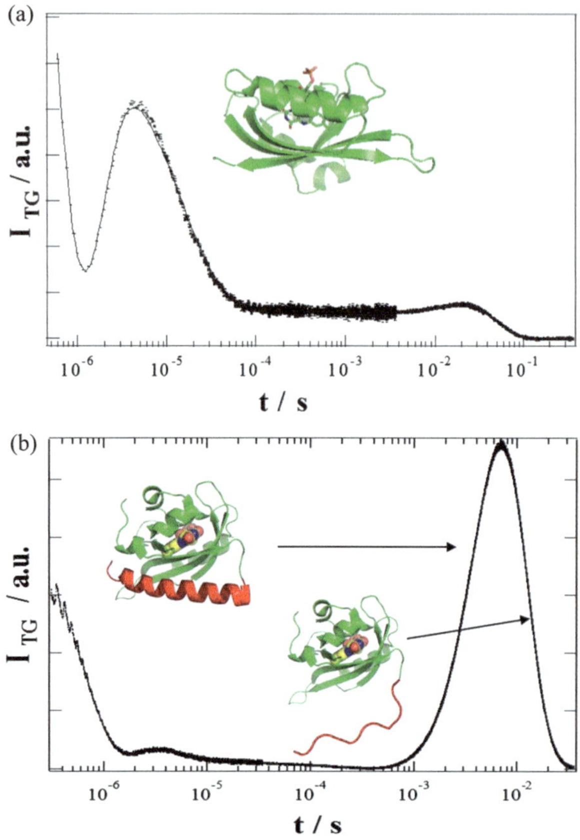

Fig. 1.9 (**a**) Observed TG signal (*dotted line*) of the LOV2 domain without the linker at $q^2 = 3.1 \times 10^{11}\ m^{-2}$. (The structure is shown in the inset.) The diffusion signal at around 10 ms is very weak compared with the thermal grating signal, which appears after a few microseconds. (**b**) Observed TG signal (*dotted line*) of the LOV2 with the linker sample at $q^2 = 5.6 \times 10^{11}\ m^{-2}$. Note the very strong diffusion signal at a few microseconds. The rising component of the diffusion signal represents the diffusion of the reactant, and the decay represents the diffusion of the product, which includes the unfolded conformation of the linker domain

curve was reproduced by the third and fourth terms of Eq. (1.17), which were assigned to the protein diffusion process of the reactant and the product. From the signs of the refractive index changes to fit the signal, the rate constants of the rise and decay components were attributed to diffusion of the reactant ($k_3 = D_R q^2$) and the product ($k_4 = D_P q^2$), respectively, and the diffusion coefficients of the reactant (D_R) and that of the product (D_P) were $10.3 \times 10^{-11}\ m^2 s^{-1}$ and $10.5 \times 10^{-11}\ m^2 s^{-1}$, respectively. The small difference between D_R and D_P suggested that the conformational change and the intermolecular interaction change between the protein and water are minor. It was found that this diffusion change was completed within 100 μs after excitation.

By using the TrL method, another reaction component was observed in a longer time range and the time constant was 9 ms at 293 K. Because there is no absorption change in this time range, this component should be attributed to an energy release process and/or volume change process. These states were denoted as S_{390}(I) and S_{390}(II). On the basis of the TG and TrL results, the reaction scheme should be expressed by Scheme 1.1.

Scheme 1.1

$$D_{447} \xrightarrow{h\nu} D^{*}_{447} \xrightarrow{\sim 30\text{ns}} L_{660} \xrightarrow{\sim 1\mu\text{s}} S_{390}(\text{I}) \xrightarrow{\sim 9\text{ms}} S_{390}(\text{II})$$

The signal intensities of the ThG and ThL components represent energy released. By comparing the ThG intensity with that of the CR sample, the reaction enthalpy, ΔH, of the adduct species (S_{390}(I): ΔH_{SI}(L)) was determined to be 80 kJ mol^{-1} at 293 K by using $\Phi_{\mathrm{f}} = 0.14$ (LOV2 from *Avena sativa* phot1 [66]), $\Phi = 0.27$ [63], and $h\nu_{\mathrm{f}} = 230$ kJ mol^{-1}. For the determination of ΔC_p for this adduct species, ΔH at various temperatures was measured and is plotted in Fig. 1.10a (open circles). This value does not depend on the temperature within experimental accuracy. This temperature independence means that the heat capacity of S_{390}(I) is the same as that of the ground state ($\Delta C_{\mathrm{p}} = 0.1 \pm 0.5$ kJ $\mathrm{mol}^{-1}\mathrm{K}^{-1}$).

ΔH of S_{390}(II) (ΔH_{SII}(L)), which is the final product in the observed time range (<100 ms), was determined from the TrL signal intensity. Comparing the ThL signal intensity with that of the CR sample, the ΔH_{SII}(L) value was 40 kJ mol^{-1} at 293 K. The temperature dependence of ΔH_{SII}(L) is shown in Fig. 1.10b (open circles). ΔC_{p} from the slope was almost zero ($\Delta C_{\mathrm{p}} = 0.2 \pm 0.5$ kJ $\mathrm{mol}^{-1}\mathrm{K}^{-1}$) within the experimental accuracy. The negligibly small change in C_{p} for LOV2 of S_{390}(I) and S_{390}(II) suggests that the energetic fluctuation of the LOV2 domain does not change during the first intermediate (adduct formation) and also for the product formation. Probably because of the small conformational change, strain could be accumulated in the protein component. This strain can explain the relatively large ΔH of this species, and this large energy could be a driving force for the subsequent reaction. The final product in our observation time frame was produced with a time constant of 9 ms at 293 K. This transition was spectrally silent. However, D of S_{390}(II) is similar to values for D_{447} and S_{390}(I).

Subtracting the thermal contribution from the observed change in the refractive index in the 9 ms dynamics of the TrL signal, the refractive index change due to the volume change was isolated. From this amplitude, the volume change contribution ΔV was determined to be 0.2 cm^3 mol^{-1} at 293 K. The small value suggests that the conformational change of the LOV2 domain is minor, which is consistent with the fact that almost no change in D was observed during this process. Furthermore, ΔV was almost temperature independent within experimental error. The slope of the plot provides the thermal expansion volume change ($V\Delta\alpha_{\mathrm{th}}$) of this species, and the value was -0.02 ± 0.02 $\mathrm{cm}^3\mathrm{mol}^{-1}\mathrm{K}^{-1}$. As stated in the above sections, the thermal expansion represents the structural fluctuation. Therefore, the negligibly small change of $V\Delta\alpha_{\mathrm{th}}$ for $S_{390}(\text{I}) \rightarrow S_{390}(\text{II})$ indicates that the fluctuation of the LOV2 domain does not change so much during this process. Similarly, ΔC_{p} represents the energetic fluctuation change, and ΔC_{p} of the second intermediate, S_{390}(II), was also small. These facts indicated that the conformational change, including the fluctuation change during this process, is not significant.

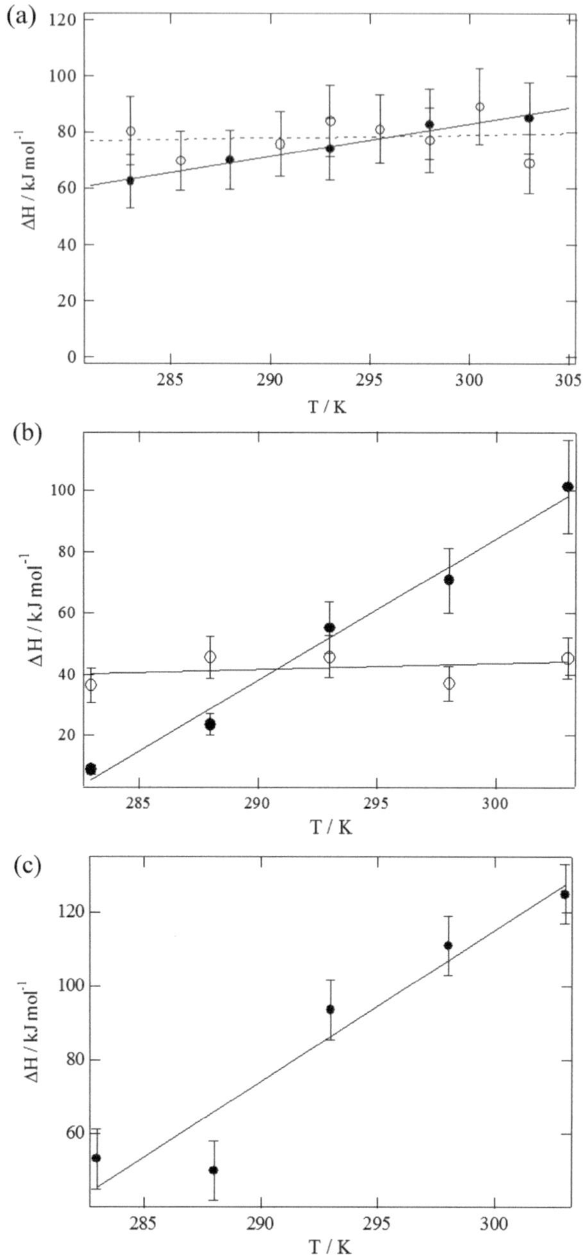

Fig. 1.10 (**a**) Temperature dependence of the enthalpy changes of the S_{390}(I) state from the ground state of LOV2 (*open circles*) and the S_{390} state of the LOV2-linker (*closed circles*). The best-fit lines by a linear function are also shown for LOV2 (*broken line*) and the LOV2-linker (*solid line*). (**b**) Temperature dependence of the enthalpy changes of the S_{390}(II) state from the ground state of LOV2 (*open circles*) and the T_{390}(II) state of the LOV2-linker (*closed circles*). The best-fit lines by a linear function are also shown for LOV2 and the LOV2-linker. (**c**) Temperature dependence of the enthalpy changes of the T_{390}(I) state of the LOV2-linker. The best-fit lines by a linear function are also shown for the LOV2-linker (*solid line*). Details of the calculation are given in the text

1.3.3.2 Reaction of the LOV2-Linker

Next, the reaction dynamics and the time-dependent fluctuation of the LOV2 domain with the linker (LOV2-linker) were revealed by the TG and TrL methods. Figure 1.9b depicts the TG signal after the photoexcitation of the LOV2-linker over a wide time range. Considering the assignment of the LOV2 signal, the initial decay with a time constant of 0.9 μs was attributed to the adduct formation process, and the next decay in 100 μs under these conditions was the decay of the ThG signal. The signal differed from that of LOV2 in the later time regions, that is, the diffusion signal of the LOV2-linker was much larger than that of LOV2. The drastic enhancement of the diffusion signal indicated that the reduction in D by the photoexcitation is much larger ($D_R = 8.8 \times 10^{-11}$ m^2s^{-1}, and $D_P = 6.8 \times 10^{-11}$ m^2s^{-1}). Furthermore, it was found that D of the photoexcited LOV2-linker was time dependent in the observed time range. D decreased from 8.8×10^{-11} m^2s^{-1} to 6.8×10^{-11} m^2s^{-1} with a time constant of 2 ms at room temperature (~293 K). This change was attributed to an enhancement of the intermolecular interactions between the protein and water molecules, which is caused by the unfolding of the α-helices in the linker region. The intermediate created by this unfolding process was referred to as the T_{390} species.

Using the TrL signal of the LOV2-linker, it was found that there were two additional phases with time constants of 140 μs and 11 ms at 293 K. The faster phase, with the time constant $\tau_f = 140$ μs, was observed only for the LOV2-linker. This is another intermediate (referred as $T_{390}{}^{pre}$) before T_{390}. Hence, the 140 μs dynamics was attributed to the formation process of the $T_{390}{}^{pre}$ state. After creation of this $T_{390}{}^{pre}$ state, the diffusion coefficient changed with a time constant of 2 ms. The following slower time constant of $\tau_s = 11$ ms was similar to $\tau_L(=9$ ms) of LOV2. These dynamics were attributed to $S_{390}(I) \rightarrow S_{390}(II)$ of the LOV2 domain. Using a similar notation for intermediates, this final state was referred as $T_{390}(II)$, because the linker region is already unfolded in this state of the LOV2-linker.

Based on these observations, the reaction scheme of the LOV2-linker may be expressed as shown in Scheme 1.2.

Scheme 1.2

$$D_{447} \xrightarrow{h\nu} D^*_{447} \xrightarrow{\sim 30ns} L_{660} \xrightarrow{\sim 1\mu s} S_{390} \xrightarrow{140\mu s} T_{390(I)pre} \xrightarrow{2ms} T_{390(I)} \xrightarrow{\sim 11ms} T_{390(II)}$$

The enthalpy of S_{390} of the LOV2-linker was measured using the ThG intensities of the LOV2-linker and CR samples. For example, at 293 K, ΔH_S(LL) was 70 kJ mol^{-1}, which is close to the corresponding value for LOV2 (ΔH_{SI}(L)). The ΔH_S(LL) values at various temperatures are plotted in Fig. 1.10a (closed circles). Although ΔH_{SI}(L) was almost temperature independent, the ΔH_S(LL) values increased slightly with increasing temperature. The heat capacity change of S_{390} was determined from the temperature dependence of ΔH_S(LL) to be 1.2 kJ $mol^{-1}K^{-1}$.

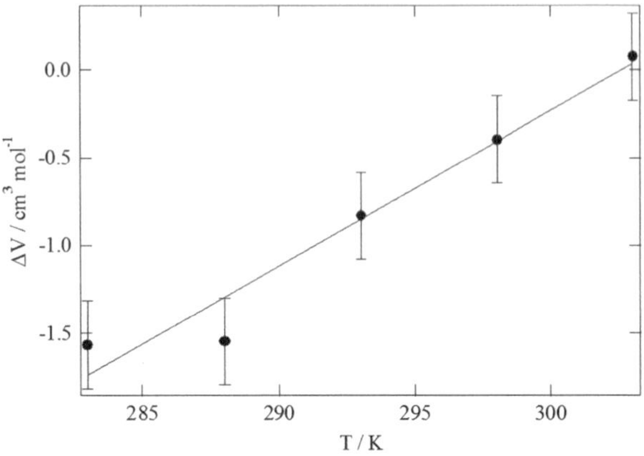

Fig. 1.11 Temperature dependence of the volume changes for the $S_{390} \rightarrow T_{390}(I)^{pre}$ transition of the LOV2-linker (*closed circles*) and the best-fit linear function

The ThL signal decayed in the 100 ms range. This decay was slower than the $T_{390}(I) \rightarrow T_{390}(II)$ transition, and hence, the ThL intensity represented the thermal energy released for the creation of $T_{390}(II)$. Using the same method as for the LOV2 sample, ΔH of $T_{390}(II)$ ($\Delta H_{TII}(LL)$) was determined to be 60 kJ mol^{-1} at 293 K. The temperature dependence of $\Delta H_{TII}(LL)$ was measured and it is plotted in Fig. 1.10b (closed circles). $\Delta H_{TII}(LL)$ decreased with decreasing temperature. From this temperature dependence, the heat capacity change ΔC_p of $T_{390}(II)$ was determined to be 4.7 kJ mol^{-1}K^{-1}. It is interesting to note that ΔC_p is negligible for LOV2 but is positive for the LOV2-linker.

The dissociation of the linker region from the LOV domain ($S_{390} \rightarrow T_{390}(I)^{pre}$) was observed as the first decay component of the TrL signal. Because there is no energy release (or uptake) during this process, ΔH of $T_{390}(I)^{pre}$ and $T_{390}(I)$ should be the same. Using this $\Delta H = 0$, the volume change during the $S_{390} \rightarrow T_{390}(I)^{pre}$ process was calculated from the amplitude of this component of the TrL signal, e.g., $\Delta V = -0.8$ cm^{3}mol^{-1} at 293 K. This volume change was sensitive to temperature (Fig. 1.11). From the slope of the plot, $V\Delta\alpha_{th}$ for $T_{390}(I)^{pre}$ was determined to be 0.08 cm^3 mol^{-1} K^{-1}. The properties of the initial adduct state S_{390} of the LOV2-linker are similar to those of $S_{390}(I)$ of LOV2. For example, the formation time constant is 0.9 μs, D is similar to the unphotolyzed state, and ΔH is relatively large. These similarities imply that the linker region does not influence the chemical bond formation between FMN and cysteine of the protein moiety. However, ΔC_p of S_{390} of the LOV2-linker is larger than the corresponding value for $S_{390}(I)$ of LOV2. This difference may be interpreted as being due to a change in the interface region between the LOV2 domain and the linker. This change also indicates the larger energetic fluctuation of the LOV2-linker and that this change may trigger the next conformational change around the linker portion and its interaction with the LOV2 core.

The enthalpy change, the heat capacity change, and the thermal expansion change of these intermediates of the LOV2-linker sample along the reaction coordinate are depicted in Fig. 1.12. It should be noted that the change in the

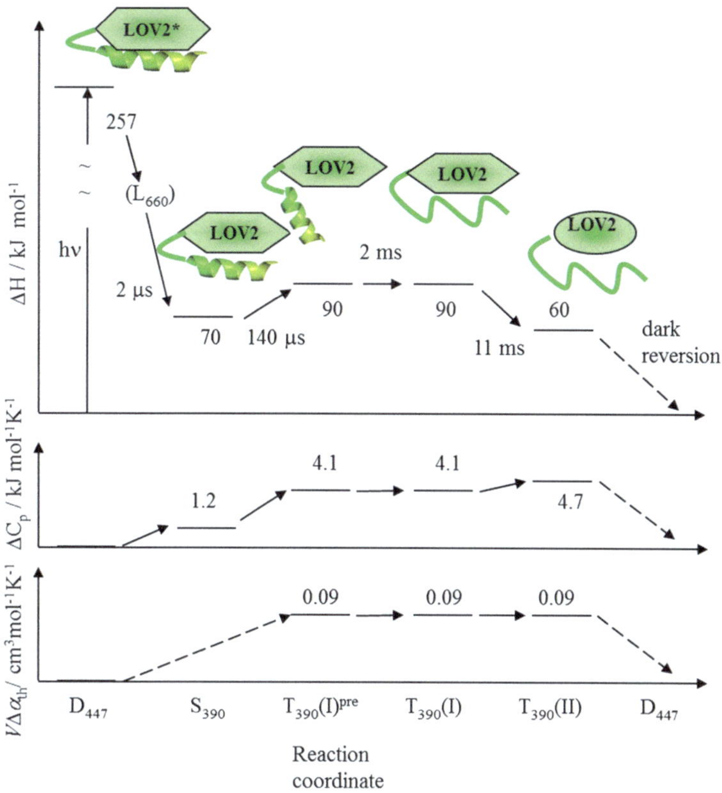

Fig. 1.12 Heat capacity change and thermal expansion coefficient change along the reaction coordinate of the Phot1LOV2-linker

thermal expansion is relatively large for the $S_{390} \rightarrow T_{390}(I)^{pre}$ step (Fig. 1.12). This fact suggests that the conformational fluctuation is enhanced at this stage and that this change may be attributed to the fluctuation in the linker region, which leads the light-dependent activation of the kinase domain. Furthermore, it is interesting to note that the enthalpy of $T_{390}(I)^{pre}$ is larger than that of S_{390} at 293 K. This fact indicates that this reaction is likely to be an entropy-driven process. Because the linker portion is dissociated from the LOV domain and this part will be flexible by this process, this enhanced entropy is reasonable.

One of the most significant observations is the large and positive ΔC_p of $T_{390}(I)$ and $T_{390}(II)$. This heat capacity change could originate from exposure of a hydrophobic area. Because a large ΔC_p was not observed for LOV2, this exposed hydrophobic area was attributed to an area in which there is light-dependent interaction between the Jα helix and the LOV2 core and the unfolded peptide at the Jα helix. This change also indicates the enhanced fluctuation in the $T_{390}(II)$ state, which should lead to the signal transduction reaction.

1.4 Summary

In this review, several time-resolved measurements of thermodynamical properties, which reflected fluctuations of proteins during reactions, were presented. In all cases, the fluctuations estimated from the thermal expansion coefficient and heat capacity increased during the reaction. These results indicate the importance of fluctuations during protein reactions. For example, because there is no route for the escape of the ligand from inside myoglobin, structural fluctuations (or conformational changes) are essential for the dissociation (or release) reaction. For some photosensor proteins, although distinct (averaged) conformational changes have not been observed after light illumination, the information of light is transferred to the downstream proteins. In such cases, fluctuations, which do not change the average conformation, could be a key factor involved in the mechanism of information transfer.

An additional property that reflects conformational fluctuations is compressibility. For measuring the compressibility, a high-pressure instrument is required. In this sense, high hydrostatic pressure combined with the time-resolved measurements by the TG and TrL methods will be a useful tool for elucidating the role of fluctuations. Research into such high-pressure systems is currently underway and will be reported in the near future.

Acknowledgments The author is deeply indebted to coauthors of the papers contained in this article. This work was supported by a Grant-in-Aid for Scientific Research on Innovative Areas (research in a proposed research area) (20107003) from the Ministry of Education, Culture, Sports, Science and Technology in Japan.

References

1. Nelson DL, Cox MM (2000) Lehninger principles of biochemistry. Worth Publishers, New York
2. Landau L, Lifsitz E (1969) Statistical physics, theoretical physics, vol 5. Pergamon Press, Oxford
3. Heremans K, Smeller L (1998) Protein structure and dynamics at high pressure. Biochim Biophys Acta 1386:353–370
4. Terazima M (2002) Molecular volume and enthalpy changes associated with irreversible photo-reactions. J Photochem Photobiol C 3:81–108
5. Terazima M (2004) Time-resolved thermodynamic properties of intermediate species during photochemical reactions. Bull Chem Soc Jpn 77:23–41
6. Terazima M (2000) Translational diffusion of organic radicals in solution. Acc Chem Res 33:687–694
7. Terazima M (2006) Diffusion coefficients as a monitor of reaction kinetics of biological molecules. Phys Chem Chem Phys 8:545–557
8. Eichler HJ, Günter P, Pohl DW (1986) Laser induced dynamic gratings. Spirnger, Berlin
9. Terazima M, Hirota N, Braslavsky SE, Mandelis A, Bialkowski SE, Diebold GJ, Miller RJD, Fournier D, Palmer RA, Tam A (2004) Quantities, terminology and symbols in photothermal and related spectroscopies. Pure Appl Chem 76:1083–1118

10. Terazima M (1998) Transient lens spectroscopy in a fast time scale. Israel J Chem 38:143–157
11. Bialkowski SE (1996) Photothermal methods for chemical analysis. Wiley, New York
12. Perman B, Anderson S, Schmidt M, Moffat K (2000) New techniques in fast time-resolved structure determination. Cell Mol Biol 46:895–913
13. Nagy AM, Raicu V, Miller RJ (2005) Nonlinear optical studies of heme protein dynamics: implications for proteins as hybrid states of matter. Biochim Biophys Acta 1749:148–172
14. Hilinski EF, Rentzepis PM (1983) Biological applications of picosecond spectroscopy. Nature 302:481–487
15. Hummer G, Schotte F, Anfinrud PA (2004) Unveiling functional protein motions with picosecond x-ray crystallography and molecular dynamics simulations. Proc Natl Acad Sci U S A 101:15330–15334
16. Lim M, Jackson TA, Anfinrud PA (1997) Ultrafast rotation and trapping of carbon monoxide dissociated from myoglobin. Nat Struct Biol 4:209–214
17. Kuczera K, Lambry JC, Martin JL, Karplus M (1993) Nonexponential relaxation after ligand dissociation from myoglobin: a molecular dynamics simulation. Proc Natl Acad Sci U S A 90 (12):5805–5807
18. Tian WD, Sage JT, Champion PM (1993) Investigations of ligand association and dissociation rates in the "open" and "closed" states of myoglobin. J Mol Biol 233:155–166
19. Hagen SJ, Hofrichter J, Eaton WA (1995) Protein reaction kinetics in a room-temperature glass. Science 269:959–962
20. Frauenfelder H, Young RD, Fenimore PW (2013) Dynamics and the free-energy landscape of proteins, explored with the Mössbauer effect and quasi-elastic neutron scattering. J Phys Chem B 117:13301–13307
21. Frauenfelder H, McMahon BH, Fenimore PW (2003) Myoglobin: the hydrogen atom of biology and a paradigm of complexity. Proc Natl Acad Sci U S A 100:8615–8617
22. Frauenfelder H, McMahon BH, Austin RH, Chu K, Groves JT (2001) The role of structure, energy landscape, dynamics, and allostery in the enzymatic function of myoglobin. Proc Natl Acad Sci U S A 98:2370–2374
23. Frauenfelder H (1995) Complexity in proteins. Nat Struct Biol 2:821–823
24. Sakakura M, Yamaguchi S, Hirota N, Terazima N (2001) Dynamics of structure and energy of horse carboxymyoglobin after photodissociation of the carbon monoxide. J Am Chem Soc 123:4286–4294
25. Sakakura M, Morishima I, Terazima M (2002) Structural dynamics of distal histidine replaced mutants of myoglobin accompanied with the photodissociation reaction of the ligand. Biochemistry 41:4837–4846
26. Sakakura M, Morishima I, Terazima M (2001) The structural dynamics and ligand releasing process after the photodissociation of sperm whale carboxymyoglobin. J Phys Chem B 105:10424–10434
27. Nishihara Y, Sakakura M, Kimura Y, Terazima M (2004) The escape process of carbon monoxide from myoglobin to solution at physiological temperature. J Am Chem Soc 126:11877–11888
28. Zipp A, Kauzmann W (1973) Pressure denaturation of metmyoglobin. Biochemistry 12:4217–4228
29. Meyer TE, Fitch JC, Bartsch RG, Tollin G, Cusanovich MA (1990) Soluble cytochromes and a photoactive yellow protein isolated from the moderately halophilic purple phototrophic bacterium, *Rhodospirillum salexigens*. Biochim Biophys Acta 1016:364–370
30. Sprenger WW, Hoff WD, Armitage JP, Hellingwerf KJ (1993) The eubacterium *Ectothiorhodospira halophila* is negatively phototactic, with a wavelength dependence that fits the absorption spectrum of the photoactive yellow protein. J Bacteriol 175:3096–3104
31. Borgstahl GE, Williams DR, Getzoff ED (1995) 1.4 A structure of photoactive yellow protein, a cytosolic photoreceptor: unusual fold, active site, and chromophore. Biochemistry 34:6278–6287

32. Hoff WD, Düx P, Hård K, Devreese B, Nugteren-Roodzant IM, Crielaard W, Boelens R, Kaptein R, van Beeumen J, Hellingwerf KJ (1994) Thiol ester-linked p-coumaric acid as a new photoactive prosthetic group in a protein with rhodopsin-like photochemistry. Biochemistry 33:13959–13962
33. Hoff WD, van Stokkum IH, van Ramesdonk HJ, van Brederode ME, Brouwer AM, Fitch JC, Meyer TE, van Grondelle R, Hellingwerf KJ (1994) Measurement and global analysis of the absorbance changes in the photocycle of the photoactive yellow protein from *Ectothiorhodospira halophila*. Biophys J 67:1691–1705
34. Meyer TE, Tollin G, Hazzard JH, Cusanovich MA (1989) Photoactive yellow protein from the purple phototrophic bacterium, *Ectothiorhodospira halophila*. Quantum yield of photobleaching and effects of temperature, alcohols, glycerol, and sucrose on kinetics of photobleaching and recovery. Biophys J 56:559–564
35. Imamoto Y, Kataoka M, Tokunaga F (1996) Photoreaction cycle of photoactive yellow protein from *Ectothiorhodospira halophila* studied by low-temperature spectroscopy. Biochemistry 35:14047–14053
36. Genick UK, Borgstahl GE, Ng K, Ren Z, Pradervand C, Burke PM, Srajer V, Teng TY, Schildkamp W, McRee DE, Moffat K, Getzoff ED (1997) Structure of a protein photocycle intermediate by millisecond time-resolved crystallography. Science 275:1471–1475
37. Brudler R, Rammelsberg R, Woo TT, Getzoff ED, Gerwert K (2001) Structure of the I1 early intermediate of photoactive yellow protein by FTIR spectroscopy. Nat Struct Biol 8:265–270
38. Rubinstenn G, Vuister GW, Mulder FA, Düx PE, Boelens R, Hellingwerf KJ, Kaptein R (1998) Structural and dynamic changes of photoactive yellow protein during its photocycle in solution. Nat Struct Biol 5:568–570
39. Takeshita K, Hirota N, Imamoto Y, Kataoka M, Tokunaga F, Terazima M (2000) Temperature dependent volume change of initial step of the photoreaction of photoactive yellow protein studied by the transient grating. J Am Chem Soc 122:8524–8528
40. Takeshita K, Hirota N, Imamoto Y, Kataoka M, Tokunaga F, Terazima M (2002) Thermodynamic and transport properties of intermediate states of photo-cyclic reaction of photoactive yellow protein. Biochemistry 41:3037–3048
41. Takeshita K, Imamoto Y, Kataoka M, Mihara K, Tokunaga F, Terazima M (2002) Structural change of site-directed mutants of PYP: new dynamics during pR state. Biophys J 83:1567–1577
42. van Brederode ME, Gensch T, Hoff WD, Hellingwerf KJ, Braslavsky SE (1995) Photoinduced volume change and energy storage associated with the early transformations of the photoactive yellow protein from *Ectothiorhodospira halophila*. Biophys J 68:1101–1109
43. Perman B, Srajer V, Ren Z, Teng T, Pradervand C, Ursby T, Bourgeois D, Schotte F, Wulff M, Kort R, Hellingwerf K, Moffat K (1998) Energy transduction on the nanosecond time scale: early structural events in a xanthopsin photocycle. Science 279:1946–1950
44. Xie A, Kelemen L, Hendriks J, White BJ, Hellingwerf KJ, Hoff WD (2001) Formation of a new buried charge drives a large-amplitude protein quake in photoreceptor activation. Biochemistry 40:1510–1517
45. Bernard C, Houben K, Derix NM, Marks D, van der Horst MA, Hellingwerf KJ, Boelens R, Kaptein R, van Nuland NAJ (2005) The solution structure of a transient photoreceptor intermediate: 25 photoactive yellow protein. Structure 13:953–962
46. Khan JS, Imamoto Y, Harigai M, Kataoka M, Terazima M (2006) Conformational changes of PYP monitored by diffusion coefficient: effect of N-terminal α-helices. Biophys J 90:3686–3693
47. Hoshihara Y, Imamoto Y, Kataoka M, Tokunaga F, Terazima M (2008) Conformational changes in the N-terminal region of photoactive yellow protein: a time-resolved diffusion study. Biophys J 94:2187–2193
48. Christie JM, Reymond P, Powell GK, Bernasconi P, Raibekas AA, Liscum E, Briggs WR (1998) Arabidopsis NPH1: a flavoprotein with the properties of a photoreceptor for phototropism. Science 282:1698–1701

49. Sakai T, Kagawa T, Kasahara M, Swartz TE, Christie JM, Briggs WR, Wada M, Okada K (2001) Arabidopsis nph1 and npl1: blue light receptors that mediate both phototropism and chloroplast relocation. Proc Natl Acad Sci U S A 98:6969–6974
50. Kagawa T, Sakai T, Suetsugu N, Oikawa K, Ishiguro S, Kato T, Tabata S, Okada K, Wada M (2001) Arabidopsis NPL1: a phototropin homolog controlling the chloroplast high-light avoidance response. Science 291:2138–2141
51. Jarillo JA, Gabrys H, Capel J, Alonso JM, Ecker JR, Cashmore AR (2001) Phototropin-related NPL1 controls chloroplast relocation induced by blue light. Nature 410:952–954
52. Kinoshita T, Doi M, Suetsugu N, Kagawa T, Wada M, Shimazaki K (2001) Phot1 and phot2 mediate blue light regulation of stomatal opening. Nature 414:656–660
53. Cheng P, He Q, Yang Y, Wang L, Liu Y (2003) Functional conservation of light, oxygen, or voltage domains in light sensing. Proc Natl Acad Sci U S A 100:5938–5943
54. Crosson S, Rajagopal S, Moffat K (2003) The LOV domain family: photoresponsive signaling modules coupled to diverse output domains. Biochemistry 42:2–10
55. Huala E, Oeller PW, Liscum E, Han IS, Larsen E, Briggs WR (1997) Arabidopsis NPH1: a protein kinase with a putative redox-sensing domain. Science 278:2120–2123
56. Swartz TE, Corchnoy SB, Christie JM, Lewis JW, Szundi I, Briggs WR, Bogomolni RA (2001) The photocycle of a flavin-binding domain of the blue light photoreceptor phototropin. J Biol Chem 276:36493–36500
57. Swartz TE, Wenzel PJ, Corchnoy SB, Briggs WR, Bogomolni RA (2002) Vibration spectroscopy reveals light-induced chromophore and protein structural changes in the LOV2 domain of the plant blue-light receptor phototropin 1. Biochemistry 41:7183–7189
58. Crosson S, Moffat K (2001) Structure of a flavin-binding plant photoreceptor domain: insights into light-mediated signal transduction. Proc Natl Acad Sci U S A 98:2995–3000
59. Crosson S, Moffat K (2002) Photoexcited structure of a plant photoreceptor domain reveals a light-driven molecular switch. Plant Cell 14:1067–1075
60. Fedorov R, Schlichting I, Hartmann E, Domratcheva T, Fuhrmann M, Hegemann P (2003) Crystal structures and molecular mechanism of a light-induced signaling switch: the Phot-LOV1 domain from *Chlamydomonas reinhardtii*. Biophys J 84:2474–2482
61. Eitoku T, Nakasone Y, Matsuoka D, Tokutomi S, Terazima M (2005) Conformational dynamics of phototropin 2 LOV2 domain with the linker upon photoexcitation. J Am Chem Soc 127:13238–13244
62. Nakasone Y, Eitoku T, Matsuoka D, Tokutomi S, Terazima M (2007) Dynamics of conformational changes of Arabidopsis phototropin 1 LOV2 with the linker domain. J Mol Biol 367:432–442
63. Kasahara M, Swartz TE, Olney MA, Onodera A, Mochizuki N, Fukuzawa H, Asamizu E, Tabata S, Kanegae H, Takano M, Christie JM, Nagatani A, Briggs WR (2002) Photochemical properties of the flavin mononucleotide-binding domains of the phototropins from Arabidopsis, rice, and *Chlamydomonas reinhardtii*. Plant Physiol 129:762–773
64. Harper SM, Neil LC, Gardner KH (2003) Structural basis of a phototropin light switch. Science 301:1541–1544
65. Eitoku T, Nakasone Y, Zikihara K, Matsuoka D, Tokutomi S, Terazima M (2007) Photochemical intermediates of *Arabidopsis* phototropin 2 LOV domains associated with conformational changes. J Mol Biol 371:1290–1303
66. Schuttrigkeit TA, Kompa CK, Salomon M, Rudiger W, Michel-Beyerle ME (2003) Primary photophysics of the FMN binding LOV2 domain of the plant blue light receptor phototropin of *Avena sativa*. Chem Phys 294:501–508

Chapter 2
Pressure Perturbation: A Prime Tool to Study Conformational Substates and Volume Fluctuations of Biomolecular Assemblies

Shobhna Kapoor and Roland Winter

Abstract Fluctuations within biomolecules dictate a plethora of biological processes and are of great importance in functional studies in molecular biophysics. From shaping the free energy landscape of biomolecules themselves to that of biomolecular interactions, they are also implicated in a number of debilitating pathological diseases, thus generating exigent issues that require in-depth investigation. In this regard, pressure perturbation serves as an important tool to mechanistically explore the causes and effects of fluctuations in biomolecules and biomolecular assemblies. Here, we review the underlying principal action of pressure on biomolecules with emphasis on lipid membranes, proteins, amyloids, and membrane-associated complexes along with some highlighted experiments. We first discuss how pressure affects the structure, phase behavior, and dynamics of lipid membranes of varying complexity. We then review the promising role of this tool to study high-energy conformational and functional substates in proteins and elaborate on the use of pressure modulation to understand protein aggregation and fibrillation phenomena. Finally, we present some recent results using pressure perturbation to explore membrane-associated biomolecular assemblies and uncover membrane-mediated conformational substates of proteins, furnishing unprecedented information on proteo-lipid interactions.

Keywords Pressure perturbation calorimetry • Lipid membranes • Amyloid • Protein aggregation • Compressibility • Volume fluctuations • Densimetry • Ultrasound velocimetry • Volume changes • Lipid phase transitions • Clapeyron equation • Bending rigidity • Protein folding • Stability diagram • Signaling processes

---

S. Kapoor • R. Winter (✉)
Department of Chemistry and Chemical Biology, Physical Chemistry I, Biophysical Chemistry, TU Dortmund University, Otto-Hahn-Strasse 4a, Dortmund 44227, Germany
e-mail: roland.winter@tu-dortmund.de

M. Terazima et al. (eds.), *Molecular Science of Fluctuations Toward Biological Functions*, DOI 10.1007/978-4-431-55840-8_2

2.1 Introduction

Fluctuations within biomolecules are instrumental in orchestrating proper physiological function of these macromolecules. Therefore, an adequate description of their molecular dynamics represents essential clues in an attempt to derive function from structure. Conformational fluctuations are an inherent property of all molecules, and their existence in proteins were already mentioned by [1] when they described the bewildering dynamic nature of proteins based on their hydrogen-deuterium exchange experiments [1]. Several years later, the work of Perutz and Mathews [2] and Weber [3] further bolstered the view of proteins as strongly fluctuating entities. Today, there is a general consensus that biomolecules, and proteins in particular, exist in a multitude of conformational states of—nearly isoenergetic—conformations (referred to as conformational substates, CS [4, 5]). CS, for the most part, have their own hierarchy and are endowed with conformational fluctuations [6]. In other words, the multidimensional conformational free energy surface of a protein in a given state manifests itself as a multitude of energy valleys or hills separated by small energy barriers (Fig. 2.1). The distribution of these states is dynamic and depends on many factors, like protein and ligand

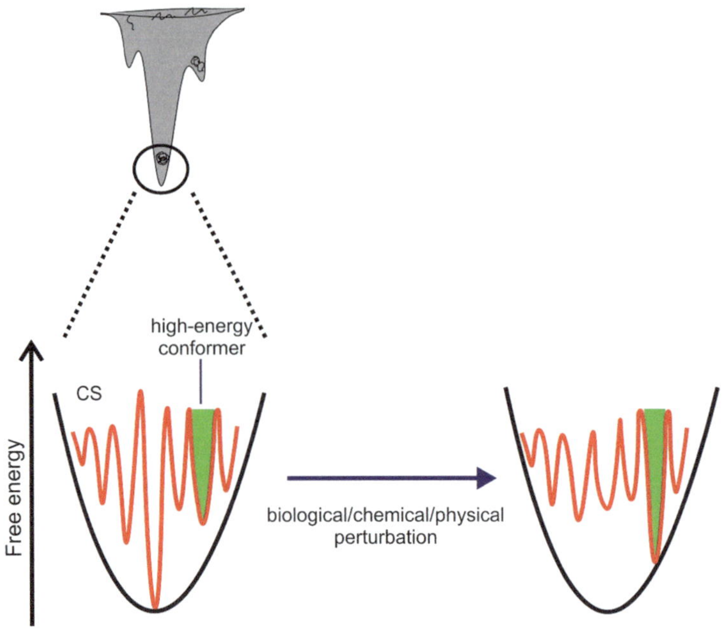

Fig. 2.1 Conformational substates (CS) and population-shift model. Schematic illustration of the free energy landscape (minima corresponds to different CS) within the native state of the protein. A population shift can be induced by different perturbation methods, whereby the protein's conformational equilibria are shifted to higher-energy (excited state) conformations. As a result, the fractional population of such states increases, thereby facilitating their experimental characterization

concentration, the protein binding state, pH, ionic strength, posttranslational modifications (phosphorylation, glycosylation, acetylation, or ubiquitination), etc. [7–9]. The distribution of these conformational ensembles is modulated to enhance functionality. One of the first experimental evidences for the existence of CS in proteins came from the laboratory of Hans Frauenfelder while interpreting fast conformational motions of proteins in terms of the *energy landscape* [6]. Based on the O_2 and CO rebinding to myoglobin, followed by flash photolysis, X-ray diffraction, and Mössbauer spectroscopy, it was concluded that the existence (up to four tiers) of CS in a given state of myoglobin generates a high level of complexity in the statistical ensemble of conformations, thus giving rise to a complex protein kinetic behavior [10, 11]. In parallel, Ansari et al. [4] articulated the use of infrared and other spectroscopies to probe internal protein motions that correspond to the interconversion of protein conformations as they move within a given protein state or visit one state after another, referred to as equilibrium fluctuations and functionally important motions, respectively. The interconnection between the conformational fluctuations and protein function came with a huge impact on the 55-year-old "induced-fit hypothesis" of Koshland [12] and led to the birth of the *conformational selection* model [6, 13]. This alternative model, implicit, for example, in the renowned Monod-Wyman-Chageux (MW) theory of allostery [14], envisions no single conformation in the native protein state but rather conformational heterogeneity. Dynamics implicates conformational ensembles, usually with low energy barriers between the substates that can be overcome, for example, by allosteric events, eventually leading to the redistribution of the relative populations of substates—now coined as the *conformational selection and population-shift mechanism.*

Bearing a similar hierarchical dynamic nature to proteins, lipid membranes are also not passive homogeneous interfaces surrounding cells but rather complex heterogeneous systems characterized by a wide variety of different lipids, whose single-particle and collective conformational dynamic fluctuations strongly influence membrane-associated biological processes. For instance, volume and area fluctuations govern the fluidity and lateral pressure profile of membranes, which in turn affect the lateral mobility of lipids and proteins within the membrane plane, ion transport through the membrane, diffusion of small molecules (like ubiquinone) in the membrane interior, and cellular uptake and release [15, 16]. Likewise, membrane thickness fluctuations (with typical amplitudes in the order of several Å) are postulated to provide a putative pore-forming mechanism [17]. A reduction in volume fluctuations (or free volume) is correlated with enhanced conformational ordering of the acyl chains of lipids, tighter packing of lipid membranes, and reduced area fluctuations. All of these are further related to the elastic properties of the lipid membranes [18], and these mechanical properties modulate the conformation and activity of membrane-bound proteins, peptides, receptors, and channels [19]. Deep-sea organisms (piezo- or barophilic) inculcate mechanisms to adapt their lipid membranes to maintain optimal membrane structural, mechanical, and dynamical properties [20, 21]. Furthermore, these organisms have proverbial means via which they alter the composition of their membrane lipids to maintain the physicochemical characteristics of the lipid matrix over incredibly wide temperature and pressure ranges [22]—the so-called homeoviscous adaptation effect.

Owing to the strong coupling between pressure, compressibility, and volume fluctuations, pressure perturbation offers key insights while probing fluctuations in biomolecular systems including proteins, lipid membranes, and membrane-associated biomolecular assemblies and will be discussed in subsequent sections in the following together with some highlighted experiments.

2.2 Elucidation of Conformational Substates and Volume Fluctuations in Biomolecules Using Pressure Perturbation Approaches

The functional relevance of high-energy, nonnative protein conformations is currently being extensively explored [23–25], and there exist many striking examples of their involvement in processes ranging from protein regulation [26], catalysis [27], and molecular recognition to signaling ([23] and refs. therein). To study the conformational subspace of proteins, identification and characterization of these high-energy conformations is indispensable. However, this task is generally limited due to the low fractional populations of the conformers at ambient conditions and hence difficult to detect by spectroscopic or other experimental means. A number of perturbation approaches (biological, chemical, and physical) are available to shift the population equilibrium, thereby being able to characterize these rare conformational substates (Fig. 2.1).

Biological perturbation entails the binding of a particular conformation of a protein to another biomolecular counterpart or ligand—thereby getting stabilized by it—thus shifting the entire equilibrium in favor of that conformation [28, 29]. This reinforces both the binding and the subsequent function with a high degree of efficiency. For example, the enzyme aspartate transcarbamoylase is suggested to exist in two conformations fully characterized by X-ray diffraction: a fully active [R] and a less active [T] state. Only recently, using NMR (nuclear magnetic resonance) spectroscopy, both these states were found to not only coexist in solution but also exhibit a population redistribution upon addition of allosteric ligands [30]. This and other studies are fully in accord with the notion that changes in conformer population that forms a generic mechanism for high-fidelity biological functions.

Chemical perturbations, on the other hand, encompass chemical denaturants (e.g., urea, GuHCl), pH, ionic strength, etc., and physical perturbation includes the thermodynamic intrinsic variable temperature and pressure. When the protein conformations or substates only differ by small energy differences, separation by purely energetic perturbation (e.g., a temperature change) is often difficult to achieve. Under such circumstances, pressure, a further fundamental physical parameter, provides an elegant and efficient means to redistribute the population via volume differences [31–35]. Thus, by favoring states, for example, of proteins, with smaller partial molar volume, pressure redistributes the equilibrium toward a system with smaller overall volume. This effect of pressure on conformation equilibria and hence conformational selection in proteins is further underscored by the fact that proteins in solution not only fluctuate in energy but also in *volume*, involving folded, unfolded, and partially unfolded conformers. The idea of using pressure perturbation to explore CS in proteins is not new [36]; however, its unmatched potential has received increased attention in recent years only, which is largely due to the coupling of this perturbation method with multidimensional NMR spectroscopy to acquire atomistic details of the lower-volume (high-energy) conformers stably trapped under pressure [37].

The quantitative description of the effect of pressure on any chemical equilibrium and reaction rate (be it protein unfolding, association reactions of protein-protein, protein-ligand, or protein-membrane) was first introduced by Planck in 1887 and is given by the equations

$$(\partial \ln K/\partial p)_T = -\Delta V/(RT) \tag{2.1}$$

$$(\partial \ln k/\partial p)_T = -\Delta V^{\#}/(RT) \tag{2.2}$$

where K is the pressure-dependent equilibrium constant, k is the rate constant of the reaction, and ΔV and $\Delta V^{\#}$ are the reaction and activation volumes, respectively. Any reaction that is accompanied by a negative $\Delta V^{\#}$, i.e., if the transition state has a smaller volume than the reactants, will be accelerated under pressure and vice versa. High-pressure mechanistic delineation based on reaction and activation volumes provides valuable information about the existence of CS and mechanistic information about the structure of the transition state of the reaction.

For an equilibrium between two (sub)states of a system, the pressure-dependent equilibrium constant $K(p)$ relative to that at atmospheric pressure (1 bar) is given to second order on pressure by

$$\ln\frac{K(p)}{K(1)} = -\frac{\Delta\overline{V}}{RT}(p-1) + \frac{\Delta\overline{\kappa}_T}{2RT}(p-1)^2 \tag{2.3}$$

where $K(p) = c_2(p)/c_1(p)$ and $K(1)$ are the equilibrium constants at pressure p and 1 bar, respectively; c_i are the concentrations of conformers i, $\Delta\overline{V} = \overline{V}_2 - \overline{V}_1$, and $\Delta\overline{\kappa}_T$ are the differences in partial molar volume and partial molar isothermal compressibility ($\overline{\kappa}_T = -\left(\partial\overline{V}/\partial p\right)_T$) of the two states at ambient conditions, respectively. For example, assuming no differences in compressibility between the two conformers, a volume difference of $-30\ \text{cm}^3\ \text{mol}^{-1}$ leads to a population shift of about an order of magnitude upon applying a pressure ramp of 2 kbar at ambient temperature. For the pressure unit, either bar or MPa is used ($1\ \text{bar} = 10^5\ \text{Pa} = 0.1\ \text{MPa} = 0.9869\ \text{atm}$).

2.2.1 Determination of Fluctuation Parameters

The partial molar volume of biomolecules along with its temperature and pressure derivatives and enthalpy and volume fluctuations are important thermodynamic factors that determine biomolecular stability, dynamics, and reactivity. These parameters can, for example, be monitored via a combination of methods including differential scanning (DSC) and pressure perturbation (PPC) calorimetry, densimetry, and ultrasound velocimetry [33, 38–40].

Molecular fluctuations of proteins and that of the surrounding solvent molecules have been a topic of particular interest in recent years. Protein fluctuations permit conformational motions, such as side-chain flips and backbone motions, thereby facilitating transitions between various substates. As the volume, enthalpy, and electric dipole moment fluctuations in the hydration shell and of the exposed amino acid residues are coupled to the surrounding solvent, the term "slaving" has been coined [41], i.e., the solvent emerges as an active participant in protein dynamics and folding.

The fluctuations of a system are given by the relevant susceptibilities, the specific heat capacity at constant pressure, c_p, and the isothermal compressibility coefficient, β_T:

$$\left\langle (\delta H)^2 \right\rangle = k_{\mathrm{B}} T^2 m\, c_p,\ \left\langle (\delta V)^2 \right\rangle = k_{\mathrm{B}} T V \beta_T \tag{2.4}$$

Here, k_{B} is the Boltzmann constant, and V and m are the volume and mass of the system, respectively. Combined enthalpy and volume fluctuations are related to the coefficient of thermal expansion, α, given by Eq. 2.5 [42]:

$$\langle (\delta H)(\delta V) \rangle = k_{\mathrm{B}} T^2 V \alpha \tag{2.5}$$

Related to the packing, void volume, and solvational properties of proteins, volume fluctuations are a yardstick of their flexibility and hence directly linked to pressure perturbation. From density measurements, the partial specific volume $\bar{v}$ (or partial molar volume $\overline{V} = \bar{v}M$) of the protein can be calculated using

$$\bar{v} = \frac{1}{c} - \frac{\rho - \rho_0}{\rho_0 c}, \tag{2.6}$$

where ρ and ρ_0 are the densities of the solution and solvent, respectively; c is the specific concentration of the protein.

The most accurate method of determining the partial molar compressibility of a solute is based on the Newton-Laplace equation [39], which generally relates the coefficient of adiabatic compressibility of a medium,

$$\beta_S = -(1(V)(\partial V/\partial p)_S = (1/V)\kappa_S, \tag{2.7}$$

with its density, ρ, and sound velocity, U:

$$U^2 = (\beta_S \rho)^{-1}. \tag{2.8}$$

For (infinitely) dilute (e.g., protein) solutions, the partial molar adiabatic compressibility is

$$\overline{K}_S^{\mathrm{o}} = \beta_{S0}\left(2\overline{V}^{\mathrm{o}} - 2[U] - M/\rho_0\right), \tag{2.9}$$

where $\overline{V} = (\partial V/\partial n)_{T,p}$ is the partial molar volume of the solute, C is the molar concentration, and M is the molar mass; β_{S0} is the coefficient of thermal expansion of the solvent; and $[U] = (U - U_0)/(U_0\ C)$ is the relative molar sound velocity (increment) of the solute, where U and U_0 are the sound velocities of the solute and solvent, respectively.

Experimental data on the isothermal compressibility of proteins are very scarce due to the technical hurdle to measure the partial specific volume as a function of pressure using densimetric techniques with high precision [43]. However, the partial molar isothermal compressibility of the solute can be obtained from the adiabatic value by [38]

$$\overline{K}_T^{\mathrm{o}} = \overline{K}_S^{\mathrm{o}} + \frac{T\alpha_0^2}{\rho_0\, c_{p,0}}\left(\frac{2\overline{E}^{\mathrm{o}}}{\alpha_0} - \frac{\overline{C}_p^{\mathrm{o}}}{\rho_0\, c_{p,0}}\right), \tag{2.10}$$

where $c_{p,0}$ is the specific heat capacity at constant pressure of the solvent, α_0 is the coefficient of thermal expansion of the solvent, and $\overline{E}^{\mathrm{o}} = \left(\partial\overline{V}^{\mathrm{o}}/\partial T\right)_p$ is the partial molar expansibility of the solute, which can be determined from PPC measurements [44]. $\overline{C}_p^{\mathrm{o}}$ is the partial molar heat capacity of the solute. The specific heat capacity of the protein at constant pressure can be obtained from

$$c_p = \Delta C_p/m + \overline{v}^{\mathrm{o}} c_{p,0}/\overline{v}_0^{\mathrm{o}}, \tag{2.11}$$

where ΔC_p is the heat capacity difference between the sample solution and solvent reference cell as obtained by a DSC calorimeter and $\overline{v}^{\mathrm{o}}$ and $\overline{v}_0^{\mathrm{o}}$ are the partial specific volumes of the solute and solvent, respectively, and $\overline{C}_p^{\mathrm{o}} \approx c_p M$ at low solute concentrations.

The partial molar volume of a solute, e.g., a protein, may be represented as the sum of the following contributions [38, 45]:

$$\overline{V}^{\mathrm{o}} = V_{\mathrm{intr}} + V_{\mathrm{therm}} + \Delta V_{\mathrm{hydr}}, \tag{2.12}$$

where V_{intr} is the intrinsic volume of the solute which originates from the van der

Waals volume of the constituent atoms plus the volume of intrinsic voids within the water-inaccessible protein interior and V_{therm} is the thermal volume that results from mutual vibrations and reorientations of solute and solvent molecules and steric and structural effects reflecting imperfect packing of solute and solvent molecules and a hydrational term (or "interaction volume") ΔV_{hydr}, which is the volume change (mostly a reduction) due to solute-water interactions, resulting from solute-solvent interactions around charged (electrostriction), polar (hydrogen bonding), and nonpolar (hydrophobic hydration) atomic groups on the protein surface.

The existing literature so far has indicated that also the compressibility of a protein can be divided into at least two components of opposite sign [38, 39]. Because of compressible cavities and voids in the protein interior, a first component is the protein's partial intrinsic compressibility coefficient, $\overline{\beta}_{S,\text{intr}}$, which is positive. The second term, $\overline{\beta}_{S,\text{hydr}}$, is the contribution to the compressibility due to hydration. As the partial compressibility of single amino acids and small peptides in solution is negative, $\overline{\beta}_{S,\text{hydr}}$ is negative. Typically, small or negative values of $\overline{\beta}_S$ are obtained for the acid-, base-, or guanidinium hydrochloride-induced unfolding processes of proteins.

Calculations of the volume fluctuations of a system by using the statistical mechanical expression

$$\left\langle \delta V^2 \right\rangle = k_{\text{B}} T V \beta_T, \tag{2.13}$$

implies that V is the actual intrinsic volume of the protein, which represents the geometric volume of its solvent-inaccessible interior, and β_T is the intrinsic coefficient of isothermal compressibility of the protein interior. Hence, Eq. 2.13 cannot be directly used for partial molar volume or compressibility data, as these still contain hydration contributions. The mean-square fluctuations of the intrinsic volume represents an effective measure of protein dynamics.

Whereas a large body of data exists on protein-folding events, only limited data have been reported on the compressibility changes and volume fluctuations of proteins upon ligand binding [45–47]. Ligand binding should manifest itself in volumetric properties, i.e., the partial molar volume and compressibility, via changes in internal atomic packing (cavities) and surface hydration of the protein since the constitutive atomic volume of the protein may be assumed to be incompressible. Generally, the cavity contribution contributes positively and the hydration contribution negatively to $\overline{\beta}_S$. $\overline{\beta}_S$ has been found to increase with increasing partial molar volume, $\overline{V}$, of the protein, and $\overline{\beta}_S$ and $\overline{V}$ have been found to be sensitively influenced by ligand binding, indicating that the conformational flexibility of the intermediates plays an important role in enzyme function [46]. A further detailed molecular-level understanding of these volume and compressibility changes must await high-pressure NMR analyses of the enzyme-ligand complexes in solution.

2.3 Conformational Fluctuations, Substates, and Phase Transitions of Biomolecular Systems Revealed by Pressure Perturbation

2.3.1 *Lipid Membranes*

Biological membranes represent one of the most pressure-sensitive cellular components [48]. Lipid membrane fluctuations, both induced directly (e.g., via thermally activated lipid conformational changes or collective excitations) and indirectly (via coupling upon interaction with other biomolecules), have been shown to modulate many membrane-associated cellular processes. Hence, to yield a deeper understanding of the structure, energetics, phase behavior, and dynamics of membranes, pressure-dependent studies—being able to fine-tune volume fluctuations—serve as a valuable tool. This section deals with the effects of pressure on lipid model membranes with varying degrees of complexity up to the level of biological membranes.

2.3.1.1 Effects of Pressure on the Structure, Phase Behavior, and Dynamics of Lipid Membranes

The matrix of biological membranes is composed of lipids, which consist of a hydrophobic and a hydrophilic portion. The propensity of the hydrophobic moieties to self-associate (which is entropically driven by water, the so-called *hydrophobic effect*), coupled with the tendency of the hydrophilic moieties to interact with the aqueous environment, forms the physical basis for the spontaneous formation of a variety of lyotropic liquid-crystalline lipid structures [49]. These can form as either type I (oil in water) or type II (water in oil) structures, indicated by subscript I or II, respectively, and include fluid lamellar phases (L_α), two-dimensional hexagonal phases (H_I, H_{II}), bicontinuous cubic phases (Q_I, Q_{II}), and micellar cubic phases. The topological structure a particular lipid system adopts depends on a number of factors, including the chemical structure of the lipid, the lipid chain length and degree of unsaturation, the cross-sectional headgroup area, and the surface charge density, all of which being affected by pressure. A qualitative understanding of the different structures can be gained by the dimensionless packing parameter, P, rationalized as $V/(A\ l)$, where V is the molecular volume, l is the molecular length, and A is the molecular area at the hydrocarbon-water interface [48, 50, 51]. Whereas values of $P = 1$ represent the formation of bilayers, i.e., lamellar structures (e.g., the fluid L_α-phase), $P > 1$ and $P < 1$ stand for molecules that are wedge shaped and designate phases as non-lamellar (e.g., inverse hexagonal (H_{II}) or cubic (Q_{II}) phases, respectively). While the lamellar phases are most biologically relevant, non-lamellar phases also participate in transient events such as pore formation, membrane fusion, or fission [52]. Membranes in eukaryotic cells have a wide repertoire of lipids including glycerophospholipids, sphingolipids, and nonpolar lipid such as sterols [53]. In addition, many possible variations in the molecule's hydrophilic headgroup and hydrophobic tails add a high level of complexity to the existing vast pool of lipid phases. Most of the phases are characterized by a different spatial arrangement and motional freedom of each lipid and are heavily influenced by their molecular structure and environmental conditions such as pH, ionic strength, hydration level, incorporated membrane proteins, temperature, and pressure [54–60]. A number of experimental methods are used to study lipid phase transitions and associated thermodynamic changes, such as calorimetry to determine the heat capacity of the system, Fourier transform infrared (FTIR) spectroscopy to monitor the conformational order of acyl chains via their internal vibrational modes, NMR spectroscopy, Laurdan fluorescence spectroscopy being sensitive to the polarity of the headgroup environment, and X-ray or neutron diffraction providing topological information about the structure adopted by the assembly of lipid molecules.

As a function of temperature, hydrated lipid bilayers generally undergo phase transitions including the main chain-melting transition. For example, in excess water, one-component saturated phospholipids generally exhibit two lamellar transitions: a gel-to-gel pre-transition ($L_{\beta'}$-$P_{\beta'}$) between ordered lipid phases of different packing and surface structure and a gel-to-liquid-crystalline main transition ($P_{\beta'}$-L_{α}). The main transition—occurring at a lipid-specific temperature T_m—is accompanied with drastic changes in the order of the system (both translational, related to the lateral diffusion coefficient in the membrane plane, and conformational, related to the *trans*/*gauche* ratio in the acyl chains) [61]. The fluidlike L_{α} (or l_d) phase is characterized by a high conformational disorder in the acyl chains of lipid molecules, where the linear *all-trans* chains (extended and ordered) present below the T_m undergo rotation and assume diverse gauche conformations to some extent. On the other hand, lipids in the L_{β} phases are arranged on a 2-D triangular lattice in the membrane phase, which is also denoted as the solid-ordered phase (s_o).

Significant thermodynamic fluctuations are known to exist near the gel-to-fluid chain-melting phase transition of pure lipid bilayers. They occur over a narrow but finite temperature range, in which the heat capacity at constant pressure displays a rather sharp maximum, and other physical membrane properties, such as the specific volume, reveal rather sharp changes, indicating a first-order transition. Obviously, the buildup of fluctuations would lead to a critical point if not preceded by the first-order transition. Therefore, the gel-to-fluid membrane transition is usually denoted as weak first order. High enthalpy fluctuations lead to high heat capacity, high volume fluctuations lead to a high volume compressibility, and high area fluctuations lead to a high area compressibility in this temperature range. In turn, area fluctuations lead to fluctuations in curvature and bending elasticity. Biological lipid membranes can also melt. Typically, such melting transitions are found about 10 °C below body or growth temperatures. Hence, it seems that biological membranes adapt their lipid compositions such that the temperature distance to the melting transition is maintained [61]. A certain extent of fluctuations may be required for optimal physiological function.

In addition to these thermotropic phase transitions, membrane lipids also conform to pressure-induced transitions, whereby they adapt to the volume restrictions by modulating their conformation and packing [60, 62, 63]. Application of pressure causes ordering of lipid acyl chains and a corresponding reduction in the hydrocarbon chain motion, resulting in a reduced cross-sectional area of the lipid hydrocarbon tail region. As the lipid headgroup is comparably little compressible, pressure only causes a decrease in the chain's splay, thus increasing the spontaneous curvature of the lipid monolayer toward the chain region [64]. Ionization of the headgroup is principally pressure sensitive due to the volume reduction accompanying electrostriction but is less pronounced due to the energy penalty required to expand and hydrate the lipid interfacial area [65]. For the flat lipid bilayers, the pressure-induced compression is thus anisotropic: lateral shrinking of the bilayer is accompanied with an increase in the membrane thickness due to the straightening of the acyl chains (e.g., in fluid 1,2-dimyristoyl-*sn*-glycero-3-phosphocholine (DMPC) bilayers using ^{2}H-NMR, an increase in the lipid thickness of around 1.3 Å $kbar^{-1}$ was obtained).

On the other hand, for inverse hexagonal and bicontinuous cubic phases, the change in the thickness may be strikingly amplified by as much as 80 Å $kbar^{-1}$ [66, 67]. This is mainly due to the reduction in the magnitude of the negative curvature following the decrease in the chain volume upon pressurization.

Generally, an increase in the lateral thickness leads to decreased area and volume fluctuations within the lipid phase and is hence accompanied with an overall reduction in the partial lipid volume. This is demonstrated by the following case: for DMPC bilayers, densimetric measurements revealed a decrease in the partial molar volume, V_L, with pressure of about $-30\ cm^3\ mol^{-1}$ during the phase transition from a liquid-crystalline Lα to the pressure-induced gel state [68]. In addition, the coefficient of isothermal compressibility (related to volume fluctuations, Eq. 2.13) of the liquid-crystalline phase is substantially higher compared to that in gel state (typically, $\beta_T(P_{\beta'}) \approx 5 \cdot 10^{-1}\ bar^{-1}$ and $\beta_T(L_\alpha) \approx 13 \cdot 10^{-5}\ bar^{-1}$) [58].

Via the pressure-induced decrease in the lipid cross-sectional area and increased lateral thickness, owing to the reduction of thermally induced volume fluctuations, a number of phase transitions can be induced by pressure as well. By scanning the hydrostatic pressure and temperature space, a full pressure-temperature (p-T) phase diagrams can be obtained and has been produced for a variety of lipid systems (e.g., see [69–72]). The phase transitions of pure one-component lipid systems (including saturated phosphatidylcholines, mono-*cis*-unsaturated phosphatidylcholines, phosphatidylserines, and phosphatidylethanolamines) exhibit fluid-to-gel transition slopes (dT_m/dp) between 20 °C and 30 °C $kbar^{-1}$ (Fig. 2.2a) [59, 73]. For single-component di-*cis*-unsaturated lipid systems (such as 1,2-dioleoyl-*sn*-glycero-3-phosphocholine (DOPC) and 1,2-dioleoyl-*sn*-glycero-3-ethanolamine (DOPE)), the slopes are smaller and can be attributed to the kinks imposed by the *cis*-double bond in the linear conformation of the lipid acyl chains. This creates significant free volume fluctuations in the bilayer and hence a lower phase transition temperature. On the other hand, for the egg-yolk sphingomyelin (EYSM), the fluid-gel phase

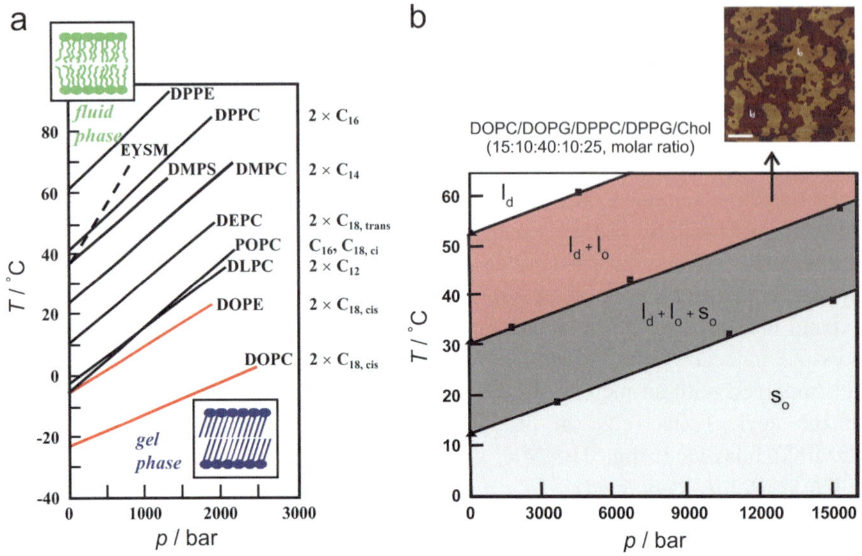

Fig. 2.2 *p-T* phase diagrams for different phospholipid bilayer systems. (**a**) The fluid-gel transition lines of one-component phospholipid bilayers where the fluidlike (L_α) phase is observed in the low-pressure, high-temperature region, and the ordered gel phases appear at low temperatures and high pressures. The length and degree of unsaturation of the acyl chains of the different lipids are denoted on the *right-hand side*. (**b**) *p-T* phase diagram of the five component raft-like lipid mixture DOPC/DOPG/DPPC/DPPG/Chol (15:10:40:10:25, molar ratio). The l_d-l_o two-phase coexistence region extends over a wide temperature and pressure range and is visualized using atomic force microscopy (AFM), shown on the *top* of the figure

boundary has a much higher slope of 40 °C kbar^{-1} (dashed line in Fig. 2.2a [74]). Sphingomyelins (SM) are a group of phospholipids based on a sphingosine backbone with an additional hydrocarbon chain attached via an amide linkage. SM-rich ordered raft-like lipid domains have been suggested as "hot spots" promoting or enabling transmembrane proteins to cluster so as to initiate cell signaling events.

The pressure dependence of the chain-melting transition temperature, T_m, can be approximately quantified using the Claperyon equation (Eq. 2.14) (valid only at the phase boundary where the free energy change for the transition is zero):

$$\frac{dT_m}{dp} = \frac{\Delta V_m}{\Delta S_m} = \frac{T_m \Delta V_m}{\Delta H_m}, \tag{2.14}$$

where ΔS_m, ΔH_m, and ΔV_m are the molar transition entropy, enthalpy, and volume changes, respectively. A positive slope can be interpreted in terms of a positive enthalpy and volume change for the gel-to-fluid transition, as observed experimentally [68]. For example, the main transition enthalpy change for 1,2-dipalmitoyl-*sn*-glycero-3-phosphocholine (DPPC) at ambient pressure is 36 kJ mol^{-1} and decreases by about 3.4 kJ mol^{-1} for every 1 kbar. The partial molar volume decreases linearly with pressure from 22.9 cm^3 mol^{-1} at 1 bar to 13 cm^3 mol^{-1} at 2 kbar [75].

As the level of complexity increases within the lipid system, the phase behavior changes according to Gibbs' phase rule. The p-T phase diagrams of binary mixtures of saturated phospholipids display a lamellar gel phase at low temperatures, a lamellar fluid phase at higher temperatures, and an intermediate fluid-gel coexistence region. The width of the coexistence region strongly depends on the type of lipids and the molar ratio of the two components. However, with increasing pressure, all systems show a ubiquitous shift of the gel-to-fluid transition to high temperatures with a transition slope on the order or 20 °C kbar^{-1}, similar to that obtained for the gel-to-fluid transition of the pure lipid components [63, 73].

High-pressure SAXS (small-angle X-ray scattering), in combination with FTIR spectroscopy, has provided valuable information about the liquid ordered-liquid disordered (l_o-l_d) lamellar phase coexistence in more complex sterol (e.g., cholesterol (Chol))-containing lipid mixtures (e.g., DPPC-Chol, DPPC:DOPC:Chol, and DPPC:DOPC:DPPG:DOPG:Chol). Sterols are known to fluidize gel-like phases and order the fluidlike phase, leading to the formation of liquid-ordered phases (l_o). Characterization of the p-T phase space of such diverse lipid mixtures has demonstrated that pressure can induce l_o-l_d phase separation starting from a fully fluid l_d phase and then induces formation of ordered gel-like structures at higher pressures [59, 69–71] (Fig. 2.2b). Usually, the overall ordered (l_o and s_o) state is generally reached within 1–3 kbar at ambient temperatures, the pressure range where membrane-protein function generally ceases for natural scenarios. In accordance with the notion that increasing lipid complexity generates a number of coexisting phases or domains with minor differences in volumetric properties and due to the possibility of lipid sorting upon changes in temperature and pressure in complex lipid mixtures, the phase transitions are rather broad and diffuse (Fig. 2.2b). This is manifested as a smaller slope of the ordered-fluid phase boundary; for example, for a five component raft-like heterogeneous lipid mixture comprising of the lipid mixture DPPC:DOPC:DPPG:DOPG:Chol, dT_m/dp of 3 °C kbar^{-1} was obtained (Fig. 2.2b), accompanied by rather smooth changes in volumetric and enthalpic properties gathered from DSC and PPC data [70].

Interestingly, high pressure can induce additional ordered phases such as partially interdigitated high-pressure gel phases ($L_{\beta i}$), which considerably differ in the tilt angle of the acyl chains and the level of hydration in the headgroup region. Such phases have been observed for phospholipid bilayers with longer acyl chains, i.e., for $C \geq 16$ [60].

Pressure also affects the dynamic properties of lipid bilayers, e.g., lateral diffusion, a process controlling, for example, lipid and membrane-protein transport and membrane-associated signaling. Jonas [76] showed that the lateral self-diffusion coefficient, D_{lat}, of DPPC in the L_α-phase decreases by 30 % from 1 bar to 300 bar at 50 °C and by 70 % upon passing from the fluid L_α to the gel phase. Notably, the rotational dynamics of the lipids in the membrane is only marginally affected by pressure, but the phospholipid flip-flop and intervesicle transfer are significantly slowed down by high pressures [55, 77].

Pressure has also a marked influence on the micromechanics of membranes, eventually involving topological changes in the lipid vesicles, which in turn comprises of a close interplay between the curvature elastic energy (g_{curv} for a lipid monolayer, Eq. 2.15), bending rigidity, and line tension [18]:

$$g_{\text{curv}} = 2\kappa_{\text{m}}\left\langle (H - H_0)^2 \right\rangle + \kappa_{\text{G}}\langle K \rangle, \tag{2.15}$$

where $H = (C_1 + C_2)/2$ and $K = C_1 C_2$ are the mean and Gaussian curvatures, respectively, C_1 and C_2 are the principal curvatures at a given point on the surface, H_0 is the spontaneous mean curvature, and κ_{m} and κ_{G} are the mean and Gaussian curvature moduli. The mean curvature and Gaussian modulus represent the energetic cost to bend an initially flat monolayer and to change the Gaussian curvature, respectively. The expression for the lipid bilayer Gaussian modulus, $\kappa_{\text{G}}^{\text{b}}$, is more complex [78, 79]:

$$\kappa_{\text{G}}^{\text{b}} = 2\left(\kappa_{\text{G}} - 4\kappa_{\text{m}}^{\text{b}} H_o l\right) \tag{2.16}$$

where l is the monolayer thickness, and the moduli $\kappa_{\text{m}}^{\text{b}}$ and $\kappa_{\text{G}}^{\text{b}}$ refer to the energetic costs of mean and Gaussian curvature deformations. Since pressure has been shown to increase the monolayer spontaneous curvature and the bending modulus [67, 80], it might also increase the bending modulus of a bilayer, as pressure tends to thicken the bilayer [81]. Lipids that form inverse structures (such as monoolein (MO), monoelaidin (ME), or DOPE) have a negative spontaneous curvature, so pressure is likely to decrease the magnitude of this negative curvature. The observation that pressure stabilizes bicontinuous cubic lipid phases [80] having negative Gaussian curvature advocates that pressure also increases the bilayer Gaussian modulus, $\kappa_{\text{G}}^{\text{b}}$ (thereby reducing the bilayer curvature elastic energy). Pressure-induced changes in the membrane micromechanics have been extensively studied both experimentally and theoretically as a function of temperature only [82], and a wide variety of equilibrium vesicle shapes have been visualized using optical microscopy. Observations of these phenomena under pressure have been quite limited, and along those lines Nicolini et al. [83], using two-photon fluorescence microscopy coupled with high pressure, visualized shape change in vesicles of different lipid compositions. They revealed that dependent on the direction of the pressure change, the pressure-jump amplitude, and the composition of the lipid vesicle, pressure-induced budding and fission may take place, even in the low-pressure regime.

It is tempting to speculate that the pressure effects on such lipid systems may have links to the physiological observations that pressure upregulates some of the genes responsible for sterol synthesis in the yeast *Saccharomyces cerevisiae*, with an aim to maintain membrane fluidity [22], a striking example of homeoviscous adaptation. Another remarkable example demonstrates that with increasing pressures, the ratio of unsaturated to saturated fatty acids increases in the barophilic deep-sea bacterium CNPT3, i.e., the ratio increases from 1.9 at 1 bar to 3.0 at 700 bar at 2 °C [84]. Direct evaluation of this phenomenon was ascertained by a pivotal study using deuterium NMR on POPC and PLPC, which showed that upon increasing the unsaturation level in the lipids, the chain-ordering effect of pressure on the lipid bilayers was mitigated [85]. Incorporation of di-*cis*-bonds within the acyl chain regions lower the fluid-to-gel transition temperature T_m, and as a result the ordering effect of high pressure is reduced, enabling the organism to maintain its membrane in an overall fluidlike state even at high pressures. The experiments highlighted above suggest that biological membranes adapt their physical state in response to changes in pressure by modulating their fatty acid synthesis pattern to incorporate appropriate lipids varying in their chain length, headgroup structure, and level of unsaturation and regulating the incorporation of sterols, with the underlying aim to prevent the T_m of the membrane from rising at alarming rates under increased pressure.

The functional adaptability of biological membranes to their local environment is also well illustrated by the study of Zhai et al. [86] working on archaeal tetraether lipid membranes. These are bipolar tetraether liposomes composed of the polar lipid fraction E (PLFE), isolated from the thermoacidophilic archaeon *Sulfolobus acidocaldarius*. Using pressure perturbation calorimetry in combination with other thermodynamic techniques and molecular acoustics, they revealed substantial differences in the isothermal and adiabatic compressibilities and relative volume and enthalpy fluctuations of PLFE membranes compared with those of commonly studied phospholipids, embarking their high thermal/chemical stability, high rigidity, and tightly packed lipid conformations. To quote an example, relative volume fluctuations in the PLFE liposome were only 0.2 %, whereas in the phospholipid DPPC, they are in the order of 3 %. Similarly, the enthalpy change associated with the gel-to-fluid phase transition is a factor of six lower in PLFE membranes. Since volume fluctuations, in particular, are closely interrelated to membrane function, this study sheds light on the mechanism by which membrane fractions from *Sulfolobus acidocaldarius*—and possibly other extremophiles—survive harsh environmental conditions to maintain the functional state of their cellular membrane.

2.3.2 *Proteins in Bulk Solution*

The era of protein X-ray crystallography had long imposed a static view of protein structures. However, with proteins becoming accessible to NMR and other spectroscopic techniques, a dynamic interpretation of biomolecular conformations is redefining the structural-functional facet of proteins and other biomolecules. In this regard, the pioneering work of Akasaka and coworkers have elegantly demonstrated that pressure perturbation coupled with NMR spectroscopy can be effectively used to modulate intrinsic fluctuations in protein by acting on the volumes of the protein conformers, thus providing novel information on the structure, thermodynamic properties, and conformational dynamics of proteins at atomic resolution [37]. Though the idea of using pressure perturbation to study conformational fluctuations in proteins dates back to a much earlier time, this field is gaining burgeoning attention due to the recent development of new pressure-resistant cells coupled to commercial NMR spectrometers [87–91]. The fundamental principle of this method deals with the analysis of the pressure-dependent ^{1}H and ^{15}N chemical shifts: linear shifts implying general compression within the folded ensembles (elastic effects) and nonlinear shifts pointing to the existence of low-lying excited states close to the folded state (conformational effects) [92].

Conformational fluctuations within the folded structure have been explored for several proteins such as melittin [93], gurmarin [94], basic pancreatic trypsin inhibitor (BPTI) [95], lysozyme [96], β-lactoglobulin [97], and apomyoglobin [98] and have provided direct evidence of the *folded macrostate in proteins to consist of microstates with different volumes*. The majority of the effects in this category arise from fluctuations in the hydrophobic core, protein backbone, torsion angles, hydrogen bond lengths, and helix orientations as well as the loss (to some extent) of void volumes [99–101]. On the other hand, a pressure-induced population of folding intermediates has been observed in various globular proteins, such as the lysozyme [96], the Ras-binding domain of RalGDS [102], the prion protein (PrP) [103] and ubiquitin [104], and the fractional populations of the intermediates varying significantly from protein to protein. The fact that these low-lying excited conformers are populated under high pressure implies that these states have a lower partial molar volume, mainly governed by hydration of water accessible to cavities within the proteins. For example, in support of this view, Kalbitzer et al. [105] have elucidated that the high-pressure-induced conformer of the signaling protein Ras in solution is accompanied with a volume change of $-17\ cm^3\ mol^{-1}$, indirectly connoting an enlarged/hydrated protein surface of this conformer. Furthermore, they unambiguously ascertained the biological function of this rare state by taking advantage of a biological perturbation (shifting the conformation equilibrium of Ras to the high-pressure-induced state by adding a potent Ras interaction partner). Likewise, in another pioneering study by Kitahara et al. [104], a series of high-energy conformers of the protein ubiquitin were characterized to reveal a unique locally disordered conformer (with a ΔV of transition of $-24\ cm^3\ mol^{-1}$). This state is postulated to be the functional protein state due to its facile interaction with other proteins. Surprisingly, the structural characterization of this pressure-stabilized conformation revealed an intimate resemblance to a kinetically trapped folding intermediate—which now appears to be a generic phenomenon for other proteins—thereby highlighting the use of pressure perturbation to gain a deeper understanding of protein-folding reactions. The same holds true for pressure-jump kinetic studies furnishing now the information on volumetric aspects of transition states of proteins (e.g., for un-/refolding or conformational interconversions) [106–108].

Proteins undergo structural fluctuations that span a wide range of time scales. Among these motions are fast backbone fluctuations on the ps-ns time scale and slower conformational fluctuations on the ms and longer time scale. Molecular flexibility on these time scales plays a central role in protein function. For example, in recognition-binding sequences, dynamic disorder on the ns-microsecond time scale may increase the efficiency of protein-protein interactions via a "fly-casting" mechanism [32]. In a recent NMR relaxation study, probing fast (sub-ns) internal motions on the protein ubiquitin at kbar pressures, it was revealed that the more compact, less dynamic, and higher-energy state populated at high pressure may be more similar to the static compact structures emphasized by low-temperature crystallographic structures of proteins. Furthermore, the dynamic response to applied pressure has been shown to be heterogeneous but spatially clustered, demonstrating localized and differential coupling of motion even within such small proteins such as ubiquitin [24].

These and further studies strongly fortify the use of pressure as a mild perturbation agent to probe conformational fluctuations. Conversely, temperature changes or chemical denaturants might alter the internal energies and chemical interactions of the protein states, eventually rendering the analysis murkier.

2.3.3 Protein Aggregation and Amyloid Formation

The ability of even the most intricate molecular structure to self-assemble with high precision and fidelity is the defining characteristics of a living system. The folding of proteins into their compact 3-D structures is the prime example of biological self-assembly. How and whether a protein folds in vivo is primarily governed by its amino acid sequence and the surrounding cellular environment [109]. Proteins fold either in the cytoplasm or inside the secretory pathway and are often assisted by chaperones, enabling the proteins to have several opportunities to fold. If folding is unsuccessful, the protein is directed to proteasomes for degradation, an intracellular quality-control machinery [110]. As only the correctly folded proteins have a long-term stability in crowed biological milieu along with the ability to selectively interact with other natural partners, the failure of protein to fold adequately, or to remain folded correctly, is the origin of a wide range of highly debilitating pathological conditions [111]. Several diseases are caused by a loss of protein function due to misfolding, such as in insulin, islet amyloid polypeptide (IAPP), β2-microglobulin, Aβ, and α-lactalbumin [112–117]. Many of these disorders are familial as the probability of misfolding is enhanced in mutational variants. Other cases encompass the following scenario: proteins with a high propensity to misfold escape the protective quality-control mechanism and thereby form intractable aggregates (amyloids) that injure and kill cells. An increasing number of disorders, including the horrifying maladies of the elderly, are associated with the deposition of such aggregates in various tissue types [118–121].

Virtually, all amyloids (in humans and animals) consist of fibrillar submicroscopic structures, i.e., bundles of long, unbranched, and often twisted rigid fibrils ranging in width from 60 to 130 Å and in length from 1000 to 16.000 Å [122], and a characteristic "cross-β" X-ray fiber diffraction pattern. The latter revealed that amyloids are ordered secondary structure motifs including a core cross-β-sheet structure, in which continuous β-sheets are formed with β-strands running perpendicular to the fibril axis [123]. In addition, it was suggested that the generic amyloid conformation—the cross-β-structure—may be a universal, free energy minimum for aggregated proteins, remarkably also found in other globular proteins that are currently not associated with protein-folding disease, e.g., muscle myoglobin [113]. The amyloid structure can be globally defined as extracellular deposit of stable, ordered, and filamentous protein aggregates, though intracellular structures sharing the same core structure have also been observed (e.g., for α-synuclein in Lewy bodies in Parkinson's disease [124]). An increasing number of diseases are linked to the deposition of amyloid fibrils and broadly fall within two kinds of amyloidosis. In each of these pathological states, a specific protein or protein fragment—ranging from intact globular proteins to largely unstructured peptide molecules—changes gradually from a highly soluble form into insoluble fibrils [125]. In the category comprising systemic amyloidosis, secreted circulating proteins produce amyloid deposits in a variety of tissues causing cell dysfunction, primarily due to a huge buildup of these deposits leading to physical compression and vascular compromise of adjacent tissues (e.g., in hemodialysis-related amyloidosis caused by β2-microglobulin). On the other side, in localized amyloidoses, such as caused by the Aβ-peptide and IAPP, the amyloid deposits are confined in a specific organ or tissue, and the protein responsible for the fibril formation is synthesized by the target cells. The localized amyloidosis is generally related to age-related diseases, such as Alzheimer's disease and diabetes mellitus type II; their amyloid deposition occurs in the neurons and β-cells, respectively. Though the amyloid structure is known to be toxic, there is a considerable discussion as to its role in the disease [113, 126]. In this regard, it has been suggested that the prefibrillar aggregates are much more toxic than the amyloid itself [127, 128]. Even if this is the case, fibrils are still likely to play an important role either as a reservoir or sink of toxic oligomers [129].

Although the structure of an amyloid fibril has not yet been atomistically determined casting shadow on the molecular mechanisms involved in their formation, a number of studies have demonstrated that stable or metastable alternative protein conformations are linked to the disease-causing misfolding transitions [125, 130]. Thus, structural characterization of these alternative conformations is currently of utmost importance and brings high-pressure perturbation approaches to the forefront. In addition, to probe the amyloid fibrils' stability, energetics, packing, and intermolecular interactions—along with pressure—temperature and cosolvent perturbations have also been applied [131]. It is now well established that high pressure can not only destabilize proteins but also dissociate the native or nonnative oligomers. Pressure effects, in general, are governed by the principle that pressure shifts the system toward the state with smallest volume. Protein structures containing a higher fraction of void volume are expected to be most pressure sensitive. In addition, pressure leads to weakening of the hydrophobic interactions, and cavities at the interface of protein oligomers represent critical nodes of pressure action, i.e., pressure-induced hydration of these regions would favor dissociation or unfolding. Moreover, volume reduction arising from electrostriction would also favor disruption of (formally favorable inter- or intramolecular) electrostatic interactions.

The work pursued by Smeller and coworkers on horse metmyoglobin ignited the interest in using pressure to study misfolding and aggregation phenomena [132]. They showed that high pressure could not only trigger the aggregate formation but also prevent or dissociate aggregates. The interesting observation that aggregation-promoting effects of pressure were sometimes observed after depressurization suggested that pressure treatment could induce formation of alternative conformations of the protein with a higher propensity of aggregation due to a relatively pronounced exposure of hydrophobic parts of the protein. Since one of the first steps in amyloid formation involves partial unfolding of the protein, the pressure-unfolded state may represent such a partially unfolded state and be the prerequisite for nonnative association and amyloid formation [133]. Moreover, stable intermediates can be trapped by carefully tuning high pressure, temperature, and cosolvent conditions, linking conformational transitions to changes in the degree of hydration and enthalpic and volumetric fluctuations.

Generally, high pressure leads to dissociation of aggregated structures, eventually leading to formation of the monomeric natively folded structures. An example represents the study by Tachibana et al. [134] who worked with amyloid fibrils of denatured lysozyme, and similar results were obtained for the tail spike protein of bacteriophage P22, myoglobin, and TTR [135–137]. Surprisingly, for TTR, after prolonged incubation of the fibrils, no sensitivity to pressure could be observed, suggesting that early aggregates underwent significant reorganization. Likewise, Grudzielanek et al. [138] showed that insulin—a well-studied model protein for aggregation studies [139, 140]—only in its early stages of aggregation involving oligomeric species dissociated upon pressurization, with no effect seen on mature insulin fibrils up to the 2 kbar pressure regime. Only at higher pressures, some degree of dissociation can be observed as well [141].

A coherent picture of a "volumetric catastrophe" accompanying the amyloidogenesis and embarking differential pressure sensitivity to the various structures formed is depicted in Fig. 2.3 for the case of insulin [142]. It encompasses the volume collapse, the reduction of area of the protein-water interface, and the simultaneous decrease of the compressibility of the protein. The apparent coefficients of thermal expansion, α, and isothermal compressibility, β_T, of insulin measured as a function of time at 60 °C (Fig. 2.3), i.e., under conditions when aggregation and fibrillation take place with time, reveal significant changes upon formation of fibrillar topologies by the orderly stacking insulin molecules: β_T decreases substantially over the first 100 min of aggregation, indicating that the volume fluctuations, which are highest at the early stage of the aggregation reaction, are drastically damped when the aggregation and fibrillation process proceeds. Upon subsequent aggregation and formation of mature fibrillar aggregates, as

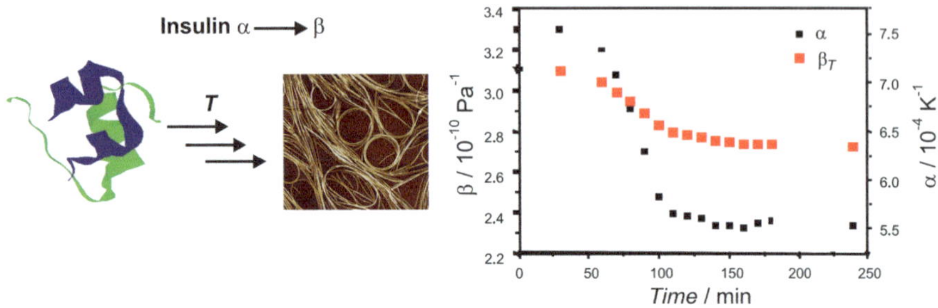

Fig. 2.3 Thermodynamic changes accompanying amyloidogenesis of insulin. The apparent coefficient of thermal expansion (α) and isothermal compressibility (β_T) of insulin at pH 1.9 and 60 °C is shown as a function of time during the progression of aggregation and fibril formation

indicated by the CD and FTIR spectroscopic data acquired under corresponding sample conditions [139], significant dehydration and compaction occur. The combined contribution of enthalpy and volume fluctuations, $\langle(\delta H)(\delta V)\rangle$, as revealed by α, decreases overall by about 30 % during the aggregation and fibrillation process, the reduction in volume fluctuations being ~7 %. The tight packing of β-sheets and depletion of internal cavities in the late fibrillar aggregates lead to a compaction and hence a ~3 % reduction of the final value of the partial specific protein volume. The decrease in solvent-accessible surface area is also reflected in the negative C_p change of -2 kJ mol − 1 K − 1 as determined by DSC. These results are thus in good agreement with the low-pressure sensitivity of the final fibrillar structure.

Taken together, the susceptibility of protein aggregates to pressure chiefly depends on the degree of the structural order of an aggregate. Whereas the fresh, amorphous aggregates are more sensitive to pressure and prone to refolding to the native state [138], in the mature fibrils, the vigor of pressure-induced dissociation depends on the particular mode of polypeptide backbone and side-chain packing that reduces the remaining void volumes. This enables the prospects of high pressure to not only differentiate between various stages of the amyloid formation but also to obtain reliable thermodynamic data, such as chemical potential and partial molar volume changes, at the early stages of the transformation—the formation of the amorphous β-pleated sheet structure. This is only possible due to the *reversibility of the process using high-pressure perturbation methodology*.

In addition, pressure-dependent studies in combination with solvation tuning were shown to be able to unravel structural and kinetic aspects of the insulin aggregation and fibrillation pathways. As an example, in the pre-aggregated regime, pressure fosters the dissociation of native insulin oligomers mainly due to a negative reaction volume of −70 mL mol − 1. Moreover, the population of monomeric but aggregation-prone species is fostered by addition of alcohols [143]. The aggregation pathways at elevated temperatures (e.g., above 60 °C) are differentially affected by pressure and in a cosolvent-dependent manner. Moderate pressures accelerate the fibrillation pathway in the presence of ethanol because both nucleus formation and elongation proceed via relatively dehydrated transition states with a negative activation volume. Alternatively, a novel and fast equilibrium pathway to distinct β-sheet-rich oligomers is accessible to partially unfolded insulin monomers at pressures up to ~200 bar in the absence of ethanol. Destabilized by moderate pressures and ethanol addition, the oligomers can reenter the slower, ultimately irreversible fibrillation pathway at higher temperatures, which can be accelerated in the absence of diffusion control. At higher pressures, above ~1000 bar, the unfavorable partial unfolding of insulin monomers, accompanied by a volumetric expansion, dominates the aggregation kinetics, which manifests in a progressive inhibition of the fibrillation. Finally, all of these studies decoded that oligomers are stabilized mainly via electrostatic and hydrophobic interactions rendering them susceptible to pressure, whereas the presence of extensive hydrogen bonding (which itself is rather pressure insensitive) and optimized side-chain packing is primarily responsible for the pressure insensitivity of mature fibrils. A similar time-dependent shift in the pressure sensitivity has also been observed for other amyloid-like aggregate of prion protein [144], IAPP [145], α-synuclein [136], and poly-L-lysine [146].

One of the most exotic applications of high pressure in protein aggregation has been the idea of destroying disease-associated fibrillar aggregates. For example, the study by Zhou et al. [147] established that moderately high pressure of 3 kbar can denature the model yeast prion protein Ure2 only in the presence of sub-denaturing doses of GdmCl; the amyloidal nature of the scrapie-type prion protein (PrP^{Sc}) with its unmatched resistance to high temperature, proteases, and many chemical denaturants has posed a significant challenge to medical and biotechnological protocols. Likewise, fresh aggregates of hamster [148] and murine [149] prion precursor proteins have been shown to be either dissociated under pressure, or the pressure treatment proved to arrest and stabilize the aggregation-prone intermediate state [103]. By contrast, work by Torrent et al. [150] showed that high pressure may actually induce amyloid formation in recombinant non-amyloidogenic PrP^{C}, which has much of the physicochemical resemblance to pathogenic PrP^{Sc} protein. Following along the same lines, the same authors have revealed that the inactivation of infectious prion seems to be dependent on the protein conformation, and this finding has been interpreted as certain β-sheet conformations being pressure sensitive (native prions) and others having more pressure stable structures (like aggregates of prions being less hydrated and with smaller cavities [151]).

To summarize this part, Fig. 2.4 shows a schematic energy landscape for protein folding and aggregation. Starting from the region of the high free energy surface, a multitude of unfolded conformations exists, "funneling" on a rugged energy surface or "landscape" toward the natively folded state via rapid intramolecular contact formation. Fluctuations within the unfolded and partially unfolded protein ensembles drive the folding reaction allowing the protein to sample different native and nonnative contacts. Folding intermediates may also be populated en route to the native state. Furthermore, via a generally slow nucleation process and subsequent autocatalytic aggregation reaction of partially folded structures (as those encountered under particular cosolvent, pH, temperature and pressure conditions, or by mutations), formation of aggregates and amyloid fibrils via intermolecular contacts

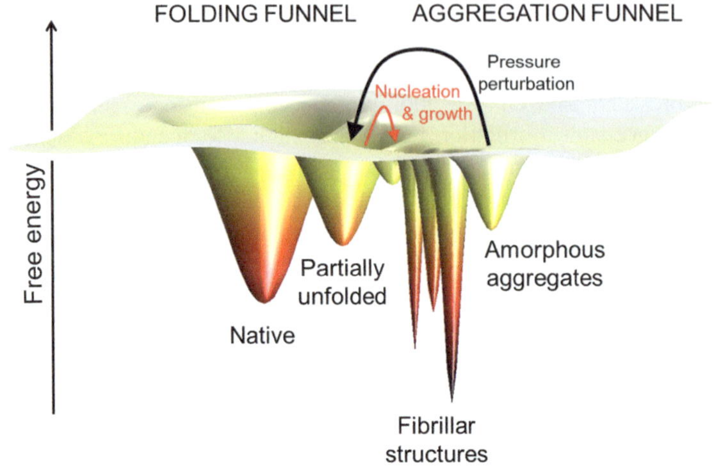

Fig. 2.4 Free energy landscape for protein folding and aggregation/amyloid formation. Under normal physiological conditions, the protein forms a well-folded 3-D native structure at its global free energy minimum. However, under amyloidogenic propensity, the protein molecules may acquire lower free energies beyond this level in the aggregation funnel (which will decrease with increasing concentration). Four kinds of energy wells are depicted: representing the native state, partially unfolded state, various fibrillar states (strains), and amorphous aggregates. Generally, pressure perturbation enables dissociation of aggregates, thus arresting aggregation

may occur, and the system enters the "aggregation funnel." Within the aggregation funnel, loosely packed oligomeric or amorphous aggregates as well as—generally several—amyloid fibrillar states (strains) with different packing properties, pressure sensitivities, and deep free energy minima may form, revealing conformational polymorphism or a multitude of substates. The cooperative character of the fibril assembly creates enormous free energy barriers for any interstrain transitions, which in the end renders the energy landscape in the aggregation funnel somewhat comblike shaped [143]. Finally, the isolation of intermediates achieved by pressure perturbation and the elucidation of particular fibrillar or prefibrillar morphologies may provide valuable targets for the development of inhibitors (drug targeting).

2.3.4 Membrane-Associated Biomolecular Assemblies

Interactions within membrane-associated assemblies are mainly stabilized by non-covalent forces such as hydrophobic and electrostatic, and the alteration of these weak bonds by pressure has the potential to procure subtle and novel insights into the mechanism of these stabilization forces and subsequently understanding their significance in biomolecular function. Specifically, membrane fluidity—among the most pressure-sensitive cellular properties—modulates many membrane-related functions such as permeability, ion transport, lateral diffusion of lipids and membrane-bound proteins, and signal transduction processes. Mechanical forces are important modulators of cellular processes, and research along those lines has revealed transmembrane signaling processes as important loci of pressure perturbation as well [152, 153]. These forces or stimuli act on top of a dynamic background of various internally generated forces, e.g., arising from cytoskeleton polymerization and molecular motors that govern the selective response of the cell. Over the last 30 years, the molecular nature of many mechanosensitive biomolecules has been identified, including mechanically gated ion channels [154], mechanosensitive receptors [155], enzymes [156], G-proteins [157], and cytoskeletons. For example, in a recent study by Petrov et al. [158], it has been shown that high pressure in the range 1–900 bar increased the channel opening probability by favoring the open state of the bacterial mechanosensitive ion channel MscL of *E. coli*, and high-pressure perturbation has been claimed to be a valuable approach toward better understanding of the gating mechanism in complex ion channels such as MscL.

A priori, pressure influences membrane-associated processes are most that accompanied by large volume changes. When the reaction volume is positive or negative, high hydrostatic pressure will stunt or foster the reaction, respectively. The contributors to negative volume changes, naming a few, are release of the void volumes (due to imperfect packing of two macromolecular surfaces), electrostriction, and lipid membrane condensation. Volume changes accompanying protein-protein associations are exhaustively studied, but those accompanying protein-membrane interactions are scarce. Membrane properties, such as surface charge, hydration, curvature, packing density, and lateral organization, present the most prominent features influencing volume changes upon protein-membrane binding.

In a detailed study from our lab, the influence of hydrostatic pressure on the membrane association, dissociation, and intervesicle transfer process of fully lipidated GDP-bound N-Ras HD/Far (hexadecyl [a C16 saturated acyl chain]/farnesyl [a C15 unsaturated acyl chain]) was explored using a FRET-based assay to obtain the kinetic parameters and volumetric properties along the reaction path of these processes [159]. Ras proteins are plasma membrane-localized small GTPases that apically control the signaling pathways regulating cell proliferation and differentiation. They function by shuttling between inactive GDP-bound and active GTP-bound forms and contain lipid-modified motifs or anchors that enable these proteins to associate with distinct membrane domains. Furthermore, oncogenic Ras is the main culprit of driving ~30 % of all human cancers. We could demonstrate that both the association of N-Ras to fluid lipid membranes and dissociation from the same were accelerated under pressure by a factor of 2 and 3.5, respectively. The pressure dependence of the rate constants (Eq. 2.2) leads to activation volume changes of $\approx -7.2\ \text{cm}^3\ \text{mol}^{-1}$ and $-9.7\ \text{cm}^3\ \text{mol}^{-1}$ for the association and the dissociation reaction, respectively. The negative activation volume of the association process disclosed a compact transition state with reduced overall volume, attributed to the higher acyl chain ordering in the membrane, invariably associated with a reduction in volume. For comparison, the pressure-induced increase in the packing density of a lipid bilayer (during a fluid-to-gel phase transition) is accompanied by a volume reduction of about $-30\ cm^3\ mol^{-1}$. A more compact lipid membrane is suggestive of increased van der Waals interactions between the lipids' acyl chains and the protein's lipid anchor at high pressure, coupled with the decrease of the conformational space in the Ras anchor region upon membrane insertion. This leads to an overall volume reduction and formation of a compact proteo-lipid transition state, favoring partitioning of Ras into membranes. By combining the activation volumes of both steps and simplistically assuming the N-Ras-DOPC interaction as a one-step process, a volume profile for this interaction could be obtained (Fig. 2.5), delineating the volume changes occurring along the N-Ras membrane interaction pathway including the tentative structure of the transition and final states of the proteo-lipid assembly. Furthermore, notable differences in activation volumes for different membrane systems were observed, suggesting that volume fluctuations in lipid membranes of varying composition strongly influence the membrane partitioning kinetic events of N-Ras and other lipidated proteins. As an example, the domain boundaries in phase-segregated heterogeneous (or raft-like) membranes are banded with high area and volume fluctuations; hence, a larger reduction in the free volume upon pressurization can be attained, leading to relatively faster association of Ras to these particular membranes [159]. Taken together, these studies using high-pressure modulation of

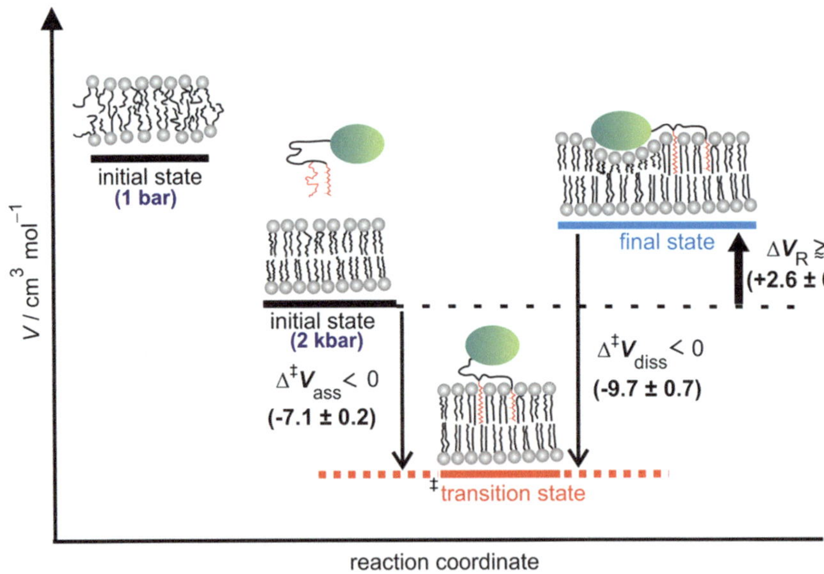

Fig. 2.5 Volume profile for the N-Ras membrane interaction process, showing relative changes in the volume of the protein-membrane system during the interaction/insertion process. Tentative structures of the initial, transition, and final proteo-lipid states are rationalized based on volumetric information obtained via high-pressure perturbation of the association and dissociation process (Reproduced with permission from Kapoor et al. [159])

Ras-membrane interactions uncovered lipidated proteins as potential mechanosensors, transducing mechanical signals to biological functions by modulating their membrane partitioning events.

Another study exploring the influence of lipid membrane's properties on the high-pressure sensitivity of lipid-protein interactions focused on multidrug resistance protein LmrA by Teichert et al. [160]. The authors demonstrated that the high-pressure-induced fluid-to-gel phase transition in lipid bilayers of various compositions had an inhibitory effect on the LmrA protein function, due to formation of an unacceptable ordered phase of the membrane. On the contrary, lipid membranes that maintained a high degree of fluidity even under pressure (e.g., for DOPC) prevented a complete loss of protein function. High order is now an accepted membrane feature induced under pressure that is not optimal for physiological activity of membrane proteins. Moreover, high-pressure-induced dissociation of the LmrA homodimer also renders the protein inactive. Pressure-induced dissociation of oligomeric protein, both membrane bound or in solution, seems to be a generic phenomenon of the pressure-related inactivation of a number of proteins [145, 161, 162]. An important mechanistic corollary gained from these high-pressure studies is that physical properties of the membrane matrix, in particular volume and area fluctuations, hydrophobic mismatch, van der Waals attractive forces, lipid chain entropy, and packing efficiency between the lipid acyl chains against the protein's hydrophobic counterpart, are critical and decisive factors in understanding protein-membrane interactions.

Pressure-dependent studies can not only improve our understanding of the mechanism or forces stabilizing membrane-protein interactions and related biological functions but also shed light on the intriguing role of lipid membranes in conformational selection and controlling the dynamics of membrane proteins. Till date, such studies are scarce, but sufficient published reports are encouraging further work in this expanding research field. An exciting example in this regard has been the Ras protein. The so-called conformational selection and population-shift mechanism is a key molecular mechanism in the regulation and functioning of Ras and other small GTPases. For instance, binding of guanine nucleotides to Ras contributes to the growing repertoire of interactions that underpins the nucleotide-induced conformational shift mechanism, predicted by simulations [163] and later on verified experimentally [164]. Distinct chemical species in the nucleotide-binding site—such as GTP, GDP, and GTP analogues—correlate with distinct global conformations. In addition, within the same nucleotide-bound state (e.g., GTP analogue), Ras sample multiple conformations among which two main conformers (substate 1 and 2) have been characterized so far [165]. The native conformer (substate 1) detectable at ambient conditions has been assigned to the effector binding state [29], whereas the nonnative high-energy conformer (substate 2) was found to interact with another class of regulating proteins in the signaling pathway, the guanine exchange factors (GEFS). The structural and functional relevance of this high-energy conformer was discovered applying high-pressure NMR spectroscopy [105].

Using high-pressure perturbation in combination with infrared and fluorescence spectroscopy, we could unravel the role of lipid membranes in inducing a marked population of a membrane-bound Ras conformational substate, here denoted as substate 3 [166]. This membrane-induced substate became clearly detectable at high pressure (up to 4 kbar) and was accompanied by a significant structural reorientation of the G-domain at the lipid interface, particularly in the GTP-bound Ras protein. At pressures higher than 4 kbar, transition from substate 3 to substate 1 was fostered, since that substate corresponds to the state with lowest free energy. In bulk solution, the volume change, ΔV, accompanying the conformational transition from substate 2 to substate 1 for Ras-GTP was found to be $-14.7\ \text{cm}^3\ \text{mol}^{-1}$, whereas in the membrane-bound form, the proteins displayed a ΔV of $-42\ \text{cm}^3\ \text{mol}^{-1}$ for the conformational transition from substate 2 to membrane-induced substate 3. A schematic representation of the pressure modulation of Ras conformational substates obtained from this study is depicted in Fig. 2.6. Efficient interaction with GEFs and effector proteins requires selection of Ras conformational substate 1 and 2, respectively, shifting the equilibrium toward these states.

This study indicates that upon membrane binding, conformational selection in Ras can additionally be modulated, preselecting substate 3. It may be speculated that this membrane-induced substate 3 may interact with particular membrane-associated interaction partners. Moreover, the same study also ascertained that temperature modulation could not reveal the high-energy conformational substates

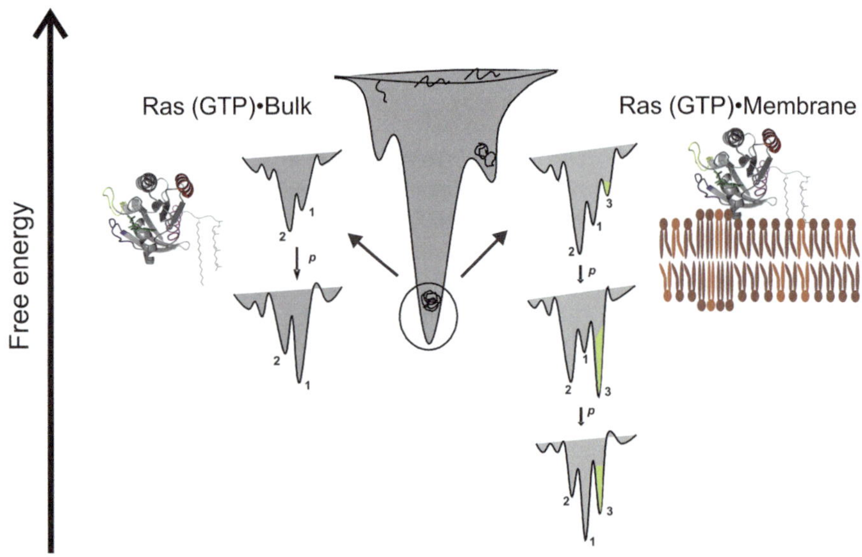

Fig. 2.6 Schematic representation of the pressure modulation of the conformational substates (CS) of the Ras protein (bound to a GTP analogue), both in bulk solution and in its membrane-bound state. The membrane-associated substate 3 is marked in *green* (Reproduced and modified with permission from [166, 167])

of Ras (both in solution and in the membrane-bound forms) and mainly resulted in irreversible unfolding and aggregation of the protein.

Another study dealt with the pressure-dependent enzymatic activity of Na^+, K^--ATPase reconstituted into the natural and model lipid membranes of diverse properties and composition [168]. The enzyme inserted into the natural plasma membrane system showed a continuous decrease in the activity with increasing pressures with an activation volume of 47 cm^3 mol^{-1}. The positive volume change was assigned to dramatic changes in the membrane fluidity under pressure, coupled with conformational changes in the protein which might proceed with a positive volume change and compensate the former effect, and finally dissociation of protein subunits and destruction of the proteo-lipid assembly at pressures exceeding 2 kbar [169]. On the other hand, embedded in model membrane systems (specifically DMPC and DOPC), the enzymatic activity first increased in the low-pressure regime, linked to the volume reduction in these lipid membranes leading to a negative activation volume and hence an increased activity. Only above 2 kbar, a decrease in the activity similar to that observed in natural membrane systems was obtained. Likewise, moderate high pressures lead to an increased ATPase activity for the H^+/ATP-driven rotary pump ($\Delta V^{\#} = -24$ cm^3 mol^{-1}), accredited to a pressure-induced enhancement in the enzymatic turnover number [170].

2.4 Outlook

Pressure acts on the structure and dynamics of biomolecular systems through changes in specific volume that are largely due to changes in hydration or packing efficiency. Thus, high hydrostatic pressure is uniquely well suited for studying the role of solvation in folding, dynamics, and interactions of proteins and other biomolecules. We have also seen that pressure is ideal for characterizing spontaneous fluctuations, because fluctuations involve a change in volume, and high-energy conformers that are often not easily accessible experimentally under ambient conditions can be stabilized by pressure. Moreover, the balance between hydrogen bonding and electrostatic and hydrophobic interactions can be tuned by pressure modulation.

Here, we focused essentially on a molecular-level-based description of pressure effects on solutions of biomolecules and the use of pressure modulation to reveal important mechanistic information on selected fundamental biomolecular processes and reactions. Grounded on accurate reference investigations of small biomolecules, water and cosolvents at high-pressure conditions, mapping of the conformational and functional substates, as well as intermolecular interactions of proteins and other biomolecular systems are feasible by pressure modulation. Pressure provides an efficient means of shifting the population distribution among fluctuating conformers of a protein and increasing the population of an otherwise marginally populated conformational substate (by a few orders of magnitude in favorable cases). This conformer may then be amenable to direct experimental observation and analysis.

A very promising but till now only sporadically explored field is the use of high hydrostatic pressure to modulate enzymatic conversions [171–175]. Given that the rate of an enzymatic reaction is often limited by the thermostability of the corresponding enzyme, it is conceivable that superimposing pressure-induced thermostabilization of the enzyme with an accelerated substrate conversion at increased temperatures could lead to an improved overall reaction rate, such as observed for amylases. For some enzymatic conversions, the catalytic activity can also directly be enhanced by high pressure, if the activation volume that is associated with the reaction is negative. In fact, based on the same thermodynamic principle, high pressure could also alter the substrate specificity of an enzyme by favoring the conversion of a substrate to a product with a smaller volume. A molecular-level understanding of pressure effects on enzymatic processes is almost completely lacking. Generally, the additional structural, thermodynamic, and dynamic information obtained by pressure-axis studies will increase our understanding of protein function, protein misfolding and disease, biological evolution, as well as adaptation, and they will help exploiting the benefits of pressure in various biotechnological applications.

Acknowledgments Financial support from the DFG Research Unit FOR 1979 is gratefully acknowledged.

References

1. Linderstrøm-Lang KU, Schellman JA (1959) Protein structure and enzyme activity. In: Boyer PD, Lardy H, Myrbäk K (eds) The enzymes vol 1, 2nd edn. Academic Press, New York, pp 443–510
2. Perutz MF, Mathews FS (1966) An X-ray study of azide methaemoglobin. J Mol Biol 21: 199–202
3. Weber G (1975) Energetics of ligand binding to proteins. Adv Protein Chem 29:1–83
4. Ansari A, Berendzen J, Bowne SF, Frauenfelder H, Iben IE, Sauke TB, Shyamsunder E, Young RD (1985) Protein states and proteinquakes. Proc Natl Acad Sci U S A 82:5000–5004
5. Frauenfelder H, Parak F, Young RD (1988) Conformational substates in proteins. Annu Rev Biophys Biophys Chem 17:451–479
6. Frauenfelder H, Sligar SG, Wolynes PG (1991) The energy landscapes and motions of proteins. Science 254:1598–1603
7. Ma BY, Kumar S, Tsai CJ, Nussinov R (1999) Folding funnels and binding mechanisms. Protein Eng 12:713–720
8. Tsai CJ, Kumar S, Ma BY, Nussinov R (1999) Folding funnels, binding funnels, and protein function. Protein Sci 8:1181–1190
9. Tsai CJ, Ma BY, Nussinov R (1999) Folding and binding cascades: shifts in energy landscapes. Proc Natl Acad Sci U S A 96:9970–9972
10. Austin RH, Beeson K, Eisenstein L, Frauenfelder H, Gunsalus IC (1975) Activation-energy spectrum in myoglobin—comment. Phys Rev Lett 34:845

11. Austin RH, Beeson KW, Eisenstein L, Frauenfelder H, Gunsalus IC (1975) Dynamics of ligand-binding to myoglobin. Biochemistry 14:5355–5373
12. Koshland DE (1958) Application of a theory of enzyme specificity to protein synthesis. Proc Natl Acad Sci U S A 44:98–104
13. Dill KA, Chan HS (1997) From levinthal to pathways to funnels. Nat Struct Biol 4:10–19
14. Monod J, Wyman J, Changeux JP (1965) On the nature of allosteric transitions: a plausible model. J Mol Biol 12:88–118
15. Marguet D, Lenne PF, Rigneault H, He HT (2006) Dynamics in the plasma membrane: how to combine fluidity and order. Embo J 25:3446–3457
16. Soderhall JA, Laaksonen A (2001) Molecular dynamics simulations of ubiquinone inside a lipid bilayer. J Phys Chem B 105:9308–9315
17. Woodka AC, Butler PD, Porcar L, Farago B, Nagao M (2012) Lipid bilayers and membrane dynamics: insight into thickness fluctuations. Phys Rev Lett 109:058102
18. Helfrich W (1973) Elastic properties of lipid bilayers—theory and possible experiments. Z Naturforsch C 28:693–703
19. Cantor RS (1997) Lateral pressures in cell membranes: a mechanism for modulation of protein function. J Phys Chem B 101:1723–1725
20. Bartlett DH (2002) Pressure effects on in vivo microbial processes. BBA-Protein Struct Mol Enzymol 1595:367–381
21. Meersmann F, Daniel I, Bartlett DH, Winter R, Hazael R, McMillan PF (2013) High-pressure biochemistry and biophysics. Rev Mineral Geochem 75:607–648
22. Fernandes PMB, Domitrovic T, Kao CM, Kurtenbach E (2004) Genomic expression pattern in Saccharomyces cerevisiae cells in response to high hydrostatic pressure. FEBS Lett 556: 153–160
23. Boehr DD, Nussinov R, Wright PE (2009) The role of dynamic conformational ensembles in biomolecular recognition. Nat Chem Biol 5:789–796
24. Fu Y, Kasinath V, Moorman VR, Nucci NV, Hilser VJ, Wand AJ (2012) Coupled motion in proteins revealed by pressure perturbation. J Am Chem Soc 134:8543–8550
25. Meinhold L, Smith JC, Kitao A, Zewail AH (2007) Picosecond fluctuating protein energy landscape mapped by pressure-temperature molecular dynamics simulation. Proc Natl Acad Sci U S A 104:17261–17265
26. Volkman BF, Lipson D, Wemmer DE, Kern D (2001) Two-state allosteric behavior in a single-domain signaling protein. Science 291:2429–2433
27. Li P, Martins IR, Amarasinghe GK, Rosen MK (2008) Internal dynamics control activation and activity of the autoinhibited Vav DH domain. Nat Struct Mol Biol 15:613–618
28. Gsponer J, Christodoulou J, Cavalli A, Bui JM, Richter B, Dobson CM, Vendruscolo M (2008) A coupled equilibrium shift mechanism in calmodulin-mediated signal transduction. Structure 16:736–746
29. Spoerner M, Nuehs A, Ganser P, Herrmann C, Wittinghofer A, Kalbitzer HR (2005) Conformational states of Ras complexed with the GTP analogue GppNHp or GppCH2p: implications for the interaction with effector proteins. Biochemistry 44:2225–2236
30. Velyvis A, Yang YR, Schachman HK, Kay LE (2007) A solution NMR study showing that active site ligands and nucleotides directly perturb the allosteric equilibrium in aspartate transcarbamoylase. Proc Natl Acad Sci U S A 104:8815–8820
31. Akasaka K (2006) Probing conformational fluctuation of proteins by pressure perturbation. Chem Rev 106:1814–1835
32. McCoy J, Hubbell WL (2011) High-pressure EPR reveals conformational equilibria and volumetric properties of spin-labeled proteins. Proc Natl Acad Sci U S A 108:1331–1336
33. Ravindra R, Winter R (2003) On the temperature–pressure free-energy landscape of proteins. ChemPhysChem 4:359–365
34. Silva JL, Foguel D, DaPoian AT, Prevelige PE (1996) The use of hydrostatic pressure as a tool to study viruses and other macromolecular assemblages. Curr Opin Struct Biol 6: 166–175

35. Silva JL, Foguel D, Royer CA (2001) Pressure provides new insights into protein folding, dynamics and structure. Trends Biochem Sci 26:612–618
36. Frauenfelder H, Alberding NA, Ansari A, Braunstein D, Cowen BR, Hong MK, Iben IET, Johnson JB, Luck S, Marden MC, Mourant JR, Ormos P, Reinisch L, Scholl R, Schulte A, Shyamsunder E, Sorensen LB, Steinbach PJ, Xie AH, Young RD, Yue KT (1990) Proteins and pressure. J Phys Chem 94:1024–1037
37. Akasaka K (2010) High pressure NMR study of proteins—seeking roots for function, evolution, disease and food applications. High Press Res 30:453–457
38. Chalikian TV (2003) Volumetric properties of proteins. Ann Rev Biophys Biomol Struct 32: 207–235
39. Gekko K, Hasegawa Y (1986) Compressibility structure relationship of globular proteins. Biochemistry 25:6563–6571
40. Gekko K, Hasegawa Y (1989) Effect of temperature on the compressibility of native globular-proteins. J Phys Chem 93:426–429
41. Fenimore PW, Frauenfelder H, McMahon BH, Parak FG (2002) Slaving: solvent fluctuations dominate protein dynamics and functions. Proc Natl Acad Sci U S A 99:16047–16051
42. Cooper A (1984) Protein fluctuations and the thermodynamic uncertainty principle. Prog Biophys Mol Biol 44:181–214
43. Seemann H, Winter R, Royer CA (2001) Volume, expansivity and isothermal compressibility changes associated with temperature and pressure unfolding of staphylococcal nuclease. J Mol Biol 307:1091–1102
44. Ravindra R, Winter R (2004) Pressure perturbation calorimetry: a new technique provides surprising results on the effects of co-solvents on protein solvation and unfolding behaviour. ChemPhysChem 5:566–571
45. Son I, Lail Shek Y, Dubins YN, Chalikian TV (2012) Volumetric characterization of tri-*N*-acetylglucosamine binding to lysozyme. Biochemistry 51:5784–5790
46. Kamiyama T, Gekko K (2000) Effect of ligand binding on the flexibility of dihydrofolate reductase as revealed by compressibility. Biochim Biophys Acta 1478:257–266
47. Fan HY, Shek YL, Amiri A, Dubins DN, Heerklotz H, Macgregor RB Jr, Chalikian TV (2011) Volumetric characterization of sodium-induced G-quadruplex formation. J Am Chem Soc 133:4518–4526
48. Winter R, Jeworrek C (2009) Effect of pressure on membranes. Soft Matter 5(17):3157–3173
49. Tanford C (1978) Hydrophobic effect and organization of living matter. Science 200: 1012–1018
50. Cevc G (ed) (1993) Phospholipids handbook. Marcel Dekker, New York
51. Lipowski R, Sackmann E (eds) (1995) Structure and dynamics of membranes, vols 1A and 1B. Elsevier, Amsterdam
52. Seddon JM (1992) Inverse cubic phases of membrane-lipids, and their relevance to the static and dynamic structure of biomembranes. Acta Pharm 42:255–262
53. van Meer G, Voelker DR, Feigenson GW (2008) Membrane lipids: where they are and how they behave. Nat Rev Mol Cell Biol 9:112–124
54. Bernsdorff C, Wolf A, Winter R (1996) The effect of temperature and pressure on structural and dynamic properties of phospholipid/sterol mixtures—a steady-state and time-resolved fluorescence anisotropy study. Z Phys Chem 193:151–173
55. Bernsdorff C, Wolf A, Winter R, Gratton E (1997) Effect of hydrostatic pressure on water penetration and rotational dynamics in phospholipid-cholesterol bilayers. Biophys J 72: 1264–1277
56. Reis O, Winter R, Zerda TW (1996) The effect of high external pressure on DPPC-cholesterol multilamellar vesicles: a pressure-tuning Fourier transform infrared spectroscopy study. Biochim Biophys Acta Biomembr 1279:5–16
57. Seddon JM, Templer RH (1993) Cubic phases of self-assembled amphiphilic aggregates. Philos Trans R Soc A 344:377–401

58. Seemann H, Winter R (2003) Volumetric properties, compressibilities and volume fluctuations in phospholipid-cholesterol bilayers. Z Phys Chem 217:831–846
59. Winter R (2001) Effects of hydrostatic pressure on lipid and surfactant phases. Curr Opin Colloid Interface Sci 6:303–312
60. Worcester D, Hammouda B (1997) Interdigitated hydrocarbon chains in C20 and C22 phosphatidylcholines induced by hydrostatic pressure. Phys B 241:1175–1177
61. Heimburg T (2007) Thermal biophysics of membranes. Wiley, Weinheim
62. Czeslik C, Reis O, Winter R, Rapp G (1998) Effect of high pressure on the structure of dipalmitoylphosphatidylcholine bilayer membranes: a synchrotron X-ray diffraction and FTIR spectroscopy study using the diamond anvil technique. Chem Phys Lipids 91:135–144
63. Winter R, Czeslik C (2000) Pressure effects on the structure of lyotropic lipid mesophases and model biomembrane systems. Z Krist 215:454–474
64. Shearman GC, Ces O, Templer RH, Seddon JM (2006) Inverse lyotropic phases of lipids and membrane curvature. J Phys Condens Matter 18:S1105–S1124
65. Scarlata SF, Rosenberg M (1990) Effect of increased lipid packing on the surface charge of micelles and membranes. Biochemistry 29:10233–10240
66. Paccamiccio L, Pisani M, Spinozzi F, Ferrero C, Finet S, Mariani P (2006) Pressure effects on lipidic direct phases: the dodecyl trimethyl ammonium chloride-water system. J Phys Chem B 110:12410–12418
67. Winter R, Erbes J, Templer RH, Seddon JM, Syrykh A, Warrender NA, Rapp G (1999) Inverse bicontinuous cubic phases in fatty acid/phosphatidylcholine mixtures: the effects of pressure and lipid composition. Phys Chem Chem Phys 1:887–893
68. Böttner M, Ceh D, Jacobs U, Winter R (1994) High pressure volumetric measurements on phospholipid bilayers. Z Phys Chem 184:205–218
69. Jeworrek C, Pühse M, Winter R (2008) X-ray kinematography of phase transformations of three-component lipid mixtures: a time-resolved synchrotron X-ray scattering study using the pressure-jump relaxation technique. Langmuir 24:11851–11859
70. Kapoor S, Werkmuller A, Denter C, Zhai Y, Markgraf J, Weise K, Opitz N, Winter R (2011) Temperature-pressure phase diagram of a heterogeneous anionic model biomembrane system: results from a combined calorimetry, spectroscopy and microscopy study. Biochim Biophys Acta Biomembr 1808:1187–1195
71. Puehse M, Jeworrek C, Winter R (2008) The temperature-pressure phase diagram of a DPPC-ergosterol fungal model membrane—a SAXS and FTIR spectroscopy study. Chem Phys Lipids 152:57–63
72. Yaghmur A, Kriechbaum M, Amenitsch H, Steinhart M, Laggner P, Rappolt M (2010) Effects of pressure and temperature on the self-assembled fully hydrated nanostructures of monoolein-oil systems. Langmuir 26:1177–1185
73. Winter R, Köhling R (2004) Static and time-resolved synchrotron small-angle X-ray scattering studies of lyotropic lipid mesophases, model biomembranes and proteins in solution. J Phys Condens Matter 16:S327–S352
74. Shaw KP, Brooks NJ, Clarke JA, Ces O, Seddon JM, Law RV (2012) Pressure-temperature phase behaviour of natural sphingomyelin extracts. Soft Matter 8:1070–1078
75. Potekhin SA, Senin AA, Abdurakhmanov NN, Khusainova RS (2008) High pressure effect on the main transition from the ripple gel P(beta)' phase to the liquid crystal (L(alpha)) phase in dipalmitoylphosphatidylcholine. Microcalorimetric study. Biochim Biophys Acta Biomembr 1778:2588–2593
76. Jonas J (ed) (1991) High pressure NMR. Springer, Berlin
77. Mantulin WW, Gotto AM, Pownall HJ (1984) Effect of hydrostatic-pressure on the transfer of a fluorescent phosphatidylcholine between apolipoprotein-phospholipid recombinants. J Am Chem Soc 106:3317–3319
78. Helfrich W, Rennschuh H (1990) Landau theory of the lamellar-to-cubic phase transition. J Phys Paris 51:C7189–C7195

79. Porte G, Appell J, Bassereau P, Marignan J (1989) L-alpha to L3—a topology driven transition in phases of infinite fluid membranes. J Phys-Paris 50:1335–1347
80. Duesing PM, Seddon JM, Templer RH, Mannock DA (1997) Pressure effects on lamellar and inverse curved phases of fully hydrated dialkyl phosphatidylethanolamines and beta-D-xylopyranosyl-sn-glycerols. Langmuir 13:2655–2664
81. Brooks NJ, Ces O, Templer RH, Seddon JM (2011) Pressure effects on lipid membrane structure and dynamics. Chem Phys Lipids 164:89–98
82. Seifert U (1997) Configurations of fluid membranes and vesicles. Adv Phys 46:13–137
83. Nicolini C, Celli A, Gratton E, Winter R (2006) Pressure tuning of the morphology of heterogeneous lipid vesicles: a two-photon-excitation fluorescence microscopy study. Biophys J 91:2936–2942
84. Delong EF, Yayanos AA (1985) Adaptation of the membrane lipids of a deep-sea bacterium to changes in hydrostatic pressure. Science 228:1101–1102
85. Skanes ID, Stewart J, Keough KMW, Morrow MR (2006) Effect of chain unsaturation on bilayer response to pressure. Phys Rev Lett E 74:051913
86. Zhai Y, Chong PLG, Taylor LJA, Erlkamp M, Grobelny S, Czeslik C, Watkins E, Winter R (2012) Physical properties of archaeal tetraether lipid membranes as revealed by differential scanning and pressure perturbation calorimetry, molecular acoustics, and neutron reflectometry: effects of pressure and cell growth temperature. Langmuir 28:5211–5217
87. Akasaka K, Yamada H (2001) On-line cell high-pressure nuclear magnetic resonance technique: application to protein studies. Methods Enzymol 338:134–158
88. Beck-Erlach M, Munte CE, Kremer W, Hartl R, Rochelt D, Niesner D, Kalbitzer HR (2010) Ceramic cells for high pressure NMR spectroscopy of proteins. J Magn Reson 204:196–199
89. Kremer W, Arnold M, Munte CE, Hartl R, Beck-Erlach M, Koehler J, Meier A, Kalbitzer HR (2011) Pulsed pressure perturbations, an extra dimension in NMR spectroscopy of proteins. J Am Chem Soc 133:13646–13651
90. Urbauer JL, Ehrhardt MR, Bieber RJ, Flynn PF, Wand AJ (1996) High-resolution triple-resonance NMR spectroscopy of a novel calmodulin peptide complex at kilobar pressures. J Am Chem Soc 118:11329–11330
91. Yamada H, Nishikawa K, Honda M, Shimura T, Akasaka K, Tabayashi K (2001) Pressure-resisting cell for high-pressure, high-resolution nuclear magnetic resonance measurements at very high magnetic fields. Rev Sci Instrum 72:1463–1471
92. Akasaka K, Li H (2001) Low-lying excited states of proteins revealed from nonlinear pressure shifts in H-1 and N-15 NMR. Biochemistry 40:8665–8671
93. Iwadate M, Asakura T, Dubovskii PV, Yamada H, Akasaka K, Williamson MP (2001) Pressure-dependent changes in the structure of the melittin alpha-helix determined by NMR. J Biomol NMR 19:115–124
94. Inoue K, Yamada H, Imoto T, Akasaka K (1998) High pressure NMR study of a small protein, gurmarin. J Biomol NMR 12:535–541
95. Li H, Yamada H, Akasaka K (1999) Effect of pressure on the tertiary structure and dynamics of folded basic pancreatic trypsin inhibitor. Biophys J 77:2801–2812
96. Kamatari YO, Yamada H, Akasaka K, Jones JA, Dobson CM, Smith LJ (2001) Response of native and denatured hen lysozyme to high pressure studied by N-15/H-1 NMR spectroscopy. Eur J Biochem 268:1782–1793
97. Kuwata K, Li H, Yamada H, Batt CA, Goto Y, Akasaka K (2001) High pressure NMR reveals a variety of fluctuating conformers in beta-lactoglobulin. J Mol Biol 305:1073–1083
98. Kitahara R, Yamada H, Akasaka K, Wright PE (2002) High pressure NMR reveals that apomyoglobin is an equilibrium mixture from the native to the unfolded. J Mol Biol 320: 311–319
99. Akasaka K, Li H, Yamada H, Li RH, Thoresen T, Woodward CK (1999) Pressure response of protein backbone structure. Pressure-induced amide N-15 chemical shifts in BPTI. Protein Sci 8:1946–1953

100. Li H, Yamada H, Akasaka K, Gronenborn AM (2000) Pressure alters electronic orbital overlap in hydrogen bonds. J Biomol NMR 18:207–216
101. Roche J, Caro JA, Norberto DR, Barthe P, Roumestand C, Schlessman JL, Garcia AE, Garcia-Moreno BE, Royer CA (2012) Cavities determine the pressure unfolding of proteins. Proc Natl Acad Sci U S A 109:6945–6950
102. Inoue K, Yamada H, Akasaka K, Hermann C, Kremer W, Maurer T, Doker R, Kalbitzer HR (2000) Pressure-induced local unfolding of the Ras binding domain of RalGDS. Nat Struct Biol 7:547–550
103. Kuwata K, Li H, Yamada H, Legname G, Prusiner SB, Akasaka K, James TL (2002) Locally disordered conformer of the hamster prion protein: a crucial intermediate to PrPSc? Biochemistry 41:12277–12283
104. Kitahara R, Akasaka K (2003) Close identity of a pressure-stabilized intermediate with a kinetic intermediate in protein folding. Proc Natl Acad Sci U S A 100:3167–3172
105. Kalbitzer HR, Spoerner M, Ganser P, Hozsa C, Kremer W (2009) Fundamental link between folding states and functional states of proteins. J Am Chem Soc 131:16714–16719
106. Kitahara R, Royer CA, Yamada H, Boyer M, Saldana JL, Akasaka K, Roumestand C (2002) Equilibrium and pressure-jump relaxation studies of the conformational transitions of P13MTCP1. J Mol Biol 320:609–628
107. Prigozhin MB, Liu Y, Wirth AJ, Kapoor S, Winter R, Schulten K, Gruebele M (2013) Misplaced helix slows down ultrafast pressure-jump protein folding. Proc Natl Acad Sci U S A 110:8087–8092
108. Woenckhaus J, Kohling R, Thiyagarajan P, Littrell KC, Seifert S, Royer CA, Winter R (2001) Pressure-jump small-angle X-ray scattering detected kinetics of staphylococcal nuclease folding. Biophys J 80:1518–1523
109. Anfinsen CB (1973) Principles that govern folding of protein chains. Science 181:223–230
110. Ellis RJ, Hemmingsen SM (1989) Molecular chaperones—proteins essential for the biogenesis of some macromolecular structures. Trends Biochem Sci 14:339–342
111. Radford SE, Dobson CM (1999) From computer simulations to human disease: emerging themes in protein folding. Cell 97:291–298
112. Bullock AN, Fersht A (2001) Rescuing the function of mutant p53. Nat Rev Cancer 1:68–76
113. Dobson CM (2001) The structural basis of protein folding and its links with human disease. Philos Trans R Soc B 356:133–145
114. Glenner GG (1980) Amyloid deposits and amyloidosis—the beta fibrilloses.1. N Engl J Med 302:1283–1292
115. Glenner GG (1980) Amyloid deposits and amyloidosis—the Beta-fibrilloses.2. N Engl J Med 302:1333–1343
116. Thomas PJ, Qu BH, Pedersen PL (1995) Defective protein folding as a basis of human disease. Trends Biochem Sci 20:456–459
117. Westermark P, Sletten K, Johnson KH (1996) Ageing and amyloid fibrillogenesis: lessons from apolipoprotein AI, transthyretin and islet amyloid polypeptide. Ciba Found Symp 199: 205–222
118. Bucciantini M, Giannoni E, Chiti F, Baroni F, Formigli L, Zurdo JS, Taddei N, Ramponi G, Dobson CM, Stefani M (2002) Inherent toxicity of aggregates implies a common mechanism for protein misfolding diseases. Nature 416:507–511
119. Horwich A (2002) Protein aggregation in disease: a role for folding intermediates forming specific multimeric interactions. J Clin Invest 110:1221–1232
120. Prusiner SB, Dearmond SJ (1995) Prion protein amyloid and neurodegeneration. Amyloid 2:39–65
121. Tan SY, Pepys MB (1994) Amyloidosis. Histopathology 25:403–414
122. Hayden MR, Tyagi SC (2001) "A" is for amylin and amyloid in type 2 diabetes mellitus. J Pancreas 2:124–139
123. Sunde M, Blake C (1997) The structure of amyloid fibrils by electron microscopy and X-ray diffraction. Adv Protein Chem 50:123–159

124. Serpell LC, Berriman J, Jakes R, Goedert M, Crowther RA (2000) Fiber diffraction of synthetic alpha-synuclein filaments shows amyloid-like cross-beta conformation. Proc Natl Acad Sci U S A 97:4897–4902
125. Kelly JW (1998) The alternative conformations of amyloidogenic proteins and their multi-step assembly pathways. Curr Opin Struct Biol 8:101–106
126. Kelly JW (1998) The environmental dependency of protein folding best explains prion and amyloid diseases. Proc Natl Acad Sci U S A 95:930–932
127. Dobson CM (2004) Experimental investigation of protein folding and misfolding. Methods 34:4–14
128. Kirkitadze MD, Bitan G, Teplow DB (2002) Paradigm shifts in Alzheimer's disease and other neurodegenerative disorders: the emerging role of oligomeric assemblies. J Neurosci Res 69: 567–577
129. Makin OS, Atkins E, Sikorski P, Johansson J, Serpell LC (2005) Molecular basis for amyloid fibril formation and stability. Proc Natl Acad Sci U S A 102:315–320
130. Kirkitadze MD, Teplow DB (2002) Conformational intermediates in amyloid beta-protein assembly. Neurobiol Aging 23:S556
131. Dzwolak W (2006) Tuning amyloidogenic conformations through cosolvents and hydrostatic pressure: when the soft matter becomes even softer. BBA-Proteins Proteom 1764:470–480
132. Smeller L, Rubens P, Heremans K (1999) Pressure effect on the temperature-induced unfolding and tendency to aggregate of myoglobin. Biochemistry 38:3816–3820
133. Silva JL, Cordeiro Y, Foguel D (2006) Protein folding and aggregation: two sides of the same coin in the condensation of proteins revealed by pressure studies. BBA-Proteins Proteom 1764:443–451
134. Niraula TN, Konno T, Li H, Yamada H, Akasaka K, Tachibana H (2004) Pressure-dissociable reversible assembly of intrinsically denatured lysozyme is a precursor for amyloid fibrils. Proc Natl Acad Sci U S A 101:4089–4093
135. Ferrao-Gonzales AD, Souto SO, Silva JL, Foguel D (2000) The preaggregated state of an amyloidogenic protein: hydrostatic pressure converts native transthyretin into the amyloidogenic state. Proc Natl Acad Sci U S A 97:6445–6450
136. Foguel D, Suarez MC, Ferrao-Gonzales AD, Porto TC, Palmieri L, Einsiedler CM, Andrade LR, Lashuel HA, Lansbury PT, Kelly JW, Silva JL (2003) Dissociation of amyloid fibrils of alpha-synuclein and transthyretin by pressure reveals their reversible nature and the formation of water-excluded cavities. Proc Natl Acad Sci U S A 100:9831–9836
137. Gorovits BM, Horowitz PM (1998) High hydrostatic pressure can reverse aggregation of protein folding intermediates and facilitate acquisition of native structure. Biochemistry 37:6132–6135
138. Grudzielanek S, Smirnovas V, Winter R (2006) Solvation-assisted pressure tuning of insulin fibrillation: from novel aggregation pathways to biotechnological applications. J Mol Biol 356:497–509
139. Dzwolak W, Ravindra R, Lendermann J, Winter R (2003) Aggregation of bovine insulin probed by DSC/PPC calorimetry and FTIR spectroscopy. Biochemistry 42:11347–11355
140. Jansen R, Grudzielanek S, Dzwolak W, Winter R (2004) High pressure promotes circularly shaped insulin amyloid. J Mol Biol 338:203–206
141. Piccirilli F, Mangialardo S, Postorino P, Baldassarre L, Lupi S, Perucchi A (2012) Sequential dissociation of insulin amyloids probed by high pressure Fourier transform infrared spectroscopy. Soft Matter 8:11863–11870
142. Smirnovas V, Winter R, Funck T, Dzwolak W (2006) Protein amyloidogenesis in the context of volume fluctuations: a case study on insulin. ChemPhysChem 7:1046–1049
143. Dzwolak W, Grudzielanek S, Smirnovas V, Ravindra R, Nicolini C, Jansen R, Loksztejn A, Porowski S, Winter R (2005) Ethanol-perturbed amyloidogenic self-assembly of insulin: looking for origins of amyloid strains. Biochemistry 44:8948–8958
144. Cordeiro Y, Kraineva J, Ravindra R, Lima LMTR, Gomes MPB, Foguel D, Winter R, Silva JL (2004) Hydration and packing effects on prion folding and beta-sheet conversion—high

pressure spectroscopy and pressure perturbation calorimetry studies. J Biol Chem 279: 32354–32359
145. Mishra R, Winter R (2008) Cold- and pressure-induced dissociation of protein aggregates and amyloid fibrils. Angew Chem Int Ed 47:6518–6521
146. Carrier D, Mantsch HH, Wong PTT (1990) Pressure-induced reversible changes in secondary structure of poly(L-Lysine)—an IR spectroscopic study. Biopolymers 29:837–844
147. Zhou JM, Zhu L, Balny C, Perrett S (2001) Pressure denaturation of the yeast prion protein Ure2. Biochem Biophys Res Commun 287:147–152
148. Alvarez-Martinez MT, Torrent J, Lange R, Verdier JM, Balny C, Liautard JP (2003) Optimized overproduction, purification, characterization and high-pressure sensitivity of the prion protein in the native (PrPC-like) or amyloid (PrPSc-like) conformation. BBA-Proteins Proteom 1645:228–240
149. Martins SM, Chapeaurouge A, Ferreira RT (2003) Folding intermediates of the prion protein stabilized by hydrostatic pressure and low temperature. J Biol Chem 278:50449–50455
150. Torrent J, Alvarez-Martinez MT, Harricane MC, Heitz F, Liautard JP, Balny C, Lange R (2004) High pressure induces scrapie-like prion protein misfolding and amyloid fibril formation. Biochemistry 43:7162–7170
151. Torrent J, Alvarez-Martinez MT, Liautard JP, Lange R (2006) Modulation of prion protein structure by pressure and temperature. BBA-Proteins Proteom 1764:546–551
152. Salvador-Silva M, Aoi S, Parker A, Yang P, Pecen P, Hernandez MR (2004) Responses and signaling pathways in human optic nerve head astrocytes exposed to hydrostatic pressure in vitro. Glia 45:364–377
153. Wu HJ, Zhang ZQ, Yu B, Liu S, Qin KR, Zhu L (2010) Pressure activates Src-dependent FAK-Akt and ERK1/2 signaling pathways in rat hepatic stellate cells. Cell Physiol Biochem 26:273–280
154. Sukharev SI, Blount P, Martinac B, Blattner FR, Kung C (1994) A large-conductance mechanosensitive channel in E. coli encoded by mscL alone. Nature 368:265–268
155. Paoletti P, Ascher P (1994) Mechanosensitivity of NMDA receptors in cultured mouse central neurons. Neuron 13:645–655
156. Lehtonen JYA, Kinnunen PKJ (1995) Phospholipase A2 as a mechanosensor. Biophys J 68: 1888–1894
157. Gudi S, Nolan JP, Frangos JA (1998) Modulation of GTPase activity of G proteins by fluid shear stress and phospholipid composition. Proc Natl Acad Sci U S A 95:2515–2519
158. Petrov E, Rohde PR, Martinac B (2011) Flying-patch patch-clamp study of G22E-MscL mutant under high hydrostatic pressure. Biophys J 100:1635–1641
159. Kapoor S, Werkmuller A, Goody RS, Waldmann H, Winter R (2013) Pressure modulation of Ras-membrane interactions and intervesicle transfer. J Am Chem Soc 135:6149–6156
160. Teichert H, Periasamy N, Winter R, Vogel RF (2009) Influence of membrane lipid composition on the activity of functionally reconstituted LmrA under high hydrostatic pressure. High Press Res 29:344–357
161. Linke K, Periasamy N, Ehrmann M, Winter R, Vogel RF (2008) Influence of high pressure on the dimerization of ToxR, a protein involved in bacterial signal transduction. Appl Environ Microbiol 74:7821–7823
162. Winter R, Lopes D, Grudzielanek S, Vogtt K (2007) Towards an understanding of the temperature/pressure configurational and free-energy landscape of biomolecules. J Non-Equilib Thermodyn 32:41–97
163. Grant BJ, McCammon JA, Gorfe AA (2010) Conformational selection in G-Proteins lessons from Ras and Rho. Biophys J 99:L87–L89
164. Meierhofer T, Rosnizeck IC, Graf T, Reiss K, Konig B, Kalbitzer HR, Spoerner M (2011) Cu2+-cyclen as probe to identify conformational states in guanine nucleotide binding proteins. J Am Chem Soc 133:2048–2051

165. Geyer M, Schweins T, Herrmann C, Prisner T, Wittinghofer A, Kalbitzer HR (1996) Conformational transitions in p21ras and in its complexes with the effector protein Raf-RBD and the GTPase activating protein GAP. Biochemistry 35:10308–10320
166. Kapoor S, Triola G, Vetter IR, Erlkamp M, Waldmann H, Winter R (2012) Revealing conformational substates of lipidated N-Ras protein by pressure modulation. Proc Natl Acad Sci U S A 109:460–465
167. Kapoor S, Weise K, Erlkamp M, Triola G, Waldmann H, Winter R (2012) The role of G-domain orientation and nucleotide state on the Ras isoform-specific membrane interaction. Eur Biophys J 41:801–813
168. Powalska E, Janosch S, Kinne-Saffran E, Kinne RKH, Fontes CFL, Mignaco JA, Winter R (2007) Fluorescence spectroscopic studies of pressure effects on Na+, K+-ATPase reconstituted into phospholipid bilayers and model raft mixtures. Biochemistry 46: 1672–1683
169. Kato M, Hayashi R, Tsuda T, Taniguchi K (2002) High pressure-induced changes of biological membrane—study on the membrane-bound Na+/K+-ATPase as a model system. Eur J Biochem 269:110–118
170. Souza MO, Creczynski-Pasa TB, Scofano HM, Graber P, Mignaco JA (2004) High hydrostatic pressure perturbs the interactions between CF(0)F(1) subunits and induces a dual effect on activity. Int J Biochem Cell B 36:920–930
171. Akasaka K, Nagahata H, Maeno A, Sakai K (2008) Pressure acceleration of proteolysis: a general mechanism. Biophysics 4:29–32
172. Eisenmenger MJ, Reyes-de-Corcuera JI (2009) High pressure enhancement of enzymes. Enzyme Microb Technol 45:331–347
173. Herve G, Tobe S, Heams T, Vergne J, Maurel MC (2006) Hydrostatic and osmotic pressure study of the hairpin ribozyme. Biochim Biophys Acta 1764:573–577
174. Kara S, Sing Long W, Berheide M, Peper S, Niemeyer B, Liese A (2011) Influence of reaction conditions on the enantioselectivity of biocatalyzed C-C bond formation under high pressure conditions. J Biotechnol 152:87–92
175. Picard A, Daniel I, Montagnac G, Oger P (2007) In situ monitoring by quantitative raman spectroscopy of alcoholic fermentation by saccharomyces cerevisiae under high pressure. Extremophiles 11:445–452

Chapter 3
Watching a Signaling Protein Function in Real Time via Picosecond Time-Resolved Laue Crystallography

Friedrich Schotte, Hyun Sun Cho, Hironari Kamikubo, Mikio Kataoka, and Philip A. Anfinrud

Abstract To understand how signaling proteins function, it is crucial to know the time-ordered sequence of events that lead to the signaling state. We recently developed on the BioCARS 14-IDB beamline at the Advanced Photon Source the infrastructure required to characterize structural changes in protein crystals with near-atomic spatial resolution and 150-ps time resolution and have used this capability to track the reversible photocycle of photoactive yellow protein (PYP) following *trans* to *cis* photoisomerization of its *p*-coumaric acid (pCA) chromophore over ten decades of time. The first of four major intermediates characterized in this study is highly contorted, with the pCA carbonyl rotated nearly 90° out of the plane of the phenolate. A hydrogen bond between the pCA carbonyl and the Cys69 backbone constrains the chromophore in this unusual twisted conformation. This novel structure, which corresponds to a strained *cis* intermediate, is short lived (~600 ps), has not been observed in prior cryo-crystallography experiments, and is the progenitor of intermediates characterized in previous nanosecond time-resolved Laue crystallography studies. The structural transitions unveiled during the PYP photocycle include *trans*/*cis* isomerization, the breaking and making of hydrogen bonds, formation/relaxation of strain, and gated water penetration into the interior of the protein. This mechanistically detailed, near-atomic resolution description of the complete PYP photocycle provides a framework for understanding signal transduction in proteins and for assessing and validating theoretical/computational approaches in protein biophysics.

Keywords Signal transduction • Time-resolved Laue crystallography • Photoactive yellow protein • Photoisomerization

---

F. Schotte • H.S. Cho • P.A. Anfinrud (✉)
Laboratory of Chemical Physics, National Institute of Diabetes and Digestive and Kidney Diseases, National Institutes of Health, Bethesda, MD 20892, USA
e-mail: anfinrud@nih.gov

H. Kamikubo • M. Kataoka
Graduate School of Materials Science, Nara Institute of Science and Technology, 8916-5 Takayama, Ikoma, Nara 630-0192, Japan

M. Terazima et al. (eds.), *Molecular Science of Fluctuations Toward Biological Functions*, DOI 10.1007/978-4-431-55840-8_3

3.1 Introduction

Proteins are not static macromolecules, as depicted by their atomic resolution structures. Rather, they are dynamic molecules whose structural fluctuations mediate conformational changes capable of modulating the activity of the protein. Although static crystal protein structures provide clues regarding how a protein might accomplish its target function, a complete mechanistic understanding requires knowledge about the time-ordered sequence of structural changes that carry out its function. When a laser pulse is used to trigger a protein's function, the ensuing structural changes can be tracked on ultrafast time scales via time-resolved Laue crystallography [1], a pump-probe methodology first demonstrated with nanosecond time resolution [2], and later extended to the picosecond time domain [3].

Light-activated signaling proteins are ubiquitous in nature and are responsible for vision in higher animals, phototropism in plants, and phototaxis in bacteria. These proteins afford the opportunity to investigate structural dynamics with near-atomic resolution on ultrafast time scales. For this study, we focused our attention on photoactive yellow protein (PYP), a 14-kD water-soluble blue-light receptor first discovered in *Halorhodospira halophila* [4], a purple sulfur bacterium that swims away from blue light and toward photosynthetically useful green light [5, 6]. Since its action spectrum for negative phototaxis matches the absorbance spectrum of PYP [6], this protein is presumed responsible for the signal that causes this extreme halophile to swim away from photons energetic enough to be genetically harmful. Since its discovery, PYP has served as a useful model system for biophysical studies of signaling proteins [7]. The chromophore in PYP is *p*-coumaric acid (pCA), which is covalently linked to the Cys69 residue via a thioester bond (Fig. 3.1). The C2=C3 double bond is *trans* in the ground state (pG) but, upon absorbing a single photon of blue light, is converted to *cis* with modest quantum efficiency. This photoisomerization event triggers a sequence of structural changes that involve spectroscopically redshifted (pR) and blueshifted (pB) intermediates, the last of which corresponds to the putative signaling state [8–11].

Using the method of picosecond time-resolved Laue crystallography, we have tracked the reversible photocycle of PYP over ten decades of time at near-atomic resolution. The time-resolved electron density maps unveiled in this study are well described by a four-state kinetic model that is both chemically and mechanistically sensible. Our comprehensive description of the structural changes occurring during the PYP photocycle helps remove some of the mystery into how proteins execute their targeted function with remarkable efficiency and extraordinary selectivity.

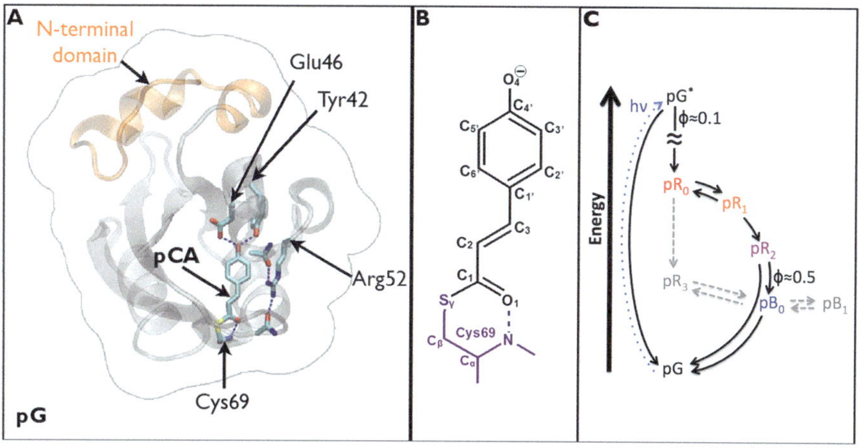

Fig. 3.1 PYP structure and photocycle. (**A**) Surface rendering (*gray*) of PYP (PDB ID: 2ZOH) in the pG state with backbone structure (*ribbon*) and atomic rendering of pCA and its hydrogen-bonding partners. *Arrows* point to the C2=C3 double bond (pCA) and the Cα atoms of key residues. Hydrogen bonds are indicated with dashed *blue lines*. Arg52 is stabilized in its "closed" state via hydrogen bonds to the protein backbone. (**B**) The C2=C3 double bond in pCA is *trans* in its ground state; it absorbs blue light and gives PYP its *yellow color*. Photoisomerization from *trans* to *cis* triggers the PYP photocycle. (**C**) The PYP photocycle is labeled and color coded according to intermediates characterized in this study. Intermediates colored *gray* and connected by *dashed lines* were not observed in this study, but have been characterized or implicated in prior studies

3.2 Experimental Methodology

3.2.1 Crystal Preparation

Wild-type PYP apoprotein was overexpressed in *E. coli*, isolated, and then reconstituted with pCA anhydride in 4 M urea buffer [12]. The holoprotein was purified by column chromatography (DEAE Sepharose CL6B, Amersham Biosciences) several times until the optical purity index (absorbance 277 nm/absorbance λ_{max}) became <0.44. Large $P6_3$ crystals of PYP were grown from microseeds via the hanging-drop vapor diffusion method [13]. The crystallization hanging-drop solution contained 24 mg-mL^{-1} PYP, 2.2 M ammonium sulfate, 1.0 M sodium chloride, and 20 mM sodium phosphate. The reservoir solution contained 2.5 M ammonium sulfate and 1.1–1.2 M sodium chloride and was maintained at 293 K. The crystallization buffers were prepared with 99.9 % heavy water (Aldrich) and titrated to pD 9.0. This method produced high-quality crystals ~2 mm in length. The crystals were mounted in 1.5-mm-diameter glass capillaries whose inner wall was texturized by bonding thin shards of crushed glass to its surface with polyvinyl formyl. Crystals mounted on the roughened surface are far less prone to slippage during data collection, which can be problematic due to the shock wave generated in the crystal after an intense, short duration laser pulse is absorbed and suddenly heats its surface.

3.2.2 *Pump-Probe Method*

Picosecond time-resolved "snapshots" of the PYP structure were acquired using the pump-probe method [3] (Fig. 3.2). Briefly, a laser pulse (pump) photoactivates a PYP crystal, after which a suitably delayed x-ray pulse (probe) passes through the crystal and records its diffraction pattern on a 2D detector. Because we use a polychromatic x-ray pulse, we capture thousands of reflections in a single image without having to rotate the crystal. This Laue approach to crystallography boosts substantially the rate at which time-resolved diffraction data can be acquired. The information needed to determine the protein's structure is encoded in the relative intensities of the diffraction spots observed. However, the structural information contained in a single diffraction image is incomplete, requiring repeated measurements at multiple crystal orientations to produce a complete set of data.

3.2.2.1 Laser Source (Pump)

The picosecond laser system used in this study employs a Spectra Physics Spitfire Pro designed to produce 5-mJ, 1.2-ps optical pulses at 780 nm. To produce ~120-ps pulses, we used a 0.8-nm band-pass interference filter to limit the bandwidth of the seed pulses delivered to the amplifier pulse stretcher and then bypassed the grating pulse compressor by inserting a large right-angle prism into the optical layout of the Spitfire Pro. The laser pulse was frequency doubled to 390 nm in a Type I BBO nonlinear crystal custom designed for these pulse characteristics. The 390-nm laser

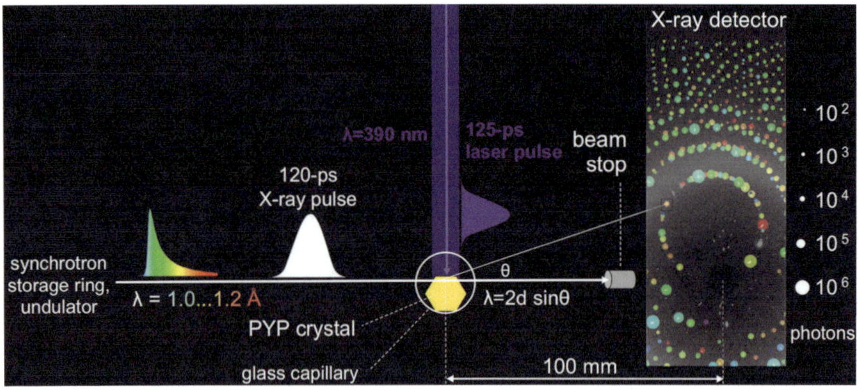

Fig. 3.2 Pump-probe geometry employed to acquire time-resolved diffraction snapshots. The PYP crystal is sealed in a thin-walled glass capillary. Because the laser penetration depth in PYP is shallow, an orthogonal pump-probe geometry is employed in which the top edge of the protein crystal is positioned at the top edge of the focused x-ray pulse. This geometry ensures optimal overlap between the laser and x-ray-illuminated volumes of the crystal. The protein crystal acts as a monochromator with various line spacings (d) and diffracts different x-ray colors (λ) in different directions (θ) according to Bragg's law ($\lambda = 2d \sin\theta$). Approximately 3000 spots are found in each time-resolved diffraction image. The *spots* in this figure are annotated according to integrated photons (spot dimension) and x-ray wavelength (spot color)

pulses were beam expanded and transported ~30 m from the laser hutch to the x-ray hutch, where beam-conditioning optics spatially filtered the beam and focused it to a 0.12 × 0.6-mm elliptical spot with the long axis aligned along the propagation direction of the x-ray beam (Fig. 3.2). The beam-conditioning optics arrangement includes a variable attenuator and a Berek polarization compensator, which were configured to deliver intense (3.5 mJ-mm^{-2}), polarized (|| to c-axis), stretched (125 ps) optical pulses tuned to the blue edge of the ground state absorption spectrum (390 nm). Under these conditions, the laser penetration depth in the PYP crystal is only ~30 μm (1/e). Use of a stretched, intense optical pulse to photoactivate the PYP crystal boosts the *trans* to *cis* isomerization yield by providing the pCA chromophore multiple opportunities to isomerize during the optical pulse. The photoisomerization yield achieved was estimated to be ~10 %.

3.2.2.2 X-Ray Source (Probe)

The data reported here were acquired on the BioCARS 14-IDB beamline [14] at the Advanced Photon Source in March 2011 when the ring was operated in hybrid mode. During this run, 12-keV polychromatic (5 % FWHM), 120-ps duration x-ray pulses with nearly 3×10^{10} photons were focused to an 80 × 40 (HxV) μm^2 spot at the sample position. A high-speed chopper, heat-load chopper, and ms shutter are capable of isolating, on demand, a single x-ray pulse from the high-frequency x-ray pulse train.

3.2.2.3 Pump-Probe Time Delay

The arrival time for the laser pump and x-ray probe pulses is monitored with noninvasive detectors, and the delay between them is controlled with a timing system based on a field-programmable gate array (FPGA). This timing system is capable of controlling the relative arrival time of the pump and probe pulses from picoseconds to tens of seconds with 10-ps precision. To calibrate the time delay, a metal-semiconductor-metal (MSM) detector that is sensitive to both laser and x-ray pulses is positioned at the sample location, and the time delay required to achieve coincident arrival is determined. The shot-to-shot timing jitter was measured to be ~8 ps rms. The time resolution achievable is not limited by this jitter but instead by the convolution of x-ray and the laser pulses, which produces an instrument response function of ~170 ps.

3.2.2.4 Data Collection Protocol

To ensure optimal overlap between the laser and x-ray pulses, it is crucial to align the top edge of the crystal with the top edge of the x-ray pulse (Fig. 3.2). To accomplish this feat, we developed an edge-finder algorithm that scans the crystal vertically through an attenuated x-ray beam while acquiring single-shot diffraction images. The extrapolated edge of the crystal is deduced from the integrated spot intensities, and the crystal is positioned to align its top edge with that of the x-ray beam.

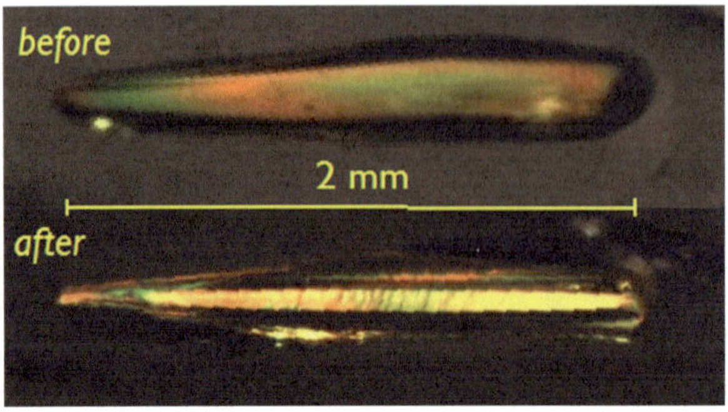

Fig. 3.3 Before and after photos of PYP7, one of nine crystals used to acquire time-resolved diffraction data. The crystal was viewed through crossed polarizers to suppress the background; variations in color are not real, but arise due to crystal birefringence. The "before" image appears homogeneous across its length. The image acquired "after" x-ray exposure reveals radiation-induced cracking of the crystal. These small cracks increase crystal mosaicity and reduce the diffraction quality. A total of 336 images were acquired from this crystal, i.e., 6 orientations with 56 time-resolved images acquired per orientation (including 12 interleaved "-1 ns" reference images)

During data collection, the crystals were chilled to 288 K with a nitrogen-gas cooling stream. Time-resolved diffraction images were acquired at 42 time delays spanning ten decades of time with four time points per decade. Interleaved between these time delays were twelve "-1 ns" reference images in which the x-ray pulse arrived 1 ns ahead of the laser pulse. A complete time series was acquired at each of six or more orientations per crystal on nine crystals. To improve the signal-to-noise ratio of each diffraction image, five consecutive pump-probe shots were integrated on the mar165 CCD detector before readout.

To mitigate the adverse effects of radiation damage during data collection, very large crystals [13] were employed. Each ~2-mm-long crystal was translated ~250 μm along its long axis after each laser shot to avoid buildup of long-lived states. The starting position for each time point in the series was offset horizontally to distribute the radiation dose uniformly over the full length of the protein crystal. This approach preserved high-quality diffraction throughout the entire time series. The number of orientations that could be obtained from each crystal was limited by radiation damage, which led to cracking of the crystals, and a concomitant loss of diffracting power (Fig. 3.3). With this protocol, the time required for crystal translation and detector readout averaged 13 s per image.

3.2.3 Data Reduction and Analysis Methodologies

3.2.3.1 Interpolated Ratio Method

To improve the accuracy of the structural information contained in the ~1.6 Å resolution Laue diffraction images, we developed an interpolated ratio method for processing the data. Briefly, the diffraction spots in each image of a time series were indexed and integrated, and the crystal orientation refined. The structural information we seek is encoded in pump-induced changes of the spot intensities, whose absolute amplitude depends on crystal volume, crystal orientation, the x-ray wavelength exciting the corresponding Bragg reflections, and the x-ray intensity at that wavelength. To minimize systematic sources of error that can arise when merging diffraction data acquired at different orientations on different crystals, we employ a variation of the ratio method developed by Coppens and coworkers [15, 16]. That method is based upon back-to-back diffraction images acquired *with* (ON) and *without* (OFF) a laser pump pulse. By merging I_{ON}/I_{OFF} ratios of integrated spot intensities for each indexed reflection (R_{hkl}), instead of intensity differences (ΔI_{hkl}), errors arising from image scaling and wavelength normalization are avoided. After merging, the ratios can be converted to structure factor amplitude differences by multiplying each ground state structure factor amplitude by its corresponding ($R_{hkl} - 1$).

Instead of using I_{ON}/I_{OFF}, we use I_{ON}/I_{REF} where I_{REF} is computed by cubic spline interpolation between integrated spot intensities from 12 reference images interleaved among the 42 time delays. From a data collection strategy point of view, this interpolated ratio method is more efficient as it requires fewer total diffraction images in each time series. For this study, the reference images correspond to "-1 ns" time points, i.e., the x-ray pulse arrives 1 ns in advance of the laser pulse.

Laser-triggered structural changes in the protein crystal cause some reflections to get brighter and some dimmer, with the average ratio for each image expected to be 1. However, the weighted average of R_{hkl} is generally found to be less than 1 due to laser-induced strain in the crystal, which increases its mosaicity and reduces its diffraction efficiency. Before processing the data further, they are put on a common scale by rescaling the ratios so the weighted average for each time point is 1.

3.2.3.2 Excitation Degree Scaling

If an R_{hkl} is sensitive to laser-induced structural changes, it will deviate from unity by a factor that is proportional to the excitation degree, which can vary due to differences in the laser pump fluence. The R_{hkl} can be corrected for differences in the excitation degree by a simple scaling procedure using

$$R_{hkl}^{s} = 1 + s_i(R_{hkl} - 1)$$

where s_i is the scale factor corresponding to time series i. We invoke excitation degree scaling by finding the set of s_i that minimizes the differences between redundant R_{hkl}^{s} acquired at different orientations and with different crystals.

3.2.3.3 R_{hkl} Merging

Instead of merging intensities time point by time point, we merge complete R_{hkl}^{s} time series with the weight for each series computed from the average uncertainty across the series, i.e., $w = 1/\sigma_R^2$. Provided the crystal orientation remains stable through the time series, the time dependence of R_{hkl} is expected to evolve smoothly and monotonically. However, laser excitation launches an acoustic shock wave in the protein crystal and occasionally causes the crystal to slip or twitch in the capillary, an event that generates a discontinuity in R_{hkl}. Time series that suffer from sudden changes in crystal orientation are excluded from the merging step. This vector-based approach to merging ensures that the completeness and weights used to calculate the electron density maps are the same throughout each time series, thereby improving the consistency of our time-resolved electron density maps and our refined kinetic rate constants.

3.2.3.4 Time-Dependent Electron Density Map Generation

The merged R^s_{hkl} vectors characterize the time-dependent change in scattering power for each observed reflection. The R^s_{hkl} are converted to structure factor amplitude differences using ground state structure factor amplitudes calculated from a high-resolution (1.25 Å) room-temperature structure of WT PYP acquired under similar conditions (PDB ID: 2ZOH). This high-resolution structure is also used to phase the experimental time-dependent structure factor amplitudes, whose Fourier transform produced the electron density maps used to generate the 100-ps time-resolved snapshots reported in this study.

3.2.3.5 Real-Space Global Analysis Method

The time-resolved electron density difference maps generated in this study can have contributions from one or more intermediates. To extract electron density difference maps for the intermediates represented in the experimental data and determine the kinetics for their interconversion, we developed a real-space global analysis approach in which the kinetic rate parameters used to define the time-dependent populations of putative intermediates are refined by nonlinear least squares, while the electron density difference maps for the corresponding intermediates are determined by linear least squares. We start with a plausible kinetic model (Fig. 3.1C) whose corresponding rate equations account not only for the first-order processes for structure interconversion but also for the rate of photoactivation, which depends upon the instrument response function, i.e., the convolution of the laser and x-ray pulses. The ability to properly account for the experimental instrument response function is crucial for accurate determination of lifetimes approaching its width, which in this case is ~170 ps.

We recover “base” electron density difference maps for the intermediates, $\mathbf{\Delta\rho}_b$, by finding the least-squares solution to

$$\mathbf{\Delta\rho}_t = \mathbf{P}_t \mathbf{\Delta\rho}_b$$

where $\mathbf{\Delta\rho}_t$ is a $42 \times n$ matrix representing 42 observed time-resolved electron density difference maps, $\mathbf{P}_t$ is a 42×4 matrix representing the time-dependent populations for each of four intermediates in our kinetic model, and $\Delta\boldsymbol{\rho}_b$ is a $4 \times n$ matrix representing four "base" electron density difference maps, one for each intermediate. The parameters in our kinetic model are refined iteratively using the Marquardt-Levenberg algorithm, which minimizes the residual found when solving this linear equation. Because structural changes are largest in the vicinity of the pCA chromophore, the residual used in the least-squares refinement was confined to $n = 4983$ voxels found within 2.5 Å of the chromophore. This restricted-range approach enhances the sensitivity of the kinetic model to structural dynamics occurring in the chromophore region and improves the accuracy of its refined parameters. Once the population matrix is known, we solve this linear equation using all voxels in the unit cell ($n = 2{,}689{,}600$) and generate complete "base" maps.

Note that this global analysis method could have been applied to structure factors of HKLs in reciprocal space rather than electron density voxels in real space. However, it is not possible to define a "region of interest" in reciprocal space, so the accuracy of the refined kinetic rate parameters would be poorer than can be achieved with our real-space approach.

Although our real-space global analysis approach recovers electron density maps for all intermediates in the model, the maps generated are not guaranteed to be chemically sensible. Therefore, structure refinement is required to assess and validate the proposed model.

3.2.3.6 Structure Refinement

Atomic structures for each intermediate in the model were refined in reciprocal space using "phenix.refine," a program that implements the restrained maximum likelihood method [17]. This program requires as input both structure factor amplitudes and their estimated uncertainties. To generate $\Delta\mathbf{F}_b$, the "base" structure factor amplitude differences, we Fourier transformed $\Delta\boldsymbol{\rho}_b$, the complete "base" difference density maps generated by our global analysis. To generate $\boldsymbol{\sigma}^2(\Delta\mathrm{F}_b)$, the variance for the "base" structure factor amplitude differences, we used linear algebra to propagate the experimental variances for $\boldsymbol{\sigma}^2(\Delta\mathrm{F}_t)$, which are derived from photon counting statistics.

To mitigate the effects of modeling error, we employed the method of difference refinement [18], using as input for the refinement program the calculated ground state structure factors, $\mathbf{F}_c$, plus our structure factor amplitude differences, $\Delta\mathbf{F}_b$ (intermediate minus ground state).

$$\mathbf{F}_b = \mathbf{F}_c + \Delta\mathbf{F}_b$$

The restraints needed for the pCA chromophore are not included in the restraint library of the PHENIX package and were instead generated by the program eLBOW (electronic Ligand Bond Builder and Optimization Workbench), a utility distributed with PHENIX. The eLBOW-generated restraints were employed with one exception: the pCA dihedral bond angle restraints were set to $0/180 \pm 4°$, as opposed to eLBOW default values of $0/180 \pm 30°$, which proved far too generous for this π-conjugated chromophore.

3.3 Results

The earliest structural event, the *trans* to *cis* photoisomerization of pCA, occurs in ~1–3 ps [19–23] and is too fast to track with 150-ps time resolution. Nevertheless, the structural changes arising from this transition are captured in the 100-ps snapshot shown in Fig. 3.4. Although structural changes are evident throughout the protein, they are most dramatic in the vicinity of the pCA chromophore. Indeed, the refined pCA structure for the first intermediate detected, which we denote pR_0, is highly contorted with its carbonyl oriented approximately 90° out of the plane of the phenolate (see curved arrows in Fig. 3.4). This novel structure has never been trapped in cryo-trapping experiments [24–26], and its lifetime is too short to be captured in prior nanosecond time-resolved Laue crystallography experiments [27–29]. It was first reported by Schotte et al. [30] and then by Jung et al. [31]. The driving force for generating this intermediate is the *trans/cis* photoisomerization of pCA, during which the distance between the Cys69 sulfur and the terminal phenolate oxygen atom (see Fig. 3.1B) contracts by about 0.7 Å. The pCA chromophore acts like a winch, pulling Glu46 and Tyr42 downward and Cys69 upward (see yellow arrows in Fig. 3.4), thereby shrinking the corresponding Cα-Cα distances between these residues and generating strain in the protein. Remarkably, all three hydrogen bonds between pCA and the protein remain intact, but are clearly strained, as evidenced by longer hydrogen-bonding distances to Glu46 and the Cys69 backbone. The pR_0 intermediate sets the stage for the transitions that follow and ultimately lead to the long-lived PYP signaling state.

Later in the PYP photocycle, multiple intermediates are simultaneously populated. To determine their structures, we developed a global data analysis methodology. Starting from a plausible kinetic model, we recover accurate structures of intermediates and accurate rate parameters for interconversion between them. Four intermediates are required to account for our experimental time-resolved electron density maps, the last of which is similar to a structure that has been assigned to a blueshifted species [27, 32]. Hence, we denote these four intermediates pR_0, pR_1, pR_2, and pB_0 (Fig. 3.5). Structures similar to pR_1, pR_2, and pB_0 have been reported before [27, 28, 32, 33]. The first intermediate, pR_0, is short lived and required picosecond time resolution to recover its structure [30]. As reported, the simplest possible kinetic model connecting these intermediates is sequential with a reversible pR_0 to pR_1 transition and with a pR_2 to pB_0 transition that short-circuits to the ground pG state approximately half the time (Fig. 3.5A). This simple model accounts for the experimental electron density maps with high fidelity, as illustrated in the calculated vs. experimental color-coded overlays shown in Fig. 3.6. The ability to assess structurally the validity of a proposed kinetic model, as has been done here, is crucial to understand a protein's mechanism and is a capability unique to time-resolved crystallography.

Not all kinetic parameters are refinable in our global least-squares analysis and must be specified independently. For example, the photoproduct yield affects the initial pR_0 population, and incorrect assumptions about this yield can lead to

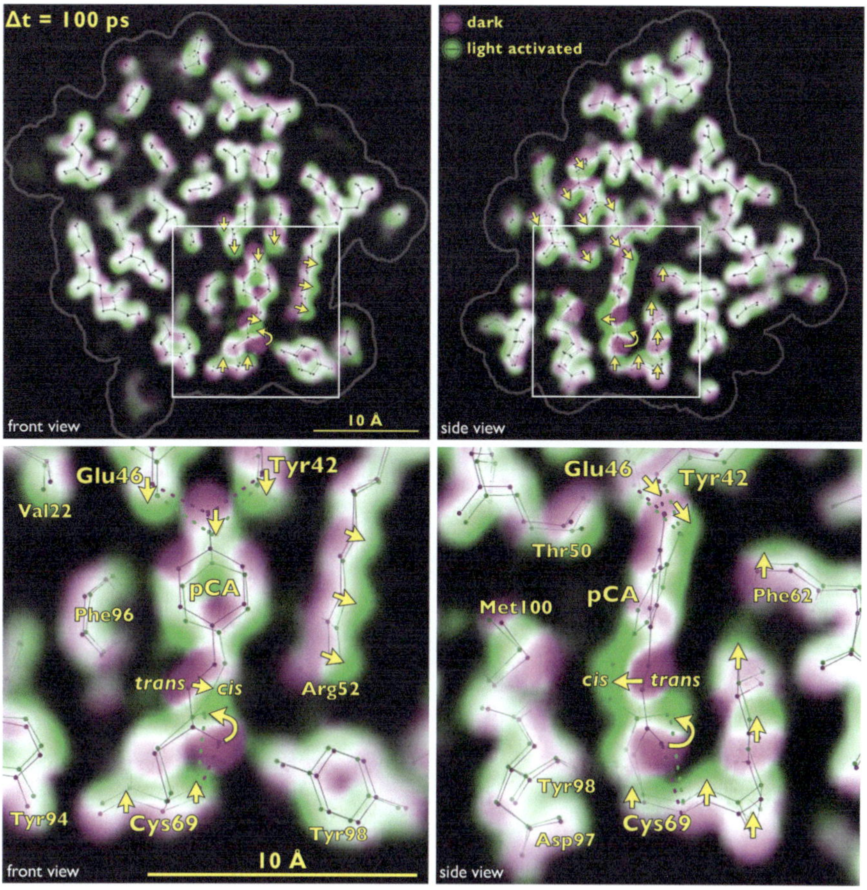

Fig. 3.4 Front and side views of time-resolved structural changes recorded 100 ps after photoexcitation of PYP. The *bottom panels* show an expanded, annotated view of the boxed region in the *top panels*. The ground state electron density map in the color-coded overlays is colored *magenta*, and the 100-ps map is colored *green*. Where *magenta* and *green* overlap, the electron density blends to *white*. The *magenta-to-green* color gradient unveils the direction of atomic motion; large-amplitude displacements are indicated with *yellow arrows*. The stick models correspond to refined structures for the ground state (pG) and the first intermediate (pR_0). Hydrogen bonds to the pCA chromophore are indicated as *dotted lines*

contamination of its base map with electron density from the ground state. This contamination was found to be negligible when the photoproduct yield was set to 10 %. When the photoproduct yield of pB_0 was assumed to be 10 % as well, the electron density map assigned to pB_0 appeared to be a 50:50 mixture of the ground and pB_0 states; when set to 5 %, the electron density map for pB_0 is well described by a single structure. Hence, about 50 % of the pR_2 population short-circuits to the ground state during the pR_2 to pB_0 transition, and our kinetic model was modified accordingly.

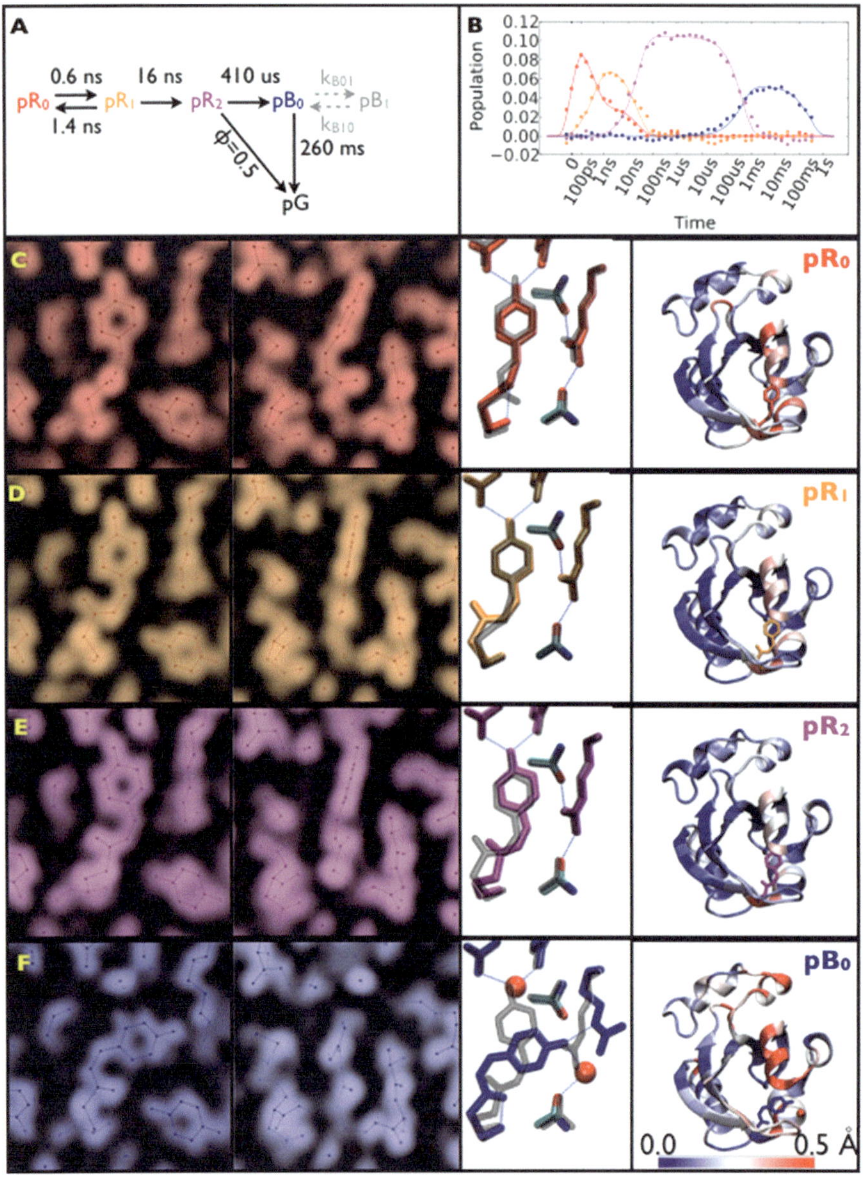

Fig. 3.5 Time-resolved population of transient intermediates and their structures. (**A**) Kinetic model used to account for structural changes spanning ten decades of time; the *arrows* are labeled with the inverse of the globally refined rate constants. Half the population short-circuits to the ground state during the pR_2 to pB_0 transition. (**B**) Time-dependent populations of each intermediate in the PYP photocycle: theoretical population predicted by the kinetic model (*solid lines*) and least-squares contributions from the four electron density base maps (*filled circles*). (**C**–**F**) Structures for pR_0, pR_1, pR_2, and pB_0 intermediates. *Left panel*: electron density base maps were derived from global analysis and phased according to their refined structures (front and side views). *Middle panel*: refined structures of pCA intermediates and their hydrogen-bonding partners. To highlight the structural

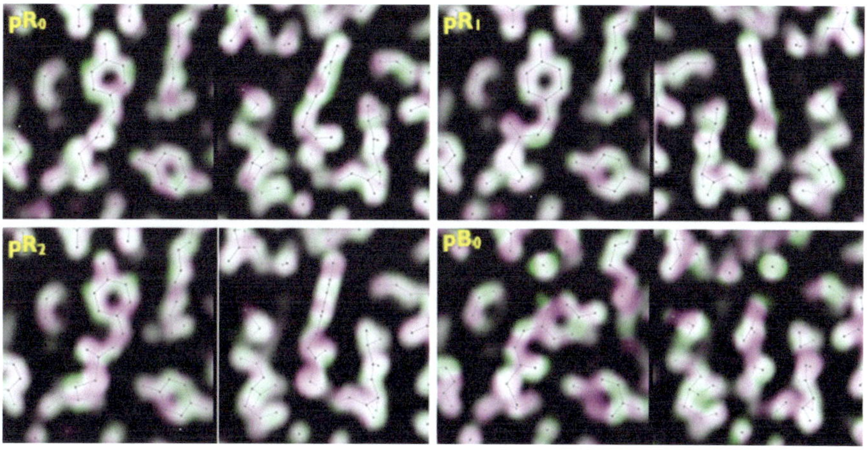

Fig. 3.6 Fit residuals are visualized via color-coded overlays of experimental (*green*) and calculated (*magenta*) electron density maps for pR_0, pR_1, pR_2, and pB_0 intermediates (front and side views). The experimental maps were obtained by global analysis of the time-resolved structure factor amplitudes. The calculated maps were generated from single structures refined against the experimental maps. The pastel coloration of the pCA electron density appears to vary randomly, demonstrating that the sequential kinetic model in Fig. 3.5A accounts for the time-dependent electron density changes with high fidelity

We assessed the validity of our kinetic model by comparing the experimental electron density maps with those computed from the refined atomic coordinates for each intermediate. If differences between the experimental and computed electron density maps rise above the noise and vary systematically in space, the kinetic model can be deemed invalid and would require modification. The aim is to find a kinetic model that accounts for the time-resolved electron density maps with the fewest possible intermediates and rate parameters, i.e., Occam's razor. The fit residuals visualized in Fig. 3.6 demonstrate that the kinetic model used in this study reproduces the experimentally determined time-resolved structures with high fidelity. This validation test failed for the first kinetic model tried: when the pR_0 to pR_1 transition was considered irreversible, the residuals between the calculated and experimental electron densities for pR_1 revealed significant contamination by pR_0 (Fig. 3.7). After making the pR_0 to pR_1 transition reversible, this contamination disappeared (Fig. 3.8).

Fig. 3.5 (continued) changes leading to the corresponding intermediate, they are overlaid with a semitransparent structure (*gray*) of the preceding state. *Right panel*: color-coded rendering of the protein backbone according to Cα displacement relative to pG (see scale in *bottom panel*)

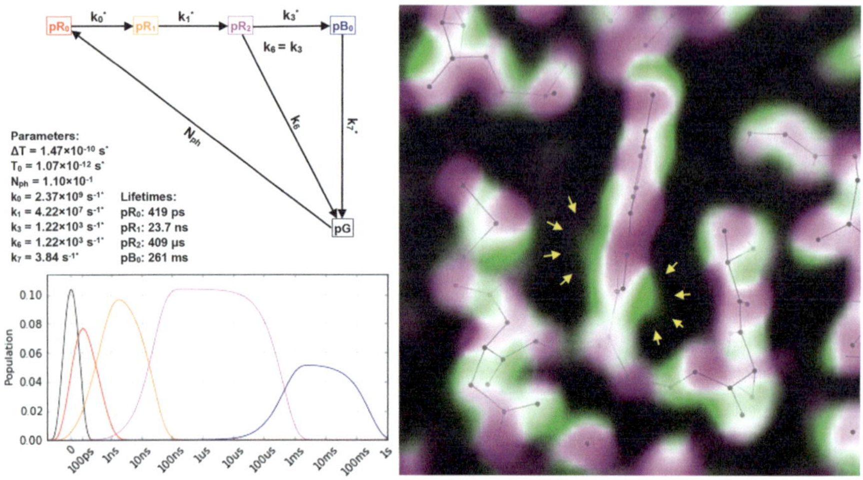

Fig. 3.7 *Left panel*: sequential kinetic model, refined parameters, and the corresponding population dynamics. *Right*: color-coded overlay of the pR_1 intermediate – atomic model density (*magenta*; F_O map phased with pR_1 model) and global analysis density (*green*; $2\,mF_O - DF_C$ map phased with pR_1 model) and ball-and-stick model of pR_1 state (*gray*). The excess green density highlighted by *yellow arrows* is found in the same location as the carbonyl oxygen in the pR_0 state, suggesting persistence of that state

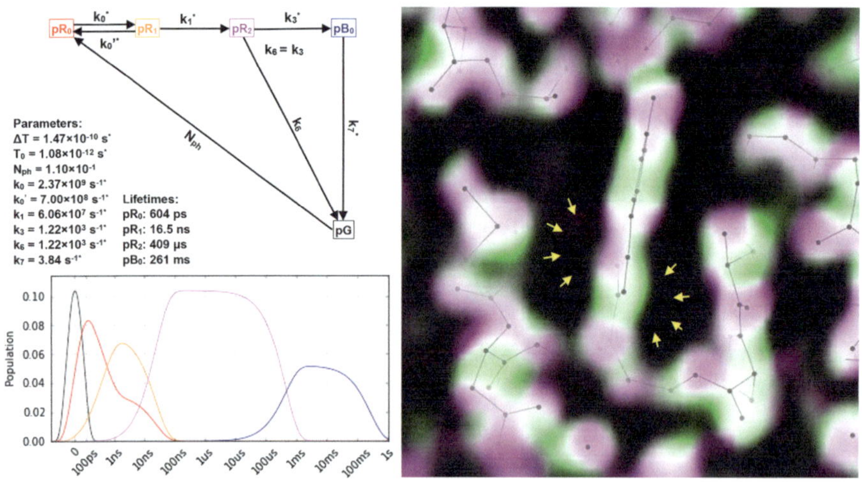

Fig. 3.8 *Left panel*: nonsequential kinetic model with equilibrium between pR_0 and pR_1, refined parameters, and the corresponding population dynamics. *Right*: color-coded overlay of the pR_1 intermediate – atomic model density (*magenta*; F_O map phased with pR_1 model) and global analysis density (*green*; $2\,mF_O - DF_C$ map phased with pR_1 model) and ball-and-stick model of pR_1 state (*gray*). The electron density differences are modest, with little evidence for pR_0 contamination

3.4 Discussion

The pR_0 structure corresponds to a strained *cis* intermediate that launches the PYP photocycle. The pCA carbonyl in pR_0 is oriented ~90° out of the plane of the phenolate and appears to be locked in this twisted conformation by the hydrogen bond between this carbonyl and the Cys69 backbone nitrogen. Characterization of pR_0 as a strained *cis* conformation is strongly supported by kinetic arguments. The equilibrium established between pR_0 and pR_1 precludes the possibility that pR_0 is a ~90° twisted, transition-state-like conformation, as the energy required to break the *cis* double bond in pR_1 and return to pR_0 would be far too high. Moreover, our finding that the quantum yield for producing pR_2 from pR_0 is near unity would be highly unlikely if pR_0 was a transition-state-like species, from which return to the *trans* pG ground state should be equally likely. According to DFT energy calculations [30], pR_0 was found to be similar in energy to pR_1, but pR_2 is about 7 kcal mol^{-1} lower. The DFT results are fully consistent with our observation that the transition between pR_0 and pR_1 is reversible, but the subsequent transition leading to pR_2 is irreversible.

The intermediate structures unveiled in this study permit a detailed examination of mechanistic issues in the PYP photocycle. To facilitate visual comparison of these intermediates, Fig. 3.5 shows their structures in three different views: color-coded electron density maps of pCA and its immediate surroundings, refined structures for pCA and relevant side chains, and color-coded renderings of protein backbone displacement.

The ultimate fate of pR_0 is dictated by the relative strength of its hydrogen bonds with the protein. In wild-type PYP, the hydrogen bond between the Cys69 backbone nitrogen and the pCA carbonyl must be weaker than the phenolate hydrogen bond to Glu46, as the pR_0 to pR_1 transition involves rupture of the hydrogen bond to Cys69. In the Glu46Gln mutant of PYP, the opposite would be true, and pR_0 would be expected to transition to the first intermediate recovered (pR_{E46Q}) in prior nanosecond time-resolved Laue studies [34], as recently observed [31]. That structure is similar to pB_0 except the phenolate is hydrogen bonded to Tyr42 instead of Arg52. To be consistent with the nomenclature presented here, we refer to that state as pR_3, as it is further along the pathway to pB_0 than is pR_2. (Jung et al. did not follow our convention and instead refer to this state as pR_1.)

The pR_0 to pR_1 transition enables the pCA to assume a planar configuration, which relieves some strain in the protein and the chromophore, but at the cost of breaking a hydrogen bond. The pR_0 to pR_1 transition is reversible, with the two populations quickly establishing an equilibrium that favors pR_1 approximately 2:1. The pR_0 to pR_1 transition leaves the C2-C1-S-Cβ dihedral angle in a sterically strained *syn* conformation. The pR_1 to pR_2 transition relieves this strain via a dihedral rotation to an *anti* conformation.

The pR_2 state is the most stable redshifted intermediate observed in this study and eventually transitions to pB_0 with a time constant of 410 μs. Compared to the first two structural transitions, which involve localized motions, the pR_2 to pB_0

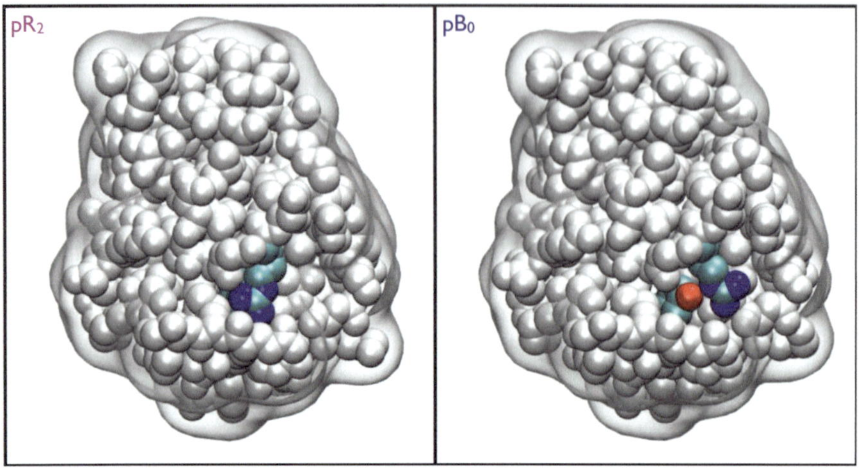

Fig. 3.9 During the pR_2 to pB_0 transition, Arg52 swings from a "closed" to an "open" conformation, exposing the phenolate oxygen (*red sphere*) to water. This open conformation facilitates penetration of a water molecule into the pCA cavity, which is stabilized by hydrogen bonds with Tyr42 and Glu46, and likely prolongs the lifetime of the pB_0 state

structural transition is complex and requires breaking two short (and strong) hydrogen bonds between the phenolate and Tyr42/Glu46, breaking two hydrogen bonds between Arg52 and the protein backbone, and ~180° rotation of one of the pCA dihedral angles. The transition state for this complex reaction pathway likely imposes much strain on the C2═C3 bond, which would account for our finding that approximately half the pR_2 population short-circuits to the ground state during this transition.

During the pR_2 to pB_0 transition, Arg52 switches to an "open" conformation that exposes the pCA phenolate to water (Fig. 3.9) and facilitates its protonation, an event that triggers a blueshift of the pCA absorbance spectrum [35]. Indeed, the pB_0 intermediate has a water molecule hydrogen bonded to the phenolate and the protein backbone. The "open" Arg52 conformation also facilitates penetration of a water molecule into the pCA cavity, which is stabilized by hydrogen bonds to Tyr42 and Glu46. Because this water molecule must escape before the pB_0 state can revert to the pG state, its presence likely prolongs the lifetime of the pB_0 state. It should be noted that the pB_0 state reported here does not have the same global structure as that produced in solution, where the N-terminal domain is free to move beyond the constraints imposed by crystal packing. In our kinetic model, the crystal-inaccessible "signaling" intermediate, in which the N-terminal domain is believed to be partially unfolded [36–38], is denoted pB_1.

Changes in backbone displacement during the PYP photocycle are clearly evident in the color-coded (blue-white-red) ribbon structures of Fig. 3.5. These maps show a diminution of red coloration as the photocycle advances from pR_0 to pR_1 to pR_2, indicating relaxation of protein strain. The pR_2 to pB_0 transition, however, exhibits large amplitude, globally distributed displacements of the protein backbone. Although the putative pB_1 "signaling" state is not achieved in the protein crystal, substantial motion in the N-terminal domain is detected during the pR_2 to pB_0 transition, demonstrating that this transition is most likely the triggering event that produces the signaling state.

The driving force for the final pB_0 to pG transition should be sufficiently high to avoid thermal activation of pB_0. This driving force appears to reside primarily in intramolecular strain in the pCA chromophore and the relative strength of its hydrogen-bonding interactions. The pB_0 to pG transition not only relieves the strain imposed by the *cis* conformation of pCA but also exchanges two weaker hydrogen bonds for two short, strong hydrogen bonds. Due to the energetic cost of reversing this transformation, thermal activation of the pG state is an improbable event, as desired for a photoactive protein.

The lifetime of a signaling state must be sufficiently long for an organism to recognize and act upon its message, but not too long to remain stuck in the "on" state. *H. halophila* responds to changes in blue-light intensity in about 0.5 s [6], which is of the same order of magnitude as the 0.26 s pB_0 lifetime reported here. This lifetime is determined by the height of the activation barrier for ground state recovery, which is likely dominated by the energy required to break the C2=C3 *cis* bond. Thus, the pCA prosthetic group, which is capable of *trans* to *cis* photoisomerization, provides an energetically and structurally attractive framework for inducing and sustaining a long-lived signaling state in PYP.

The PYP photocycle found here differs from what was reported in earlier nanosecond time-resolved Laue crystallography studies [27, 29]. Those studies lacked the time resolution to capture pR_0, but managed to characterize pR_1, which was believed to bifurcate to both pR_2 and pR_3 with pR_2 favored 3:2. If pR_3 is present in this study, its population is too small to be characterized. Moreover, those studies did not report water penetration into the pCA cavity [27, 29], except at pH 4 [29], and found the ground state recovery from pB to be biexponential, with most of the population converting to pG on a 10–20 ms time scale [27, 29]. Here, the ground state recovery from pB_0 is exponential with a much longer, 260-ms time constant. The differences between this and earlier studies cannot be ascribed solely to improvements in data quality and the data analysis procedure, but appear to be due to differences in sample preparation. Extreme halophiles tolerate a high intracellular NaCl concentration. For example, *Halobacterium salinarum* grown in 4 M NaCl has been shown to have an intracellular NaCl concentration of 3.6 M [39]. Nevertheless, most studies of PYP have been performed under conditions with relatively little or no NaCl present. In contrast, this study was carried out with $P6_3$ PYP crystals grown with 1.1 M NaCl in D_2O, pD 9.0. Under these conditions, which are arguably more physiologically relevant compared to prior crystallography studies, the PYP photocycle can be described by a simple, sequential model with four well-defined structural intermediates, whose lifetimes correlate favorably with those measured in solution studies. Similar results have been obtained at pH 7.0, demonstrating that the differences between this and earlier work are not due to pH or H/D isotope differences. Evidently, the presence of 1.1 M NaCl mediates penetration of water into the pCA cavity, which accounts for the long-lived pB_0 state reported here. Interestingly, the longer-lived pB population reported in earlier studies [27, 29] has a lifetime similar to what we find here and may arise from water occupancy in the pCA cavity that is too low to be detected in their electron density maps.

The intermediates found in this work provide a structural framework for interpreting the wealth of prior time-resolved spectroscopic studies of PYP in solution, a few examples of which are presented here. (i) Transient optical absorbance spectra of PYP in solution [20, 22] unveiled a short-lived intermediate peaked around 510 nm that relaxed to a longer-lived feature peaked around 460 nm. We associate the short-lived 510-nm intermediate with pR_0 and the longer-lived 460-nm intermediate with pR_1 and pR_2, which are similar in structure and would be expected to have similar optical absorbance spectra. The lifetimes reported for the short-lived intermediate, 3.0 ns for $I_0^{\ddagger}$ [22] and 1.3 ns for PYP_B [20], are longer than the 600-ps lifetime we find for pR_0; however, the lifetime recovered in this study is expected to be shortened by the unavoidable adiabatic temperature jump that arises when photoactivating PYP crystals. This temperature jump is largest at the protein surface and decreases as a function of depth as the laser pulse is absorbed. Given the conditions used in this study, its magnitude decreases from about 30–7 K across the 40-μm vertical cross section of the x-ray beam. Because the thermal conductivity in a protein crystal is quite high and its heat capacity large, the temperature in the x-ray-illuminated volume of the crystal rapidly cools to the cooling-stream temperature. Consequently, the early-time dynamics are recorded at an elevated temperature, which would be expected to shorten the lifetimes of the earliest detected intermediates, while the longtime dynamics are recorded at a near-normal temperature. In solution-phase optical studies, the protein concentration is comparatively low, and the adiabatic temperature jump is negligible. (ii) Prior studies uncovered an equilibrium between red- and blueshifted intermediates [40, 41]. We find no evidence for an equilibrium between pR_2 and pB_0, nor would we expect it for reasons already discussed; however, if pR_3 were formed in solution, it would be expected to coexist in equilibrium with pB_0 and offer a possible structural explanation for this equilibrium and for the biexponential kinetics reported for the conversion of pR to pB. (iii) Hellingwerf and coworkers characterized the PYP photocycle over a wide range of pH and pD [41] and in Table 1 of ref 30 reported relaxation times at pH 8.1 and pH 9.55, whose averages are remarkably similar to what we find in PYP crystals at pD 9.0: 420 (410) μs for pR to pB and 220 (260) ms for pB to pG. The fact that the pR and pB lifetimes in crystals grown in ~1.1 M NaCl are similar to those found in solution at a similar pH/pD [41] suggests that the pCA photocycle in our crystals is authentic and is only loosely connected to the large-amplitude displacement that accompanies the putative pB_0 to pB_1 signaling transition.

The time-resolved structural evolution presented here, spanning ten decades of time, unveils a simple, stepwise structural progression of PYP conformations toward a long-lived pB_0 state, with each transition characterized at an unprecedented level of detail. The highly contorted structure of the pR_0 state provides a visual clue regarding the conformational gymnastics that must accompany *trans*/*cis* isomerization in a highly constrained protein environment. These results illustrate how a protein can employ hydrogen-bonding networks, gated water penetration, and strain to steer the direction of structural transitions in a fashion that facilitates its target function. Such detailed information is crucial to properly assess and validate theoretical and computational approaches in biophysics and to develop a detailed understanding of the relations between protein structure, dynamics, and function.

Acknowledgments We thank Gerhard Hummer and Ville Kaila for useful discussions. Structures in several figures were generated using VMD: http://www.ks.uiuc.edu/Research/vmd/. Use of the BioCARS Sector 14 was supported by grants from the National Center for Research Resources (5P41RR007707) and the National Institute of General Medical Sciences (8P41GM103543) from the National Institutes of Health. Use of the Advanced Photon Source, an Office of Science User Facility operated for the US Department of Energy (DOE) Office of Science by Argonne National Laboratory, was supported by the US DOE under Contract No. DE-AC02-06CH11357. The time-resolved setup at Sector 14 was funded in part through a collaboration with Philip Anfinrud (NIH/NIDDK). This research was supported by the Intramural Research Program of the National Institute of Diabetes and Digestive and Kidney Diseases and National Institutes of Health.

References

1. Eaton WA, Henry ER, Hofrichter J (1996) Nanosecond crystallographic snapshots of protein structural changes. Science 274(5293):1631–1632
2. Srajer V, Teng TY, Ursby T, Pradervand C, Ren Z, Adachi S, Schildkamp W, Bourgeois D, Wulff M, Moffat K (1996) Photolysis of the carbon monoxide complex of myoglobin: nanosecond time-resolved crystallography. Science 274(5293):1726–1729
3. Schotte F, Lim MH, Jackson TA, Smirnov AV, Soman J, Olson JS, Phillips GN, Wulff M, Anfinrud PA (2003) Watching a protein as it functions with 150-ps time-resolved X-ray crystallography. Science 300(5627):1944–1947
4. Meyer TE (1985) Isolation and characterization of soluble cytochromes, ferredoxins and other chromophoric proteins from the halophilic phototrophic bacterium ectothiorhodospira-halophila. Biochim Et Biophys Acta 806(1):175–183
5. Hustede E, Liebergesell M, Schlegel HG (1989) The photophobic response of various sulfur and nonsulfur purple bacteria. Photochem Photobiol 50(6):809–815
6. Sprenger WW, Hoff WD, Armitage JP, Hellingwerf KJ (1993) The eubacterium ectothiorhodospira-halophila is negatively phototactic, with a wavelength dependence that fits the absorption-spectrum of the photoactive yellow protein. J Bacteriol 175(10):3096–3104
7. Hellingwerf KJ, Hendriks J, Gensch T (2003) Photoactive yellow protein, a new type of photoreceptor protein: will this "yellow lab" bring us where we want to go? J Phys Chem A 107(8):1082–1094. doi:10.1021/Jp027005y
8. Hoff WD, Vanstokkum IHM, Vanramesdonk HJ, Vanbrederode ME, Brouwer AM, Fitch JC, Meyer TE, Vangrondelle R, Hellingwerf KJ (1994) Measurement and global analysis of the absorbance changes in the photocycle of the photoactive yellow protein from ectothiorhodospira-halophila. Biophys J 67(4):1691–1705
9. Meyer TE, Yakali E, Cusanovich MA, Tollin G (1987) Properties of a water-soluble, yellow protein isolated from a halophilic phototrophic bacterium that has photochemical activity analogous to sensory rhodopsin. Biochemistry 26(2):418–423
10. Meyer TE, Tollin G, Hazzard JH, Cusanovich MA (1989) Photoactive yellow protein from the purple phototrophic bacterium, ectothiorhodospira-halophila – quantum yield of

photobleaching and effects of temperature, alcohols, glycerol, and sucrose on kinetics of photobleaching and recover. Biophys J 56(3):559–564
11. Meyer TE, Tollin G, Causgrove TP, Cheng P, Blankenship RE (1991) Picosecond decay kinetics and quantum yield of fluorescence of the photoactive yellow protein from the halophilic purple phototrophic bacterium, ectothiorhodospira-halophila. Biophys J 59 (5):988–991
12. Mihara K, Hisatomi O, Imamoto Y, Kataoka M, Tokunaga F (1997) Functional expression and site-directed mutagenesis of photoactive yellow protein. J Biochem 121(5):876–880
13. Yamaguchi S, Kamikubo H, Shimizu N, Yamazaki Y, Imamoto Y, Kataoka M (2007) Preparation of large crystals of photoactive yellow protein for neutron diffraction and high resolution crystal structure analysis. Photochem Photobiol 83(2):336–338. doi:10.1562/2006-07-23-Ra-977
14. Graber T, Anderson S, Brewer H, Chen YS, Cho HS, Dashdorj N, Henning RW, Kosheleva I, Macha G, Meron M, Pahl R, Ren Z, Ruan S, Schotte F, Rajer VS, Viccaro PJ, Westferro F, Anfinrud P, Moffat K (2011) BioCARS: a synchrotron resource for time-resolved X-ray science. J Synchrotron Radiat 18:658–670. doi:10.1107/S0909049511009423
15. Coppens P, Pitak M, Gembicky M, Messerschmidt M, Scheins S, Benedict J, Adachi S, Sato T, Nozawa S, Ichiyanagi K, Chollet M, Koshihara S (2009) The RATIO method for time-resolved Laue crystallography. J Synchrotron Radiat 16:226–230. doi:10.1107/S0909049508040892
16. Vorontsov I, Pillet S, Kaminski R, Schmokel MS, Coppens P (2010) LASER – a program for response-ratio refinement of time-resolved diffraction data. J Appl Crystallogr 43:1129–1130. doi:10.1107/S0021889810029900
17. Adams PD, Afonine PV, Bunkoczi G, Chen VB, Davis IW, Echols N, Headd JJ, Hung LW, Kapral GJ, Grosse-Kunstleve RW, McCoy AJ, Moriarty NW, Oeffner R, Read RJ, Richardson DC, Richardson JS, Terwilliger TC, Zwart PH (2010) PHENIX: a comprehensive Python-based system for macromolecular structure solution. Acta Crystallogr D Biol Crystallogr 66 (Pt 2):213–221. doi:10.1107/S0907444909052925
18. Terwilliger TC, Berendzen J (1996) Bayesian difference refinement. Acta Crystallogr D Biol Crystallogr 52(Pt 5):1004–1011. doi:10.1107/S0907444996006725
19. Groot ML, van Wilderen LJGW, Larsen DS, van der Horst MA, van Stokkum IHM, Hellingwerf KJ, van Grondelle R (2003) Initial steps of signal generation in photoactive yellow protein revealed with femtosecond mid-infrared spectroscopy. Biochemistry 42 (34):10054–10059. doi:10.1021/Bi034878p
20. Imamoto Y, Kataoka M, Tokunaga F, Asahi T, Masuhara H (2001) Primary photoreaction of photoactive yellow protein studied by subpicosecond-nanosecond spectroscopy. Biochemistry 40(20):6047–6052. doi:10.1021/Bi002437p
21. Larsen DS, van Stokkum IHM, Vengris M, van der Horst MA, de Weerd FL, Hellingwerf KJ, van Grondelle R (2004) Incoherent manipulation of the photoactive yellow protein photocycle with dispersed pump-dump-probe spectroscopy. Biophys J 87(3):1858–1872. doi:10.1529/Biophysj.104.043794
22. Ujj L, Devanathan S, Meyer TE, Cusanovich MA, Tollin G, Atkinson GH (1998) New photocycle intermediates in the photoactive yellow protein from ectothiorhodospira halophila: picosecond transient absorption spectroscopy. Biophys J 75(1):406–412
23. van Wilderen LJGW, Van der Horst MA, van Stokkum IHM, Hellingwerf KJ, van Grondelle R, Groot ML (2006) Ultrafast infrared spectroscopy reveals a key step for successful entry into the photocycle for photoactive yellow protein. Proc Natl Acad Sci U S A 103 (41):15050–15055. doi:10.1073/Pnas.0603476103
24. Anderson S, Srajer V, Moffat K (2004) Structural heterogeneity of cryotrapped intermediates in the bacterial blue light photoreceptor, photoactive yellow protein. Photochem Photobiol 80 (1):7–14
25. Genick UK, Soltis SM, Kuhn P, Canestrelli IL, Getzoff ED (1998) Structure at 0.85 A resolution of an early protein photocycle intermediate. Nature 392(6672):206–209. doi:10.1038/32462

26. Kort R, Hellingwerf KJ, Ravelli RBG (2004) Initial events in the photocycle of photoactive yellow protein. J Biol Chem 279(25):26417–26424. doi:10.1074/Jbc.M311961200
27. Ihee H, Rajagopal S, Srajer V, Pahl R, Anderson S, Schmidt M, Schotte F, Anfinrud PA, Wulff M, Moffat K (2005) Visualizing reaction pathways in photoactive yellow protein from nanoseconds to seconds. Proc Natl Acad Sci U S A 102(20):7145–7150. doi:10.1073/Pnas.0409035102
28. Perman B, Srajer V, Ren Z, Teng T, Pradervand C, Ursby T, Bourgeois D, Schotte F, Wulff M, Kort R, Hellingwerf K, Moffat K (1998) Energy transduction on the nanosecond time scale: early structural events in a xanthopsin photocycle. Science 279(5358):1946–1950
29. Tripathi S, Srajer V, Purwar N, Henning R, Schmidt M (2012) pH dependence of the photoactive yellow protein photocycle investigated by time-resolved crystallography. Biophys J 102(2):325–332. doi:10.1016/J.Bpj.2011.11.4021
30. Schotte F, Cho HS, Kaila VR, Kamikubo H, Dashdorj N, Henry ER, Graber TJ, Henning R, Wulff M, Hummer G, Kataoka M, Anfinrud PA (2012) Watching a signaling protein function in real time via 100-ps time-resolved Laue crystallography. Proc Natl Acad Sci U S A 109 (47):19256–19261. doi:10.1073/pnas.1210938109
31. Jung YO, Lee JH, Kim J, Schmidt M, Moffat K, Srajer V, Ihee H (2013) Volume-conserving trans-cis isomerization pathways in photoactive yellow protein visualized by picosecond X-ray crystallography. Nat Chem 5(3):212–220. doi:10.1038/nchem.1565
32. Genick UK, Borgstahl GE, Ng K, Ren Z, Pradervand C, Burke PM, Srajer V, Teng TY, Schildkamp W, McRee DE, Moffat K, Getzoff ED (1997) Structure of a protein photocycle intermediate by millisecond time-resolved crystallography. Science 275(5305):1471–1475
33. Anderson S, Crosson S, Moffat K (2004) Short hydrogen bonds in photoactive yellow protein. Acta Crystallogr D Biol Crystallogr 60:1008–1016. doi:10.1107/S090744490400616x
34. Rajagopal S, Anderson S, Srajer V, Schmidt M, Pahl R, Moffat K (2005) A structural pathway for signaling in the E46Q mutant of photoactive yellow protein. Structure 13(1):55–63. doi:10.1016/J.Str.2004.10.016
35. Hendriks J, Hoff WD, Crielaard W, Hellingwerf KJ (1999) Protonation deprotonation reactions triggered by photoactivation of photoactive yellow protein from Ectothiorhodospira halophila. J Biol Chem 274(25):17655–17660
36. Harigai M, Yasuda S, Imamoto Y, Yoshihara K, Tokunaga F, Kataoka M (2001) Amino acids in the N-terminal region regulate the photocycle of photoactive yellow protein. J Biochem 130 (1):51–56
37. Harigai M, Imamoto Y, Kamikubo H, Yamazaki Y, Kataoka M (2003) Role of an N-terminal loop in the secondary structural change of photoactive yellow protein. Biochemistry 42 (47):13893–13900. doi:10.1021/Bi034814e
38. van der Horst MA, van Stokkum IH, Crielaard W, Hellingwerf KJ (2001) The role of the N-terminal domain of photoactive yellow protein in the transient partial unfolding during signalling state formation. Febs Lett 497(1):26–30
39. Christian JH, Waltho JA (1962) Solute concentrations within cells of halophilic and non-halophilic bacteria. Biochim Biophys Acta 65:506–508
40. Borucki B, Otto H, Joshi CP, Gasperi C, Cusanovich MA, Devanathan S, Tollin G, Heyn MP (2003) pH dependence of the photocycle kinetics of the E46Q mutant of photoactive yellow protein: protonation equilibrium between I-1 and I-2 intermediates, chromophore deprotonation by hydroxyl uptake, and protonation relaxation of the dark state. Biochemistry 42(29):8780–8790. doi:10.1021/Bi034315d
41. Hendriks J, van Stokkum IHM, Hellingwerf KJ (2003) Deuterium isotope effects in the photocycle transitions of the photoactive yellow protein. Biophys J 84(2):1180–1191

Chapter 4
NMR Explorations of Biomolecular Systems with Rapid Conformational Exchanges

Maho Yagi-Utsumi, Takumi Yamaguchi, Ryo Kitahara, and Koichi Kato

Abstract Biomolecules such as proteins and oligosaccharides undergo dynamic conformational exchanges, which are relevant to regulation of biologically functional processes as typified by molecular recognition. Nuclear magnetic resonance (NMR) spectroscopy provides useful approaches to characterize the conformational dynamics of biomolecules over a broad range of time scales. However, detailed characterizations of individual conformers are inherently challenging for those biomolecules that exhibit rapid conformational interconversions. Here we describe several NMR strategies to deal with dynamic conformational equilibria and ensembles using monomeric and dimeric ubiquitin (Ub) and the oligosaccharide moieties of gangliosides as model molecules. A specific Ub conformer could be enriched using high pressure combined with a single amino acid substitution. Introducing a covalent linkage constrained the conformational states of a Ub dimer. NMR spectroscopy was also useful for validating molecular dynamics simulations of highly flexible oligosaccharides. These methods provided for detailed determinations of dynamic conformational exchanges that involve minor conformational species.

M. Yagi-Utsumi
Okazaki Institute for Integrative Bioscience and Institute for Molecular Science, National Institutes of Natural Sciences, 5-1 Higashiyama, Myodaiji, Okazaki 444-8787, Japan

Graduate School of Pharmaceutical Sciences, Nagoya City University, 3-1 Tanabe-dori, Mizuho-ku, Nagoya 467-8603, Japan

Department of Chemistry, University of Cambridge, Lensfield Road, Cambridge CB2 1EW, UK

T. Yamaguchi • K. Kato (✉)
Okazaki Institute for Integrative Bioscience and Institute for Molecular Science, National Institutes of Natural Sciences, 5-1 Higashiyama, Myodaiji, Okazaki 444-8787, Japan

Graduate School of Pharmaceutical Sciences, Nagoya City University, 3-1 Tanabe-dori, Mizuho-ku, Nagoya 467-8603, Japan
e-mail: kkatonmr@ims.ac.jp

R. Kitahara
College of Pharmaceutical Sciences, Ritsumeikan University, 1-1-1 Noji-higashi, Kusatsu 525-8577, Japan

M. Terazima et al. (eds.), *Molecular Science of Fluctuations Toward Biological Functions*, DOI 10.1007/978-4-431-55840-8_4

Keywords Nuclear magnetic resonance spectroscopy • Conformational exchange • Ubiquitin • Oligosaccharides • Molecular dynamics simulation

4.1 Introduction

As a part of living systems, biomacromolecules are characterized by their various degrees of freedom with regard to internal motions, which provide them with conformational plasticity that is required during binding and reaction processes and also for conformational adaptability to their binding partners. For example, many proteins exhibit multiple conformations in the absence of their cognate ligands, which select and stabilize specific conformers of the target proteins [1, 2]. Conformational dynamics during molecular recognition processes are exemplified by intrinsically disordered proteins that are devoid of stable secondary or tertiary structures and are often folded into specific secondary structures upon interacting with their binding partners [3, 4].

Nuclear magnetic resonance (NMR) spectroscopy provides potentially useful tools for characterizing the conformational dynamics of biomolecules over a wide range of time scales (Fig. 4.1) [5, 6]. If the conformational interconversion is sufficiently slow on the NMR time scale, then NMR signals originating from individual conformers are observable and provide detailed information for each conformational state. However, under circumstances when conformational transitions occur faster, the NMR peaks converge at a position that corresponds to a population-weighted average of chemical shifts. Even under these conditions, NMR relaxation dispersion experiments can provide information on a chemical shift and a population of conformers as well as the exchange rate if the interconversion occurs on the micro-to-millisecond time scale [7–9].

Conformational fluctuations on the subnanosecond time scale are characterized based on measurements of longitudinal and transverse relaxation rates (R_1 and R_2, respectively) and the heteronuclear Overhauser effect (NOE). For example, a model-free formalism is widely used to describe protein backbone dynamics with order parameters and correlation times of individual amino acid residues on the basis of observations of ^{15}N R_1, R_2, and {^{1}H}-^{15}N NOE [10, 11]. However, it is challenging to characterize individual conformers that form such a dynamic ensemble of biomolecules.

In this chapter, we present several examples of strategies for obtaining detailed information on specific conformers under dynamic equilibrium, particularly minor conformational species under physiological conditions.

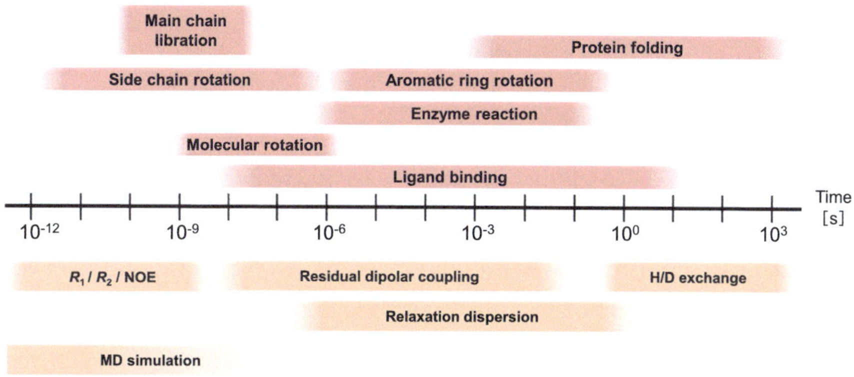

Fig. 4.1 Biomolecular NMR experiments to characterize protein dynamics on different time scales

4.2 Trapping Conformational States

4.2.1 *High-Pressure NMR Combined with a Mutational Approach*

In many biomolecular NMR studies, spectral measurements are made under equilibrium solution conditions in which the molecules undergo dynamic conformational interconversions. Among pre-existing conformers, major conformers are not always biologically active. For detailed structural analyses of their minor conformational states, the equilibria are perturbed by changing solution conditions, such as the pH, salt concentration, and temperature, so as to make the originally minor species more populated.

High-pressure NMR spectroscopy is one of the most powerful methods for capturing low-population conformational species of proteins, which have higher Gibbs free energies than their basic folded state (native state). Because the conformational fluctuations of a protein in solution are closely related to the fluctuations of its partial molar volume [12, 13], applying high pressure causes a population shift of conformers within the ensemble, which results in NMR spectral changes [14]. Here we illustrate the utility of the high-pressure NMR approach using ubiquitin (Ub) as a model.

Ub is a small protein that consists of 76 amino acid residues and is a posttranslational protein modifier, which plays crucial regulatory roles in diverse cellular events [15]. A Ub modification can be made through the formation of isopeptide linkages at its C-terminus. Although Ub can be covalently attached to target proteins in a monomeric form, it can also be linked to another Ub to generate various types of polyUb chains. PolyUb chains that are mediated through a Lys48 isopeptide linkage provide a signal that is recognized by the 26S proteasome for selective protein degradation in cells. The formation of Ub isopeptide linkages is catalyzed by three enzymes, Ub-activating enzyme E1, Ub-conjugating enzyme E2, and Ub-protein ligase E3, in sequential order.

Previously reported high-pressure NMR investigations demonstrated that Ub existed in equilibrium among its natively folded state N_1, a totally unfolded state U, and at least two high-energy states, an alternatively folded state N_2 and a locally disordered state I [16–18]. Based on spectral data acquired at low (30 bar) and high (3 kbar) pressures, it was concluded that the N_2 state accounted for 77 % of the population at high pressure, which involved reorientation of its C-terminal β_5-strand and α_1-helix and provided an open conformation in the proximity of its C-terminus [18]. The structural changes of pressure-stabilized N_2 were supported by residual dipolar coupling (RDC) experiments [19] and 1 μs molecular dynamics (MD) simulations [20] of the protein under high pressure.

We demonstrated that the high-pressure NMR approach provided a new strategy for amplifying particular fluctuations in proteins by rational mutation designed based on the conformational states observed under different pressure conditions [21]. With this strategy, the N_1 (30 bar) and N_2 (3 kbar) structures of wild-type (WT) Ub were first compared to identify the key intramolecular interactions. Subsequently, the hydrogen bond or salt bridge thus identified was systematically broken by single amino acid replacements to control conformational fluctuations. A series of Ub mutants (K11A, E34A, Q41A, and Q41N) were subjected to heteronuclear single-quantum coherence (HSQC) experiment-based chemical shift comparisons (Fig. 4.2), which indicated that the Q41N mutant preferentially (70 %) adopted the N_2 conformation at atmospheric pressure. The NMR structure of the N_2 conformer of Q41N-mutated Ub determined at atmospheric pressure confirmed that the hydrogen bond between the Ile36 backbone carbonyl group and the Gln41 side chain amide group was critical for controlling the N_1-N_2 conformational interconversion in WT Ub.

In the Q41N mutant, the N_2 state accounted for 97 % of the population at high pressure (2.5 kbar), which enabled detailed conformational characterizations of the "pure" N_2 state of Ub [22]. The N_2 state exhibited an altered orientation of β_5-strand as compared with that of the natively folded N_1 state (Fig. 4.3). Interestingly, the conformation of N_2 was quite similar to the crystal structure of Ub in complex with the Ub-activating enzyme E1 (Fig. 4.4), which indicated that Ub recognition by E1 was best explained by conformational selection rather than induced-fit motion. Similar high-energy, biologically active states are shared by NEDD8, a Ub-like posttranslational modifier that has its own E1–E2–E3 reaction cascade [23]. These findings suggest that both the major ground state and the minor active conformational state are conserved among the Ub-like modifiers.

Thus, high-pressure NMR measurements combined with suitably chosen point mutations could be a useful strategy for conformational trapping of natively low-populated but functionally relevant protein states.

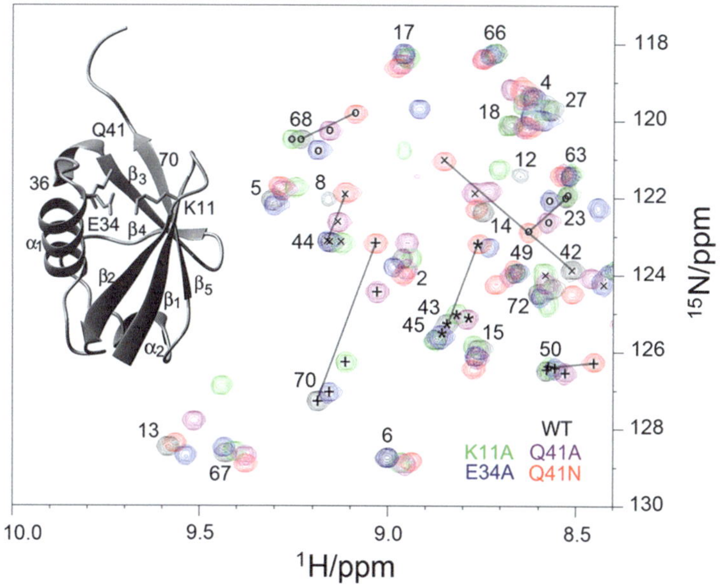

Fig. 4.2 Overlay of the HSQC spectra of WT (*black*) and mutated Ubs (K11A, *green*; E34A, *blue*; Q41A, *purple*; Q41N, *red*). The substituted side chains are shown in the inset Ub structure. All mutant peaks lie approximately in a straight line that connects the WT and Q41N peaks (Adapted with permission from Ref. [21])

4.2.2 Covalent Constraints of Biomolecular Conformations

Protein functions are derived primarily through the cooperative interplay between two or more domains. The plasticity in spatial arrangements of these domains is a key feature of multidomain proteins endowed with sophisticated molecular mechanisms typified by allosteric regulation. Interconversion of domain arrangements takes place over a wide range of temporal and spatial time scales. When domain rearrangements are rapid, it is again not possible to describe individual conformational states based on NMR data obtained as population-weighted averages unless specific conformers can be *frozen*. Here we present our results for characterizing the conformational dynamics of a Lys48-linked Ub dimer (diUb) by NMR spectroscopy in conjunction with X-ray crystallography [24].

The previously reported crystal structure of Lys48-linked diUb exhibited a compact closed conformation in which the hydrophobic surfaces in both Ub units were packed against each other and shielded from the solvent [25]. The Ub hydrophobic surface is involved in interactions with the Ub-binding proteins that function in the Ub-/proteasome-mediated protein degradation pathway, including the proteasome subunits Rpn10 and Rpn13 [26, 27]. Therefore, one may wonder how Ub-binding proteins can efficiently access the functionally important but sequestered hydrophobic surface of the Lys48-linked polyUb chains. We addressed

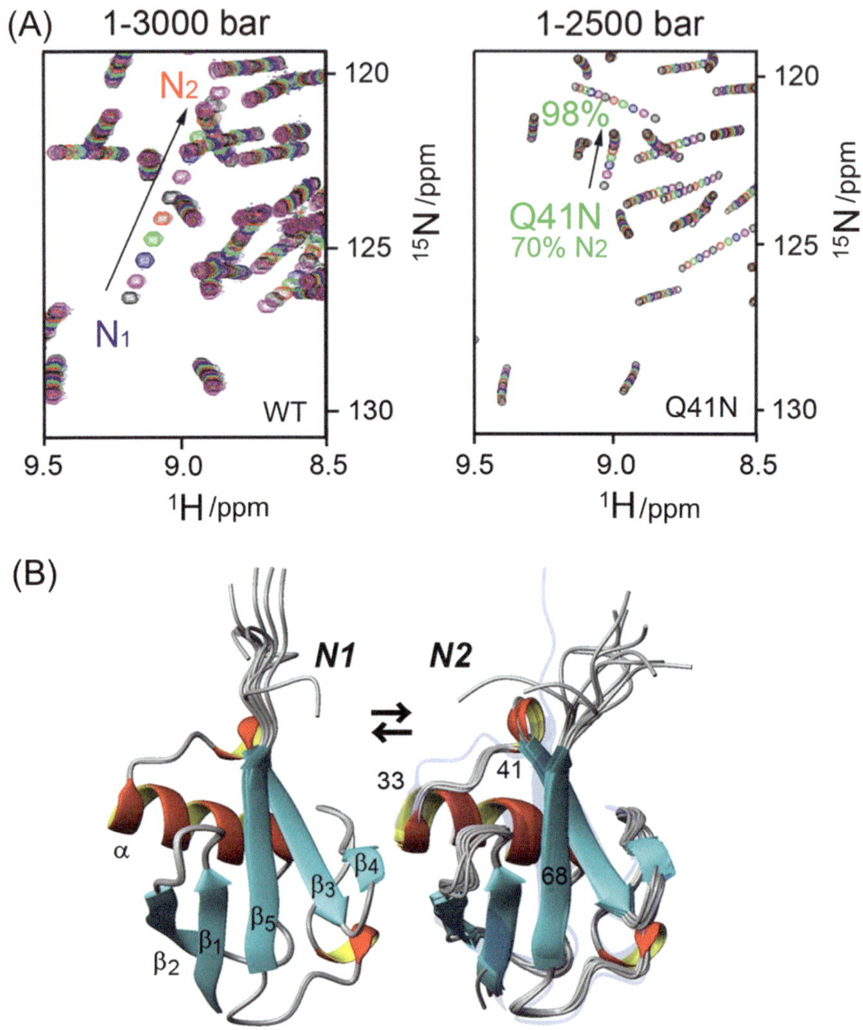

Fig. 4.3 (**a**) Superposition of HSQC spectra of WT (*left*) and Q41N (*right*) Ub at different pressures. The Q41N mutant preferentially (70 %) adopts the N_2 conformation at atmospheric pressure. (**b**) Solution structure of WT ubiquitin at 1 bar, a model for the basic folded conformation N_1 [Protein Data Bank (PDB) entry 1D3Z] (*left*), and the Q41N variant at 2500 bar, model for the alternatively folded conformation N_2 (PDB entry 2RU6) (*right*) (Adapted with permissions from Refs. [21] and [22])

this using Lys48-linked diUb prepared by in vitro E1–E2 (E2-25 K) enzymatic reactions [28], which generated WT Lys48-linked polyUb chains as well as their cyclic forms, in which each Ub unit was linked *via* two isopeptide linkages (Fig. 4.5). Chromatographically isolated WT and cyclic Lys48-linked diUb were then subjected to structural analyses.

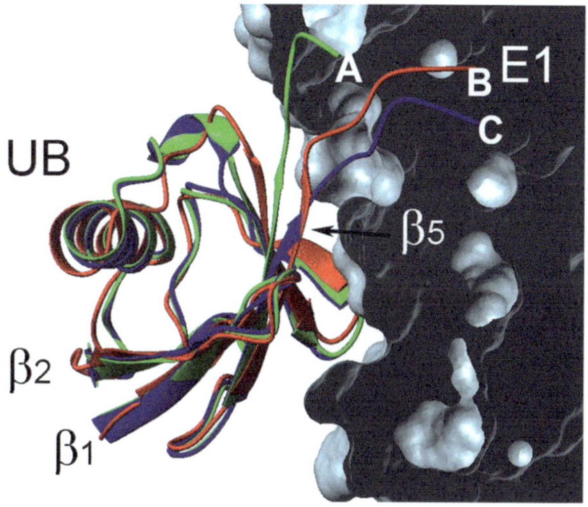

Fig. 4.4 Superposition of NMR-derived structures of WT Ub at 1 bar (*green*, PDB entry 1D3Z) and Q41N at 2500 bar (*red*, PDB entry 2RU6) on the crystal structure of the Ub–E1 complex (*blue*, PDB entry 3CMM). E1 is shown as a molecular surface (*gray*) (Adapted with permission from Ref. [22])

Intriguingly, our resolved crystal structure of WT Lys48-linked diUb exhibited an open conformation in contrast to the previously reported crystal structures and suggested that the crystallographic data provided snapshots of the diUb molecule under dynamic equilibrium in solution (Fig. 4.6a). Furthermore, our NMR-derived 3D structure of the cyclic diUb form closely resembled the closed conformation observed in the crystals (Fig. 4.6b). Thus, the closed conformer could be successfully *isolated* by introducing additional isopeptide linkages. Interestingly, endogenous cyclic Lys48-linked diUb was detected in skeletal muscle and cultured mammalian cells [29], although its physiological role remains unknown.

HSQC spectral data of WT Lys48-linked diUb were compared with those of the cyclic form and monomeric Ub, which mimics an open form of diUb in terms of its hydrophobic surface exposure (Fig. 4.7). In WT diUb, each amino acid residue located on the hydrophobic surface provided a single HSQC peak between the peaks that originated from the corresponding sites in monomeric Ub and cyclic diUb in the same straight line with an identical dividing ratio. These data demonstrated that Lys48-linked diUb interconverted between open and closed conformations, which were mimicked by monomeric Ub and cyclic diUb, respectively, in the fast-exchange regime. In this system, the populations of open and closed conformers were determined to be 75 % and 25 %, respectively, based simply on the division ratio of the chemical shift difference. The open conformation accounted for a greater population under lower pH conditions and almost completely predominated at pH 4.5 (Fig. 4.8). These data suggested that the intrinsic properties of Lys48-linked Ub chains to adopt the open conformation may be advantageous for their interactions with Ub-binding proteins.

This study demonstrates the utility of conformational trapping of multidomain proteins by introducing covalent linkage constraints.

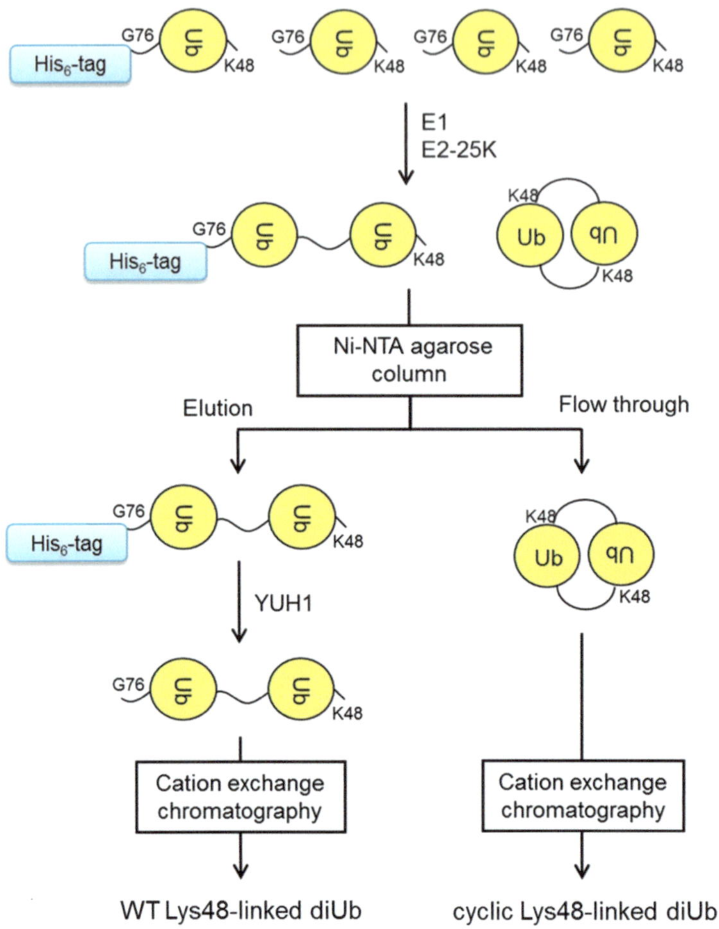

Fig. 4.5 Schematic diagram for preparing WT and cyclic Lys48-linked Ub chains (Adapted with permission from Ref. [24])

4.3 Exploration of Conformational Spaces of Flexible Biomolecules

In the previous sections, we described protein conformational dynamics as interchanges among a few conformers. However, in general, biomolecular dynamics involve not only conformational transitions from a low-energy minimum to another in the energy landscape but also harmonic motions in each energy minimum. This is particularly true for highly flexible biomolecules best represented by oligosaccharides and intrinsically disordered proteins. Oligosaccharides have branched structures and have numerous degrees of freedom of motion, which provide conformational spaces with a number of shallow and broad energy minima

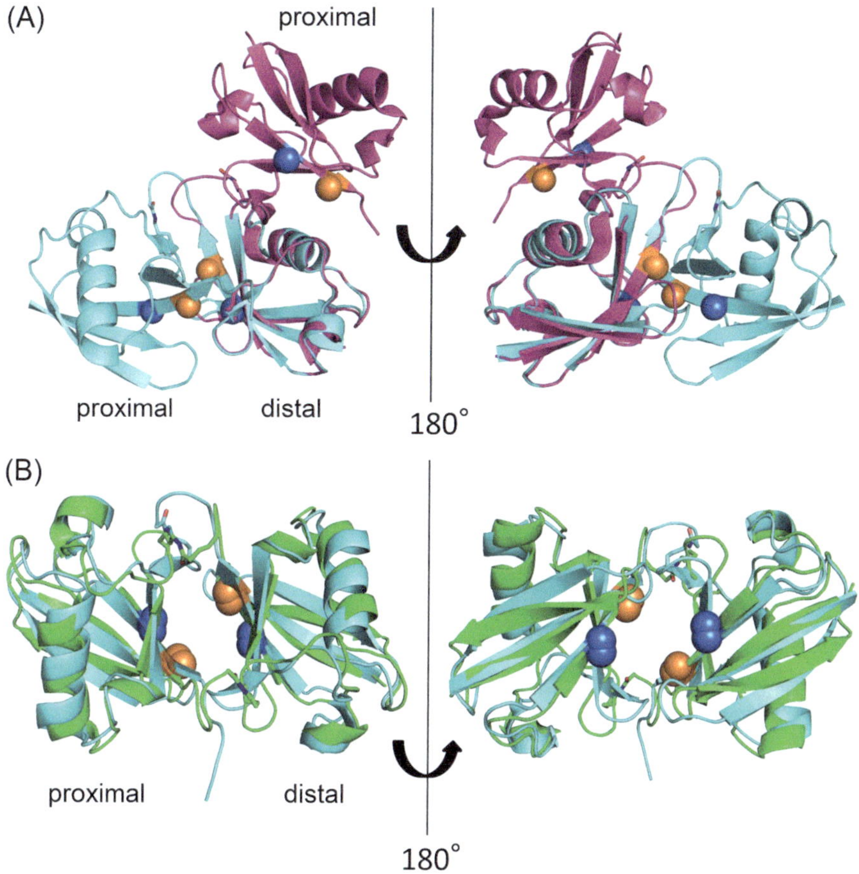

Fig. 4.6 (**a**) Superposition of the crystal structures of WT Lys48-linked diUb in the open (present structure; PDB code, 3AUL) (*magenta*) and closed forms (PDB code, 1AAR) (*cyan*). (**b**) Superposition of the NMR-derived model of cyclic Lys48-linked diUb (*green*) and the crystal structure of WT Lys48-linked diUb showing the closed conformation (PDB code, 1AAR) (*cyan*). The C-terminal group of Ub is linked to another Ub (designated the distal and proximal moieties, respectively) through its Lys48 side chain (Adapted with permission from Ref. [24])

[30]. Thus, the three-dimensional (3D) structure of such a flexible oligosaccharide should be interpreted as a vast ensemble of dynamic conformations, rather than only a limited number of conformational states.

The biological functions of oligosaccharides are exerted primarily through molecular recognition processes that are mediated by carbohydrate-binding proteins collectively designated lectins [31, 32]. For example, various high-mannose-type oligosaccharides attached to proteins serve as quality indicators in the early secretory pathway, which are recognized by a series of intracellular lectins that function during protein folding, transport, and degradation processes [33–35]. On cell surfaces, various glycoconjugates mediate a number of physiological and

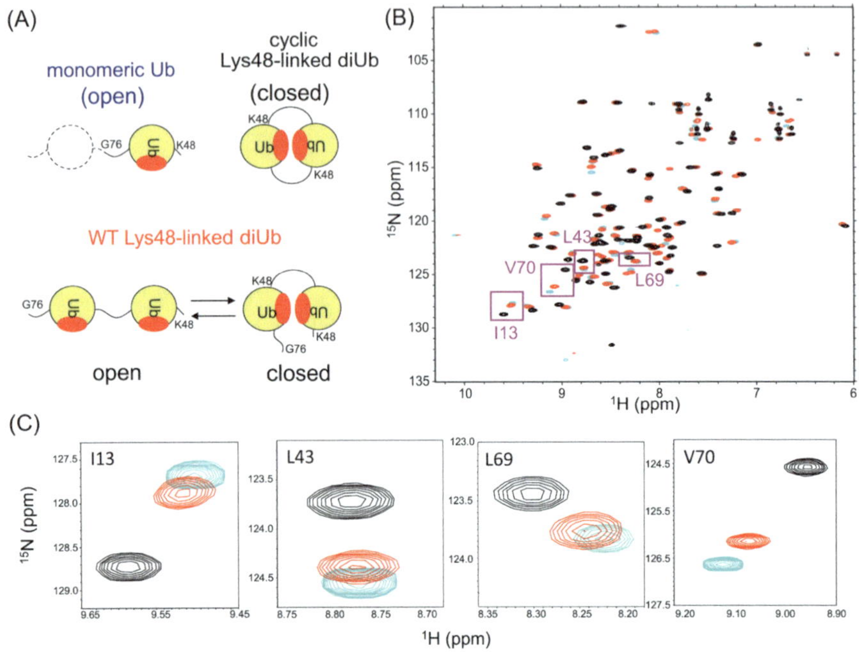

Fig. 4.7 (**a**) Cartoon model of the conformational equilibrium of WT Lys48-linked diUb. Cyclic Lys48-linked diUb was used as a mimic of the closed state, and monomeric Ub was used as a mimic of the open state. The hydrophobic surface is colored red. (**b**) ^{1}H–^{15}N HSQC spectra of monomeric Ub (*cyan*), WT Lys48-linked diUb (*red*), and cyclic Lys48-linked diUb (*black*) at pH 7.0. (**c**) Close-up views of the spectral regions (*boxed*), including the peaks for Ile13, Leu43, Leu69, and Val70, are displayed at the bottom (Adapted with permission from Ref. [24])

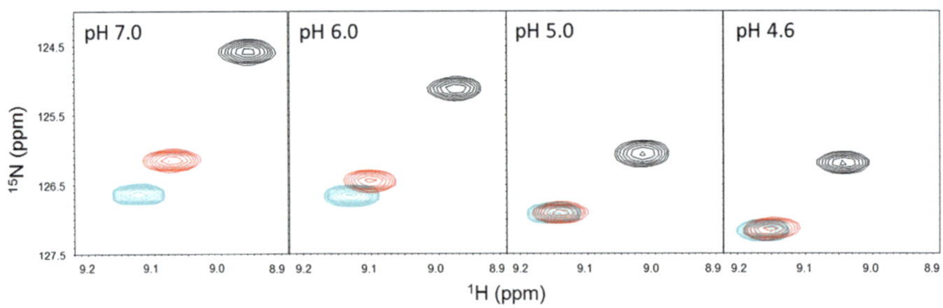

Fig. 4.8 pH-dependent chemical shift changes of the HSQC peaks for monomeric Ub (*cyan*), WT Lys48-linked diUb (*red*), and cyclic Lys48-linked diUb (*black*), which shows that the open conformation of WT Lys48-linked diUb becomes populated upon lowering the pH (Adapted with permission from Ref. [24])

pathological processes, as exemplified by gangliosides, glycosphingolipids containing sialyl residue(s), which are involved in cellular communications and development, cancer metastasis, viral infections, and the onset and progression of neurodegenerative disorders [36–39].

To better understand the mechanisms underlying oligosaccharide functions, their 3D structural information should be acquired with quantitative descriptions of their dynamic, rapid conformational fluctuations. NMR methods combined with computational simulation techniques provide useful tools to explore the conformational spaces of flexible biomolecules. For example, MD simulations can be used to predict the dynamic conformational motions of highly flexible oligosaccharides in solution [40, 41]. However, a frequent problem with this approach is its dependence on simulation protocols, including initial structures, computational times, and the choice of force fields. Therefore, it is critically important to validate the simulation results by comparisons with experimental data for biomolecular fluctuations on time scales ranging from subnanoseconds to microseconds. We developed a methodology to evaluate MD simulation results using paramagnetism-assisted NMR techniques to explore the conformational spaces occupied by flexible oligosaccharides in solution [42].

Introducing a paramagnetic center, such as lanthanide ions, into a target biomolecule results in significant modifications of its NMR spectra due to the generated magnetic dipole–dipole interactions between nuclei and an unpaired electron [43, 44]. For example, pseudocontact shift (PCS), a well-known paramagnetic effect observed in the presence of lanthanide ions with anisotropic magnetic susceptibility, can modulate the chemical shift of each NMR signal according to the spatial arrangements of individual nuclei with respect to the paramagnetic center [45]. Because of the large magnetic moment of unpaired electrons, through-space dipole interactions perturb not only the signals from the nuclei in proximity to the paramagnetic center but also those that are distal from it. Thus, PCS measurements have the potential to provide valuable atomic long-distance information for characterizing biomolecular conformations.

We applied this method to evaluate the conformational dynamics of a GM3 trisaccharide (NeuAcα2-3Galβ1-4Glc), the common core oligosaccharide structure shared among gangliosides [46]. Attaching a paramagnetic lanthanide tag to this trisaccharide allowed us to measure PCS effects (Fig. 4.9a). Using two-dimensional ^{1}H–^{13}C HSQC spectra, the PCS values of each ^{1}H and ^{13}C nucleus were acquired as the differences in chemical shifts between the oligosaccharide tagged with a paramagnetic ion, such as Tm^{3+}, and that modified with a diamagnetic reference La^{3+} ion (Fig. 4.9b). Because the 3D structure of the oligosaccharide rapidly fluctuates in solution, the observed PCS data should be interpreted as averages over the dynamic conformational ensemble. Thus, we performed MD simulations in explicit water to create a vast conformational ensemble of this oligosaccharide. The simulation results showed wide distributions of torsion angles in this trisaccharide, particularly around a glycosidic linkage between the NeuAc and Gal residues (Fig. 4.10). From the computed trajectory of the trisaccharide over a total of 120 ns, 2000 conformers were sampled, which involved harmonic motions

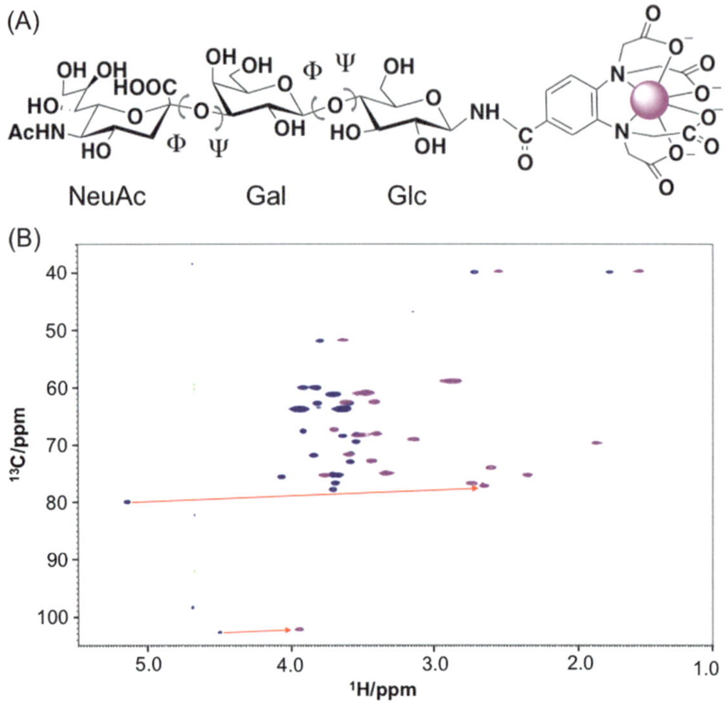

Fig. 4.9 PCS measurements for the GM3 trisaccharide. (**a**) Introducing a paramagnetic lanthanide tag. (**b**) ^{1}H–^{13}C HSQC spectra of the tagged oligosaccharide in complex with Tm^{3+} (*magenta*) and La^{3+} (*blue*). *Arrows* indicate chemical shift differences for the anomeric CH groups of Gal and Glc residues (Adapted with permission from Ref. [46])

in a local minimum as well as transitions from one low-energy region to another in the energy landscape.

Using the atomic coordinates generated from this dynamic ensemble model, the conformational space of the GM3 trisaccharide derived from the MD simulations was quantitatively validated. With this approach, the experimentally obtained PCS values were compared with those that were back calculated from the ensemble model. The results for the back-calculated PCS values were in excellent agreement with the experimental data with a small Q value (used as criteria for agreement between experimental and calculated values), which indicated successful exploration of the conformational spaces of this oligosaccharide (Fig. 4.11). Using most single conformers or using a combination of selected low-energy conformers gave compromised Q values. Furthermore, such a low Q value was not estimated from a conformational ensemble of this trisaccharide without considering its low-populated (around 2 %) conformational cluster (Figs. 4.10 and 4.11). These results underscore the existence of the minor conformers in solution and their significant contributions to the observed NMR data.

A unique structural feature of oligosaccharides is their branching with multiple modes of linkage, which provide diverse 3D structures. To examine the impact of

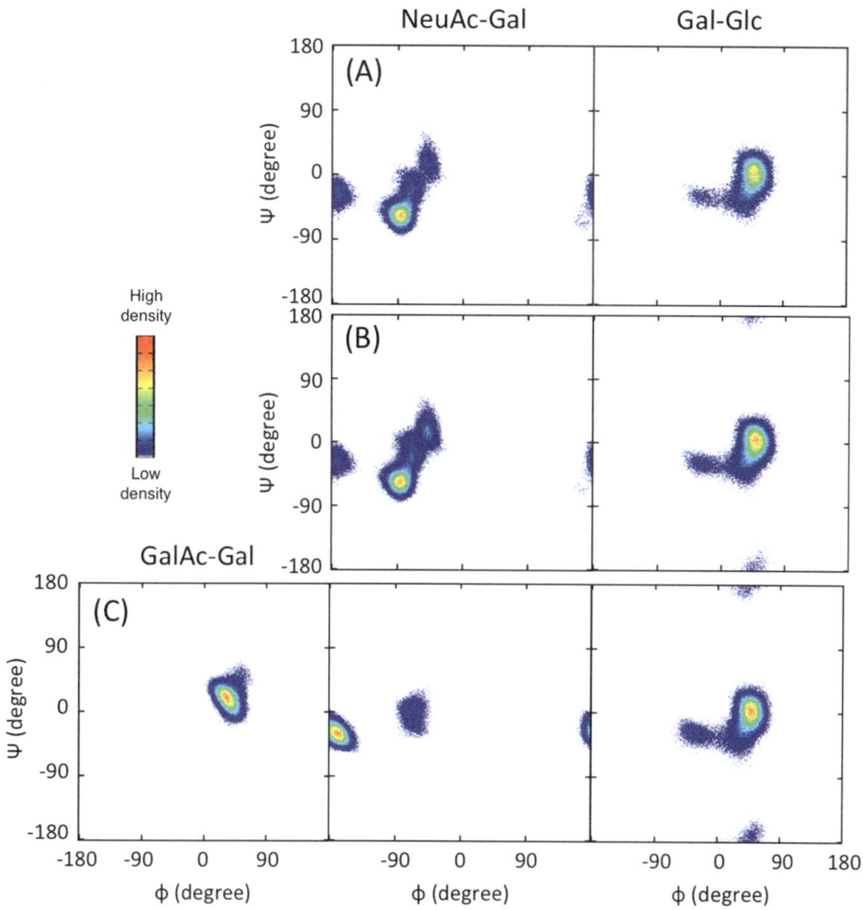

Fig. 4.10 Torsion angle density maps of the MD trajectory for the GM3 trisaccharide with a total simulation time of (**a**) 60 ns and (**b**) 120 ns and (**c**) the GM2 tetrasaccharide with a total simulation time of 120 ns (Adapted with permissions from Refs. [46] and [47])

these branches on the conformations and dynamics of oligosaccharides, paramagnetism-assisted NMR analyses in conjunction with MD simulations were also used to conformationally characterize a branched oligosaccharide, GM2 tetrasaccharide (GalNAcβ1-4[NeuAcα2-3]Galβ1-4Glc), which has an additional GalNAc branch as compared with the GM3 trisaccharide [47].

The PCS-validated conformational spaces of the GM2 and GM3 oligosaccharides showed similarities in their Gal–Glc glycosidic linkage conformations. In contrast, significant differences were observed between the GM2 and GM3 oligosaccharides in terms of the conformational spaces around the NeuAc–Gal glycosidic linkages (Fig. 4.10c). A corresponding cluster of this linkage to the most populated conformations in the GM3 trisaccharide was missing in the

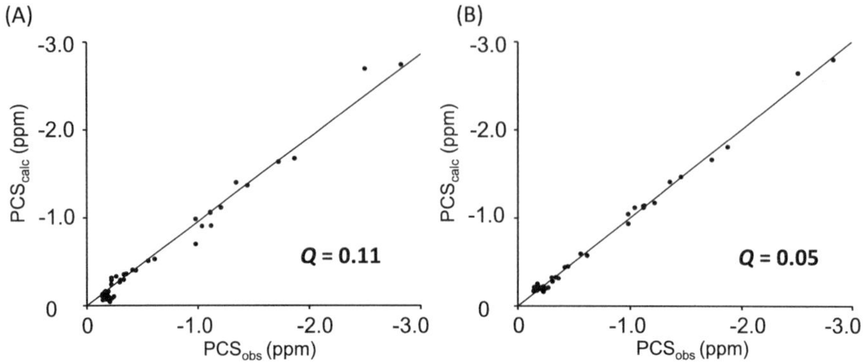

Fig. 4.11 Correlations between experimentally observed PCS values with Tm^{3+} and computationally calculated PCS data for the GM3 trisaccharide. The PCS values were back calculated from ensemble models derived from (**a**) the trajectory with a total simulation time of 60 ns and (**b**) those with a simulation time of 120 ns (Adapted with permission from Ref. [46])

conformational space of the GM2 tetrasaccharide. This suggests that in the biantennary GM2 tetrasaccharide, the additional GalNAc residue controls the conformational fluctuation of another branch. The corresponding NeuAc–Gal glycosidic linkage has conformational freedom in the linear GM3 oligosaccharide.

4.4 Concluding Remarks

In this chapter, we outlined two contrasting types of NMR approaches for characterizing individual conformers of biomolecules that rapidly interchange with one another. In one approach, a specific conformer is enriched or even isolated for detailed NMR analyses. Our NMR studies of mono- and diUb illustrated this conformational trapping by applying high pressure, point mutations, pH optimization, and covalent constraints. Crystallographic snapshots can also be used to interpret NMR data for a dynamic conformational ensemble. In another more systematic approach, the dynamic behaviors of biomolecules in solution can be directly observed by NMR-validated MD simulations. This method can be applied to describe dynamic pictures of flexible biomolecules, including intrinsically disordered proteins. Thus, sophisticated NMR spectroscopic techniques combined with X-ray crystallography and MD simulations provide versatile tools for characterizing the conformational dynamics of biomolecules and exploring their conformational spaces.

In living systems, these flexible biomolecules are self-organized into highly ordered supramolecular structures and, more importantly, are autonomously disassembled to create higher functions. For these systems, NMR spectroscopy provides valuable microscopic information on the local conformational dynamics of biomolecules during the processes involved in biomolecular organization, which can be complemented by macroscopic information on the overall structures of these supramolecular complexes provided by solution scattering, electron microscopy, and other biophysical methods. In this context, NMR has important roles for exploring the dynamic ordering of biomolecular systems for the generation of integrated functions.

Acknowledgments This work was partly supported by the JSPS/MEXT Grants in Aid for Scientific Research on Innovation Areas (23107729, 20107004, and 25102008), Young Scientists (B) (25840025 and 24750170), and Challenging Exploratory Research (26560451) and by the Okazaki ORION project. M. Y. U. is a recipient of the Naito Foundation Grant for Studying Overseas.

References

1. Boehr DD, Nussinov R, Wright PE (2009) The role of dynamic conformational ensembles in biomolecular recognition. Nat Chem Biol 5:789–796
2. Lange OF, Lakomek NA, Fares C, Schroder GF, Walter KF, Becker S et al (2008) Recognition dynamics up to microseconds revealed from an RDC-derived ubiquitin ensemble in solution. Science 320:1471–1475
3. Mittag T, Kay LE, Forman-Kay JD (2010) Protein dynamics and conformational disorder in molecular recognition. J Mol Recognit 23:105–116
4. Uversky VN, Dunker AK (2010) Understanding protein non-folding. Biochim Biophys Acta 1804:1231–1264
5. Kleckner IR, Foster MP (2011) An introduction to NMR-based approaches for measuring protein dynamics. Biochim Biophys Acta 1814:942–968
6. Mittermaier AK, Kay LE (2009) Observing biological dynamics at atomic resolution using NMR. Trends Biochem Sci 34:601–611
7. Hansen DF, Vallurupalli P, Kay LE (2008) An improved ^{15}N relaxation dispersion experiment for the measurement of millisecond time-scale dynamics in proteins. J Phys Chem B 112:5898–5904
8. Loria JP, Rance M, Palmer AG (1999) A relaxation-compensated Carr–Purcell–Meiboom–Gill sequence for characterizing chemical exchange by NMR spectroscopy. J Am Chem Soc 121:2331–2332
9. Sugase K, Dyson HJ, Wright PE (2007) Mechanism of coupled folding and binding of an intrinsically disordered protein. Nature 447:1021–1025
10. Lipari G, Szabo A (1982) Model-free approach to the interpretation of nuclear magnetic resonance relaxation in macromolecules. 1. Theory and range of validity. J Am Chem Soc 104:4546–4559
11. Lipari G, Szabo A (1982) Model-free approach to the interpretation of nuclear magnetic resonance relaxation in macromolecules. 2. Analysis of experimental results. J Am Chem Soc 104:4559–4570
12. Gekko K, Hasegawa Y (1986) Compressibility-structure relationship of globular proteins. Biochemistry 25:6563–6571
13. Chalikian TV, Breslauer KJ (1998) Thermodynamic analysis of biomolecules: a volumetric approach. Curr Opin Struct Biol 8:657–664
14. Akasaka K (2003) Highly fluctuating protein structures revealed by variable-pressure nuclear magnetic resonance. Biochemistry 42:10875–10885
15. Hershko A, Ciechanover A (1998) The ubiquitin system. Annu Rev Biochem 67:425–479

16. Kitahara R, Yamada H, Akasaka K (2001) Two folded conformers of ubiquitin revealed by high-pressure NMR. Biochemistry 40:13556–13563
17. Kitahara R, Akasaka K (2003) Close identity of a pressure-stabilized intermediate with a kinetic intermediate in protein folding. Proc Natl Acad Sci U S A 100:3167–3172
18. Kitahara R, Yokoyama S, Akasaka K (2005) NMR snapshots of a fluctuating protein structure: ubiquitin at 30 bar-3 kbar. J Mol Biol 347:277–285
19. Fu Y, Wand AJ (2013) Partial alignment and measurement of residual dipolar couplings of proteins under high hydrostatic pressure. J Biomol NMR 56:353–357
20. Imai T, Sugita Y (2010) Dynamic correlation between pressure-induced protein structural transition and water penetration. J Phys Chem B 114:2281–2286
21. Kitazawa S, Kameda T, Yagi-Utsumi M, Sugase K, Baxter NJ, Kato K et al (2013) Solution structure of the Q41N variant of ubiquitin as a model for the alternatively folded N2 state of ubiquitin. Biochemistry 52:1874–1885
22. Kitazawa S, Kameda T, Kumo A, Yagi-Utsumi M, Baxter NJ, Kato K et al (2014) Close identity between alternatively folded state N2 of ubiquitin and the conformation of the protein bound to the ubiquitin-activating enzyme. Biochemistry 53:447–449
23. Kitahara R, Yamaguchi Y, Sakata E, Kasuya T, Tanaka K, Kato K et al (2006) Evolutionally conserved intermediates between ubiquitin and NEDD8. J Mol Biol 363:395–404
24. Hirano T, Serve O, Yagi-Utsumi M, Takemoto E, Hiromoto T, Satoh T et al (2011) Conformational dynamics of wild-type Lys-48-linked diubiquitin in solution. J Biol Chem 286:37496–37502
25. Cook WJ, Jeffrey LC, Carson M, Chen Z, Pickart CM (1992) Structure of a diubiquitin conjugate and a model for interaction with ubiquitin conjugating enzyme (E2). J Biol Chem 267:16467–16471
26. Varadan R, Assfalg M, Raasi S, Pickart C, Fushman D (2005) Structural determinants for selective recognition of a Lys48-linked polyubiquitin chain by a UBA domain. Mol Cell 18:687–698
27. Zhang N, Wang Q, Ehlinger A, Randles L, Lary JW, Kang Y et al (2009) Structure of the s5a: k48-linked diubiquitin complex and its interactions with rpn13. Mol Cell 35:280–290
28. Yao T, Cohen RE (2000) Cyclization of polyubiquitin by the E2-25K ubiquitin conjugating enzyme. J Biol Chem 275:36862–36868
29. Sokratous K, Strachan J, Roach LV, Layfield R, Oldham NJ (2012) Cyclisation of Lys48-linked diubiquitin in vitro and in vivo. FEBS Lett 586:4144–4147
30. Wormald MR, Petrescu AJ, Pao YL, Glithero A, Elliott T, Dwek RA (2002) Conformational studies of oligosaccharides and glycopeptides: complementarity of NMR, X-ray crystallography, and molecular modelling. Chem Rev 102:371–386
31. Gabius HJ, Andre S, Jimenez-Barbero J, Romero A, Solis D (2011) From lectin structure to functional glycomics: principles of the sugar code. Trends Biochem Sci 36:298–313
32. Kamiya Y, Yagi-Utsumi M, Yagi H, Kato K (2011) Structural and molecular basis of carbohydrate-protein interaction systems as potential therapeutic targets. Curr Pharm Des 17:1672–1684
33. Aebi M, Bernasconi R, Clerc S, Molinari M (2010) N-glycan structures: recognition and processing in the ER. Trends Biochem Sci 35:74–82
34. Kamiya Y, Satoh T, Kato K (2012) Molecular and structural basis for N-glycan-dependent determination of glycoprotein fates in cells. Biochim Biophys Acta 1820:1327–1337
35. Kato K, Kamiya Y (2007) Structural views of glycoprotein-fate determination in cells. Glycobiology 17:1031–1044
36. Regina Todeschini A, Hakomori SI (2008) Functional role of glycosphingolipids and gangliosides in control of cell adhesion, motility, and growth, through glycosynaptic microdomains. Biochim Biophys Acta 1780:421–433
37. Lopez PH, Schnaar RL (2009) Gangliosides in cell recognition and membrane protein regulation. Curr Opin Struct Biol 19:549–557

38. Ariga T, McDonald MP, Yu RK (2008) Role of ganglioside metabolism in the pathogenesis of Alzheimer's disease – a review. J Lipid Res 49:1157–1175
39. Matsuzaki K, Kato K, Yanagisawa K (2010) Abeta polymerization through interaction with membrane gangliosides. Biochim Biophys Acta 1801:868–877
40. Fadda E, Woods RJ (2010) Molecular simulations of carbohydrates and protein-carbohydrate interactions: motivation, issues and prospects. Drug Discov Today 15:596–609
41. Re S, Nishima W, Miyashita N, Sugita Y (2012) Conformational flexibility of N-glycans in solution studied by REMD simulations. Biophys Rev 4:179–187
42. Zhang Y, Yamaguchi T, Kato K (2013) New NMR tools for characterizing the dynamic conformations and interactions of oligosaccharides. Chem Lett 42:1455–1462
43. Rodriguez-Castaneda F, Haberz P, Leonov A, Griesinger C (2006) Paramagnetic tagging of diamagnetic proteins for solution NMR. Magn Reson Chem 44 Spec No:S10–S6
44. Bertini I, Luchinat C, Parigi G, Pierattelli R (2008) Perspectives in paramagnetic NMR of metalloproteins. Dalton Trans 3782–3790
45. Otting G (2010) Protein NMR, using paramagnetic ions. Annu Rev Biophys 39:387–405
46. Yamamoto S, Zhang Y, Yamaguchi T, Kameda T, Kato K (2012) Lanthanide-assisted NMR evaluation of a dynamic ensemble of oligosaccharide conformations. Chem Commun (Camb) 48:4752–4754
47. Zhang Y, Yamamoto S, Yamaguchi T, Kato K (2012) Application of paramagnetic NMR-validated molecular dynamics simulation to the analysis of a conformational ensemble of a branched oligosaccharide. Molecules 17:6658–6671

Chapter 5
Site-Specific Incorporation of Fluorescent Nonnatural Amino Acids into Proteins and Its Application to Fluorescence Analysis of Proteins

Takahiro Hohsaka

Abstract Nonnatural amino acid mutagenesis using expanded genetic codes allows us to introduce artificial functional groups such as fluorophores at specific sites of proteins. By applying this technique, site-specific fluorescence labeling has been achieved. This method enables to generate fluorescent protein probes that show ligand-dependent fluorescence change based on fluorescence quenching of the incorporated fluorophores by endogenous Trp residues. This strategy has been proved to be effective using ligand-binding proteins and antibodies. Moreover, site-specific double-labeling of ligand-binding proteins with two fluorophores using two expanded genetic codons has achieved ratiometric detection of the ligand binding based on FRET and fluorescence quenching. These achievements demonstrate the usefulness of the site-specific incorporation of fluorescent nonnatural amino acids and their potential utility as research and diagnostic tools.

Keywords Nonnatural amino acid • Protein modification • FRET • Fluorescence quenching

5.1 Introduction

Proteins are biosynthesized from 20 types of naturally occurring amino acids according to the genetic information on DNA. Living organisms have been improving proteins by altering amino acid sequences of proteins. Recent development on genetic engineering techniques such as site-directed mutagenesis allows us to alter amino acid sequence of proteins. Amino acid substitution has been used to improve protein properties including thermostability, catalytic and binding activity, substrate specificity, etc. In addition, structure-function relationship of proteins can be analyzed by replacing amino acids predicted to be essential for proteins. However,

T. Hohsaka (✉)
School of Materials Science, Japan Advanced Institute of Science and Technology, 1-1 Asahidai, Nomi, Ishikawa 923-1292, Japan
e-mail: hohsaka@jaist.ac.jp

M. Terazima et al. (eds.), *Molecular Science of Fluctuations Toward Biological Functions*, DOI 10.1007/978-4-431-55840-8_5

the replacement of amino acid residues is usually restricted within the 20 types of naturally occurring amino acids.

On the other hand, it has been achieved to introduce nonnatural amino acids into proteins in a site-specific manner using a nonnatural amino acid mutagenesis technique. The nonnatural amino acids are applicable to protein analysis as well as engineering proteins with artificial functions. A large variety of nonnatural amino acids carrying artificial functional groups such as photo and electronic functional groups have been successfully incorporated into proteins. Amino acids analogous to naturally occurring ones are also available to alter local environments of proteins for structure-function relationship studies.

In this chapter, the principle of nonnatural amino acid mutagenesis is first described. Applications of the nonnatural amino acid mutagenesis to fluorescence analysis of proteins and to design and synthesis of protein probes based on fluorescence resonance energy transfer (FRET) and fluorescence quenching are then reviewed.

5.2 Incorporation of Nonnatural Amino Acids

The incorporation of nonnatural amino acids has been achieved by modifying the genetic codon table [1–7]. In a natural ribosomal translation system, amino acids are incorporated into polypeptide chains through decoding triplet codons on mRNAs with the use of aminoacyl-tRNAs having the complementary triplet anticodons. There are 61 codons for amino acids and three stop codons on the genetic code table; therefore, no codons are available exclusively for nonnatural amino acids.

We have developed a four-base codon method for adding nonnatural amino acids to the genetic codon table [7–9]. A nonnatural aminoacyl-tRNA, which has a four-base anticodon in place of triplet anticodon and is aminoacylated with a nonnatural amino acid, is used to introduce a nonnatural residue into a protein in response to a four-base codon. Figure 5.1 illustrates the scheme of the translation in the case of CGGG four-base codon. When an mRNA containing a CGGG codon is translated in the presence of a nonnatural aminoacyl-tRNA containing a CCCG anticodon, the CGGG codon is decoded by the four-base anticodon tRNA, and as a consequence, a full-length protein containing the desired nonnatural amino acid is successfully obtained. On the other hand, when the three-base CGG is recognized by a naturally occurring arginyl-tRNA that contains a CCG anticodon, arginine is incorporated instead of nonnatural amino acid. In this case, however, the reading frame shifts incorrectly, and a downstream stop codon terminates the peptide elongation. As a result, a full-length protein must contain the desired nonnatural amino acid at the desired position.

Nonnatural aminoacyl-tRNA having a four-base codon, a key molecule in the nonnatural amino acid mutagenesis, can be synthesized by chemical or enzymatic method. In a chemical method [10, 11], a tRNA lacking 3′ C and A nucleotides is

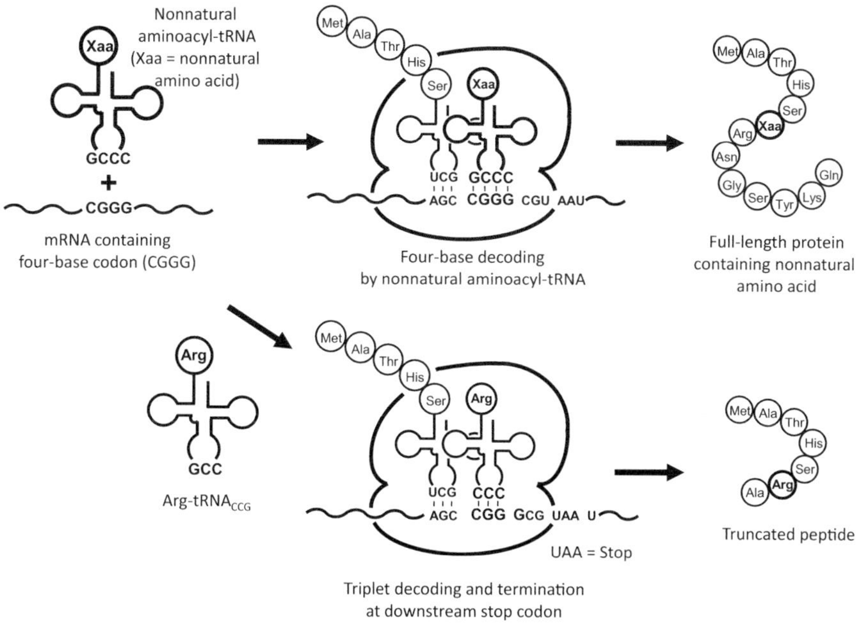

Fig. 5.1 Schematic illustration of the incorporation of nonnatural amino acids into proteins. Four-base codon (CGGG) is recognized by nonnatural aminoacyl-tRNA having the complementary four-base anticodon, and as a consequence, the nonnatural amino acid is site-specifically incorporated into the full-length protein. On the other hand, if the three bases (CGG) are recognized as a triplet codon by endogenous triplet tRNA, the reading frame shifts and downstream stop codon terminates the peptide elongation

prepared by T7 RNA polymerase from a synthetic tRNA gene having a four-base anticodon. On the other hand, a nonnatural amino acid is chemically esterified to a CA dinucleotide. The truncated tRNA and aminoacyl dinucleotide are linked together by T4 RNA ligase to obtain a full-length nonnatural aminoacyl-tRNA. The aminoacyl-tRNA is added to a cell-free translation system together with an expression gene mutated to contain a four-base codon.

We have demonstrated that a nonnatural amino acid is successfully incorporated into a protein in response to four-base codons. Various four-base codons have been examined and some of them are effective to introduce nonnatural amino acids [9]. The incorporation efficiency is also dependent on the type of side chains. We have observed that nonnatural amino acids carrying linearly extended side chains such as *p*-substituted phenylalanine derivatives are good substrates for the ribosomal translation system, but widely expanded side chains are hardly incorporated into proteins (Fig. 5.2) [8].

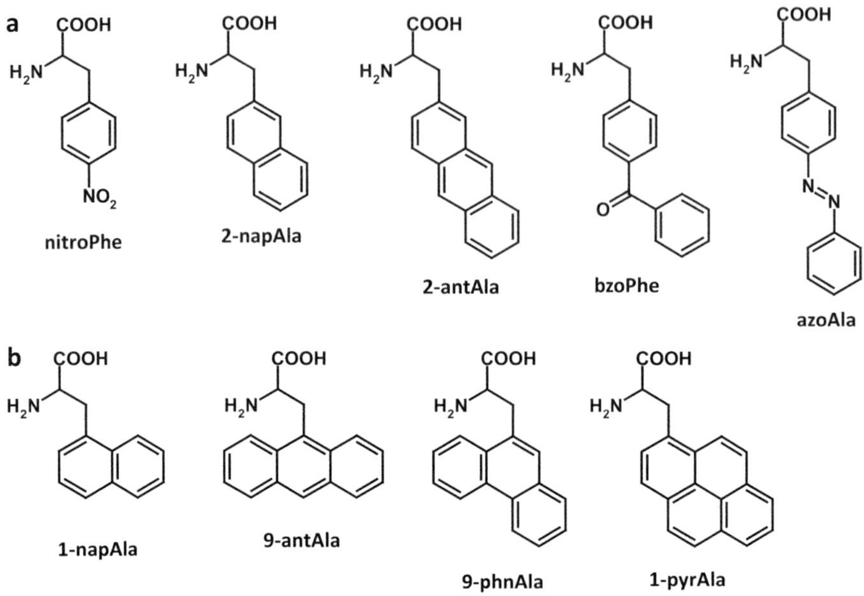

Fig. 5.2 (**a**) Nonnatural amino acids carrying linearly extended side chains are efficiently incorporated into proteins through the ribosomal translation. (**b**) Widely expanded side chains are poor substrates for the ribosomal translation

5.3 Incorporation of Fluorescent Nonnatural Amino Acids

The nonnatural amino acid mutagenesis has been applied to site-specific fluorescence labeling of proteins. Fluorescence labeling of biomolecules is a useful method for probing structure and function of biomolecules. In particular, fluorescence labeling of DNA greatly contributes to the recent progress in DNA analytical technologies including DNA sequencing and microarray analysis. Fluorescence labeling of proteins is also useful for protein researches. Fluorescence labeling of proteins is usually carried out by chemical modification. However, conventional chemical methods cannot achieve site-specific and quantitative labeling because reactive residues such as lysine exist multiply on the protein surface and the labeling yields are usually below 100 % (Fig. 5.3a). Alternatively, proteins can be labeled by the fusion with green fluorescent protein (GFP) derivatives, although the steric hindrance of the large GFP molecules may affect structural and functional properties of proteins (Fig. 5.3b). It is also disadvantageous that the fusion is limited to be at N- or C-terminus of proteins.

The incorporation of nonnatural amino acids carrying fluorescent dyes overcomes these problems. Based on the findings that *p*-substituted phenylalanine derivatives are good substrates for the ribosomal translation [8], we have designed visible-wavelength fluorescent nonnatural amino acids in which fluorophores are linked at the *p*-position of phenylalanine [12] (Fig. 5.3c). As a fluorophore,

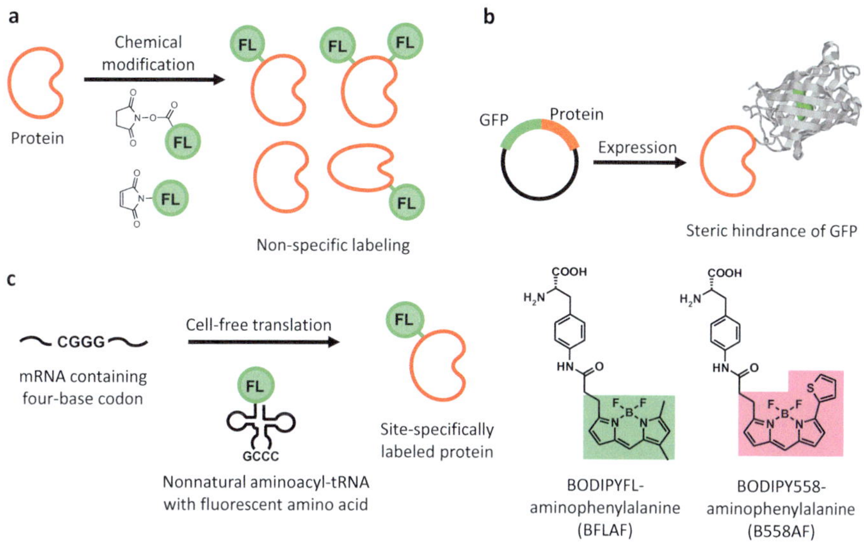

Fig. 5.3 Methods for fluorescence labeling of proteins. (**a**) Chemical modification of proteins with fluorescent-labeling reagents toward Lys or Cys residue. (**b**) Fusion expression of proteins with *green* fluorescent protein (GFP) derivatives. (**c**) Site-specific incorporation of fluorescent nonnatural amino acids into proteins in response to the four-base codon

BODIPYFL dye showing fluorescein-like green fluorescence was first used, because the relatively small molecular size of BODIPY dye can be helpful in the efficient incorporation through the ribosomal translation.

The incorporation was examined using a streptavidin gene containing a CGGG four-base codon at Tyr83 position and confirmed by fluorescence image of SDS-PAGE gel. The result showed that BODIPYFL-linked aminophenylalanine derivative was incorporated in good efficiency [12]. On the other hand, BODIPY-linked lysine derivatives were poorly incorporated, suggesting that the translation system may prefer the aromatic amino acid framework as substrates. Other BODIPY dyes that have additional π-conjugating systems and emit red-shifted fluorescence were also examined. Although the incorporation efficiency was lower than the case of BODIPYFL, BODIPY558-linked aminophenylalanine was successfully incorporated [12]. The pair of the green and red fluorescent amino acids is useful for FRET pair as described below.

Incorporation of fluorescent nonnatural amino acids carrying more large fluorophores such as fluorescein and rhodamine dyes (Fig. 5.4a) was also examined [13]. These fluorophores were poorly incorporated at the Tyr83 position of streptavidin. The incorporation at other positions of streptavidin was then investigated for TAMRA-linked amino acid, but negligible incorporation was observed. At the N-terminal region, however, these fluorophores were successfully incorporated. This result suggests that relatively large fluorophores can be incorporated downstream of short peptides but be rejected downstream of long peptides. This

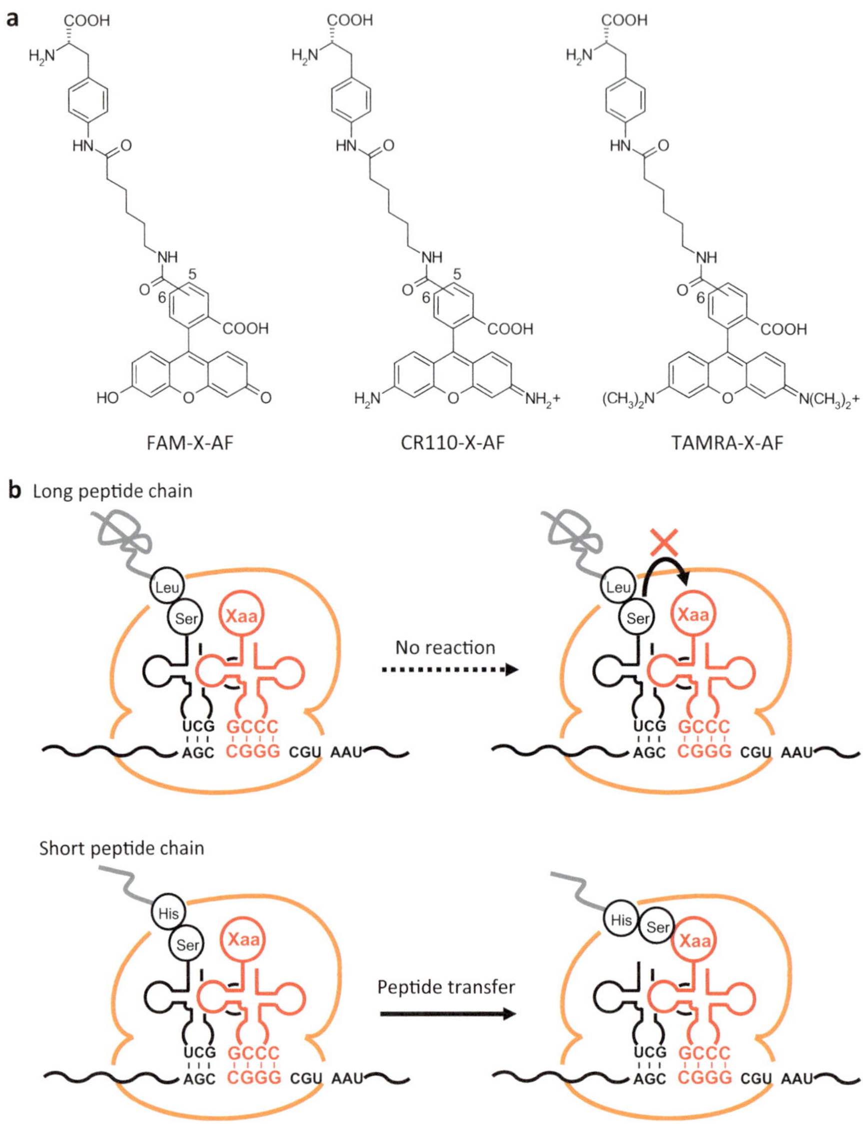

Fig. 5.4 Incorporation of fluorescent nonnatural amino acids carrying large fluorophores at the side chain. (**a**) Structure of large fluorescent nonnatural amino acids. (**b**) Schematic illustration of peptide transfer on ribosome. Long polypeptide on ribosomal P-site cannot react with nonnatural amino acid carrying large fluorophore, possibly because molecular fluctuation of these reactants is not enough to promote the peptide bond formation. Short polypeptide on ribosomal P-site increases the molecular fluctuation of the peptidyl-tRNA and can promote the peptide bond formation

phenomenon may be explained by considering the fluctuation of polypeptide chain on ribosomal P-site (Fig. 5.4b). During the translation, amino acid on ribosomal A-site reacts with growing polypeptide on P-site to form a peptide bond. Large fluorophores at the side chain of nonnatural amino acids may interfere with the peptide bond formation by suppressing the fluctuation of the amino acid main chain, and as a consequence, the fluorescent nonnatural amino acids fail to react with the growing polypeptide. In case that the peptide on the P-site is short, the peptide may be highly fluctuating and be able to react with the nonnatural amino acids even though they have less fluctuating large fluorophores.

5.4 Application of Fluorescent Nonnatural Amino Acids to Protein Analysis

Site-specific incorporation of fluorescent nonnatural amino acids is available as a useful tool for various protein analyses such as protein-protein interaction, conformational change, and oligomerization. Because the fluorescent label can be introduced in response to a four-base codon, proteins can be labeled at positions in which the fluorescence labeling doesn't affect the structure and activity of proteins.

We have demonstrated the utility of the BODIPYFL-linked aminophenylalanine. A single-chain antibody was labeled with BODIPYFL at the N-terminal region, and the resulting labeled antibody was utilized for the antigen-binding measurement by monitoring fluorescence polarization. DNA-binding activity of fluorescently labeled lambda Cro repressor protein was applied to gel shift assay and fluorescence imaging of the gel.

The present method is more advantageous when applied to FRET. For intramolecular FRET analysis of proteins, energy donor and acceptor pairs must be introduced into two specific positions of proteins. Conventional chemical synthetic routes such as modification of two cysteine residues with two different fluorophores are difficult to produce site-specifically and quantitatively double-labeled proteins. Fusion with two GFP derivatives (e.g., cyan and yellow fluorescent proteins) is another strategy for double labeling of proteins [14, 15], although only N- and C-termini can be labeled, and moreover, two large GFP molecules may affect the labeled proteins.

Incorporation of two different fluorescent nonnatural amino acids has been achieved using two four-base codons or a pair of four-base and amber codons [9, 12, 16, 17]. In the case of calmodulin, BODIPYFL- and BODIPY558-linked aminophenylalanine derivatives were incorporated in response to the GGGU and CGGG four-base codons, respectively (Fig. 5.5a). The fluorescence spectrum of BODIPYFL is overlapped with absorption spectrum of BODIPY558, and therefore, FRET from BODIPYFL to BODIPY558 can occur. Förster distance (R_0) was calculated to be 59 Å with assumption that the orientation factor is 2/3. BODIPYFL and BODIPY558 were incorporated at the Gly40 position and the N-terminal region

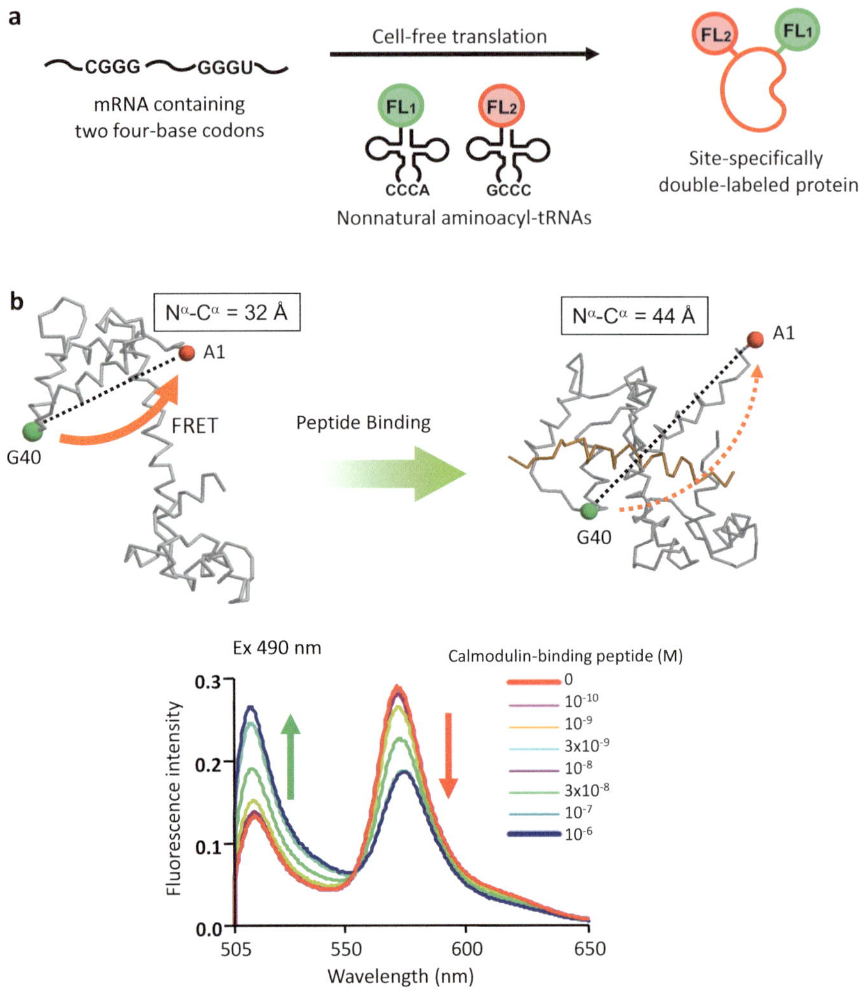

Fig. 5.5 FRET analysis of conformational change of calmodulin upon the binding of calmodulin-binding peptide. (**a**) Synthesis of double-labeled calmodulin containing BODIPYFL (donor) and BODIPY558 (acceptor) in response to the GGGU and CGGG four-base codons. (**b**) Fluorescence spectral change of the double-labeled calmodulin upon the addition of calmodulin-binding peptide in the presence of Ca^{2+}

of calmodulin. The double-labeled calmodulin expressed in the cell-free translation system was purified using C-terminal histidine tag, and the fluorescence spectra were measured in the presence of Ca^{2+} ion and various concentration of calmodulin-binding peptide (Fig. 5.5b). In the absence of the binding peptide, weak green fluorescence and strong red fluorescence were observed. Upon the addition of the binding peptide, green fluorescence increased with decrease in red fluorescence. This fluorescence change indicates that FRET from BODIPYFL to

BODIPY558 was suppressed upon the binding of calmodulin-binding peptide. Calmodulin changes its conformation and the distance between the N-terminus and Gly40 increases upon the ligand binding from 32 to 44 Å. The observed FRET change well reflects the conformational change of calmodulin.

5.5 Synthesis of Fluorescent Protein Probes by Introducing Fluorescent Nonnatural Amino Acids

Fluorescent amino acids are available not only as fluorescent labels but as fluorescent probes. We have observed that the fluorescent amino acids incorporated at the particular positions of ligand-binding proteins show fluorescence change depending on the surrounding Trp residues. It has been reported that various fluorophores can be quenched by Trp in solution and in polypeptide chains [18]. The alteration of fluorescence quenching upon the ligand binding allows us to detect the ligand binding as a fluorescence change.

In the case of maltose-binding protein (MBP), BODIPYFL-linked aminophenylalanine was incorporated at 15 tyrosine residues of MBP, respectively, and fluorescence spectral changes upon the ligand binding were measured [19]. Among these positions, MBP containing BODIPYFL at Tyr210 showed remarkable increase in the fluorescence intensity (by tenfold) upon the addition of maltose (Fig. 5.6). According to the crystallographic data of MBP [20], Tyr210 locates near the ligand-binding pocket in MBP. The fluorescence of BODIPYFL could be quenched by tryptophan residues around the binding pocket in the absence of the ligand. In the presence of the ligand, however, the tryptophan residues hydrophobically interact with the ligand in place of BODIPYFL, and as a consequence, the fluorescence quenching of BODIPYFL is eliminated. The mutation of the respective tryptophan residues to phenylalanine revealed that Trp341 dominantly quenched the fluorescence of BODIPYFL in a ligand-dependent manner [19].

The single-labeled protein probe was further derivatized to double-labeled FRET protein probes that show fluorescence ratio change upon the ligand binding. While single-labeled protein probes allow quantitative detection of ligand binding under constant concentration of protein probes, double-labeled protein probes achieve quantitative detection without depending on the protein concentration. This property is advantageous especially when used in living cells.

In order to derivatize BODIPY-labeled MBP to double-labeled one, BODIPYFL-linked aminophenylalanine was introduced at the N-terminal Lys1, and BODIPY558-aminophenylalanine was introduced at Tyr210 position as FRET donor and acceptor, respectively [19] (Fig. 5.7). The distance between Lys1 and Tyr210 (43 Å) in the crystallographic data [20] is shorter than Förster distance for BODIPYFL and BODIPY558, and therefore, FRET from BODIPYFL to BODIPY558 can occur efficiently. The double-labeled MBP showed significant

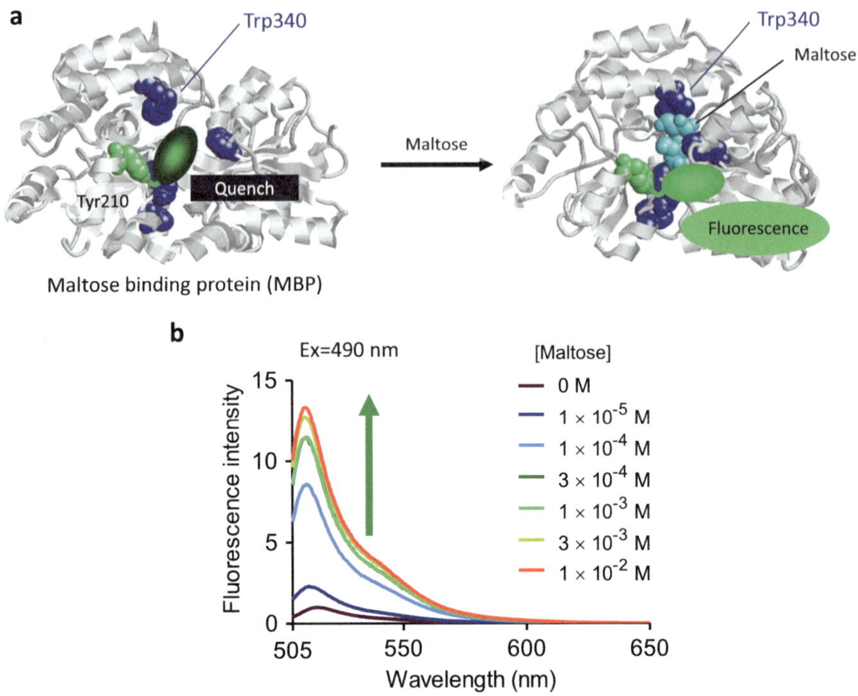

Fig. 5.6 Fluorescence detection of the ligand binding of maltose-binding protein (*MBP*) based on fluorescence quenching. (**a**) Schematic illustration of BODIPYFL-labeled MBP at Tyr210 showing ligand-dependent fluorescence quenching. Trp340 mainly interacts with BODIPYFL and quenches the fluorescence of BODIPYFL in the absence of the ligand; however, the quenching is eliminated by the ligand binding due to the interaction of Trp with the ligand. (**b**) Fluorescence spectral change of MBP containing BODIPYFL at Tyr210 upon the addition of maltose

increase in the BODIPY558 fluorescence with decrease in the BODIPYFL fluorescence in a ligand-dependent manner under excitation of BODIPYFL (Fig. 5.7b). The increase in BODIPY558 fluorescence can be explained due to FRET from BODIPYFL at Lys1 to BODIPY558 at Tyr210 and fluorescence quenching of BODIPY558 by Trp residues only in the absence of maltose (Fig. 5.7a). The decrease in the fluorescence of BODIPYFL upon the ligand binding suggests that the conformational change of MBP may decrease the FRET efficiency. The fact that the distance between Lys1 to Tyr210 decreases from 43 to 35 Å upon the binding of maltose in the crystallographic data supports this FRET change. The fluorescence intensity ratio at 515 and 575 nm significantly increased from 0.14 to 1.33. This remarkable ratio change is not usually achieved by FRET alone and is obviously due to the contribution of ligand-dependent fluorescence quenching.

The double labeling of proteins can also be achieved by combining the fusion with fluorescent or bioluminescent proteins and the incorporation of fluorescent nonnatural amino acids. A variety of FRET protein probes carrying two fluorescent proteins such as CFP and YFP have been developed including those for Ca^{2+}

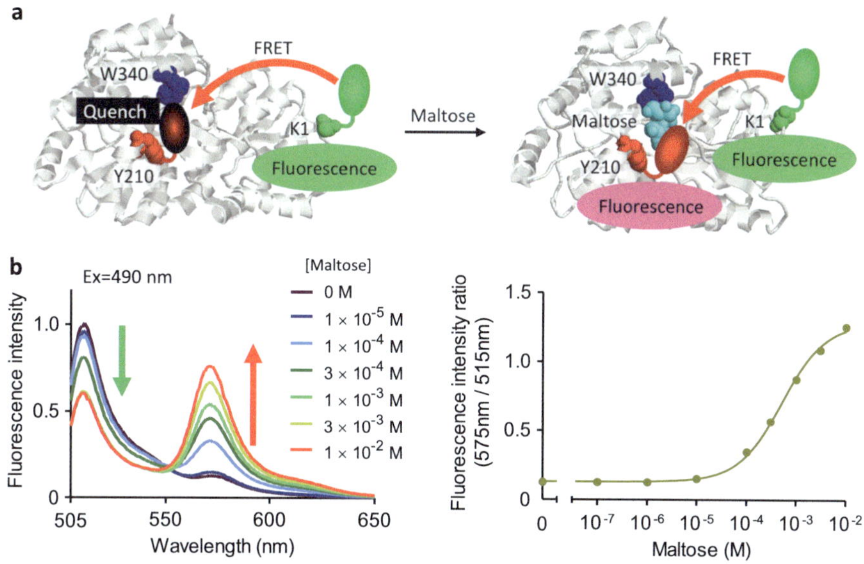

Fig. 5.7 (**a**) Double-labeled MBP containing BODIPYFL at the N-terminus and BODIPY558 at Tyr210 shows significant change in the fluorescence ratio. FRET from BODIPYFL to BODIPY558 occurs regardless of the absence and presence of the ligand, but the fluorescence of BODIPY558 at Tyr210 is quenched in the absence of the ligand. Consequently, the fluorescence ratio (BODIPY558/BODIPYFL) increases significantly. (**b**) Fluorescence spectra and fluorescence ratio of the double-labeled MBP in the absence and presence of maltose

binding [14] and protein phosphorylation [21]. In the case of double-labeled bioluminescent proteins, luciferase and fluorescent proteins were fused to target proteins to generate luminescent protein probes based on bioluminescence resonance energy transfer (BRET). Luminescent substrates for sequence-specific proteases have been reported using luciferase, and fluorescent protein fusion linked each other with specific substrate peptides [22, 23]. A ligand-binding protein, β-arrestin, which shows agonist-induced conformational rearrangements, was successfully derivatized to a luminescent probe by fusing Renilla luciferase and yellow fluorescent protein [24]. These BRET protein probes allow us to detect biomolecular processes with very low background without using external excitation light. However, it is difficult to design suitable BRET protein probes responding to the protein structural changes, because two light-emitting proteins are limited to be attached at the N- and C-termini of proteins. Moreover, fusion of both luciferase and fluorescent proteins may cause steric hindrance due to their large molecular size.

We have demonstrated that the combination of the incorporation of fluorescent nonnatural amino acids and the fusion with luciferase is an effective strategy to achieve BRET protein probes (Fig. 5.8a) [25]. As a BRET donor, *Gaussia* luciferase (GLuc) [26] from a marine copepod *Gaussia princeps* was fused at the C-terminus of MBP. A BRET acceptor, BODIPY558AF, was incorporated at

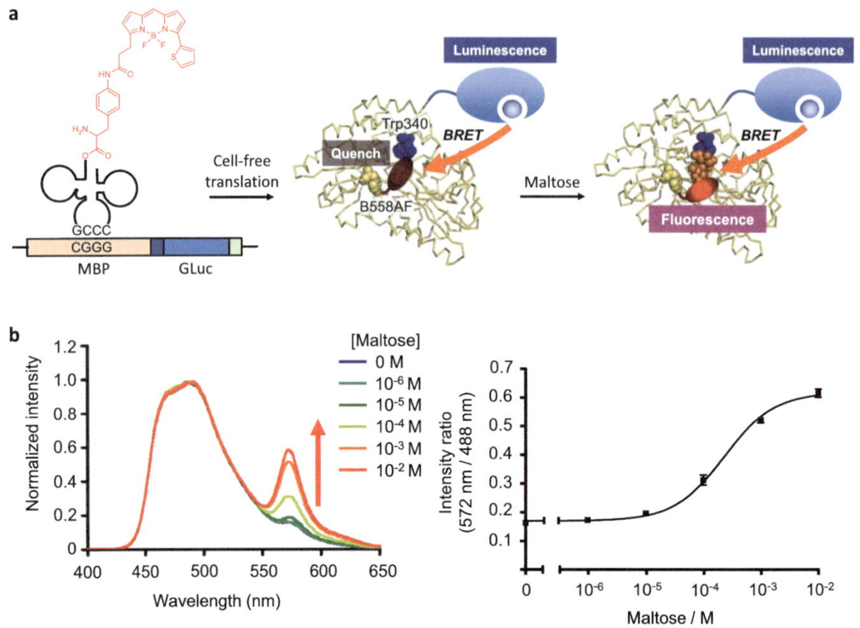

Fig. 5.8 (**a**) BRET and fluorescence quenching of luciferase-fused MBP containing BODIPY558 at Tyr210. (**b**) Luminescence spectra and emission ratio of the BODIPY558-containing luciferase-MBP fusion in the absence and presence of maltose. BRET occurs regardless of the absence and presence of the ligand, but the fluorescence of BODIPY558 at Tyr210 is quenched in a ligand-dependent manner

Tyr210 of MBP-GLuc fusion in response to the four-base codon CGGG. GLuc shows bright luminescence in the 450–550 nm range, which is overlapped with the absorption spectrum of BODIPY558. Therefore, BRET from GLuc to BODIPY558 is expected to occur efficiently. BODIPY558 at Tyr210 of MBP is not only BRET acceptor but also fluorescent probe that is quenched by Trp residues in a ligand-dependent manner. As expected, the BODIPY558-containing MBP-GLuc fusion showed luminescence of GLuc and fluorescence of BODIPY558 due to BRET, but the BODIPY558 fluorescence significantly increased upon the ligand binding (Fig. 5.8b). Correspondingly, the intensity ratio of luminescence of GLuc and fluorescence of BODIPY558 significantly increased upon the ligand binding. This result supports the idea that BRET occurs from GLuc to BODIPY558 and the resulting excited BODIPY558 is quenched only in the presence of the ligand by adjacent Trp residues.

As in the case of the luciferase fusion, BODIPY558-containing MBP was fused with GFP in place of GLuc [25]. The BODIPY558-MBP-GFP showed remarkable increase in BODIPY558 fluorescence but negligible change in GFP fluorescence with excitation at 490 nm for GFP. This phenomenon can be explained that FRET occurs from GFP to BODIPY558, and BODIPY558 is quenched in a ligand-dependent manner. The result that the ligand binding did not affect the fluorescence intensity of GFP suggests that FRET efficiency is identical in the absence and presence of ligand, possibly because the conformational change of MBP is not enough to alter the distance between GFP chromophore and BODIPY558 at Tyr210. In general, fluorescent protein-based FRET protein probes require relatively large conformational change upon the ligand binding or other modifications. The present strategy combining FRET from fluorescent proteins to fluorescent nonnatural amino acids and subsequent ligand-dependent fluorescence quenching allows us to design and generate FRET protein probes even though the conformational change of the target proteins is insufficient to cause measurable FRET change.

The fluorescence quenching mechanism has been applied to other ligand-binding proteins. Lectins, a family of glycan-binding proteins, are attractive target proteins because lectins are important not only as biological functional molecules but also as useful probes for glycan analysis [27, 28]. Fluorescent-labeled lectins that can detect the ligand binding as a fluorescence change will become useful research and diagnostic tools. We used a sialic acid-binding lectin [29] which had been developed from an earthworm galactose-binding lectin by random mutagenesis and in vitro selection. The sialic acid-binding lectin has one binding site for α2,6-sialyllactose. Trp31 residue, which locates at the binding site for hydrophobic interaction with the ligand, is expected to be available as a quencher when a fluorescent nonnatural amino acid is introduced near the binding site (Fig. 5.9a).

We introduced BODIPYFL-aminophenylalanine at several residues near the ligand-binding site in response to the amber codon using a highly efficient amber suppressor tRNA derived from *Mycoplasma capricolum* [30]. Fluorescence spectral measurements revealed that fluorescence intensity of BODIPYFL-containing lectins increased upon the addition of sialyllactose, while the extent of the fluorescence change depends on the incorporation position [31]. The most significant increase (3.7-fold) was observed for BODIPYFL-containing lectin at Thr38 (Fig. 5.9b). To examine the possibility of the quenching by Trp31, Trp31 was substituted with phenylalanine. The extent of the increase in the fluorescence intensity of BODIPYFL-containing lectin at Thr38 with the W31F mutation was smaller than that without the mutation, suggesting that the Trp31 residue may contribute to the ligand-dependent quenching. A variety of lectins have Trp residues at the ligand-binding site for hydrophobic interaction with hydrophobic moieties in glycans; therefore, the present strategy will be utilized for other types of lectins. A series of fluorescent lectin probes will be a powerful tool for systematic glycan analysis as in the case of lectin microarray [32].

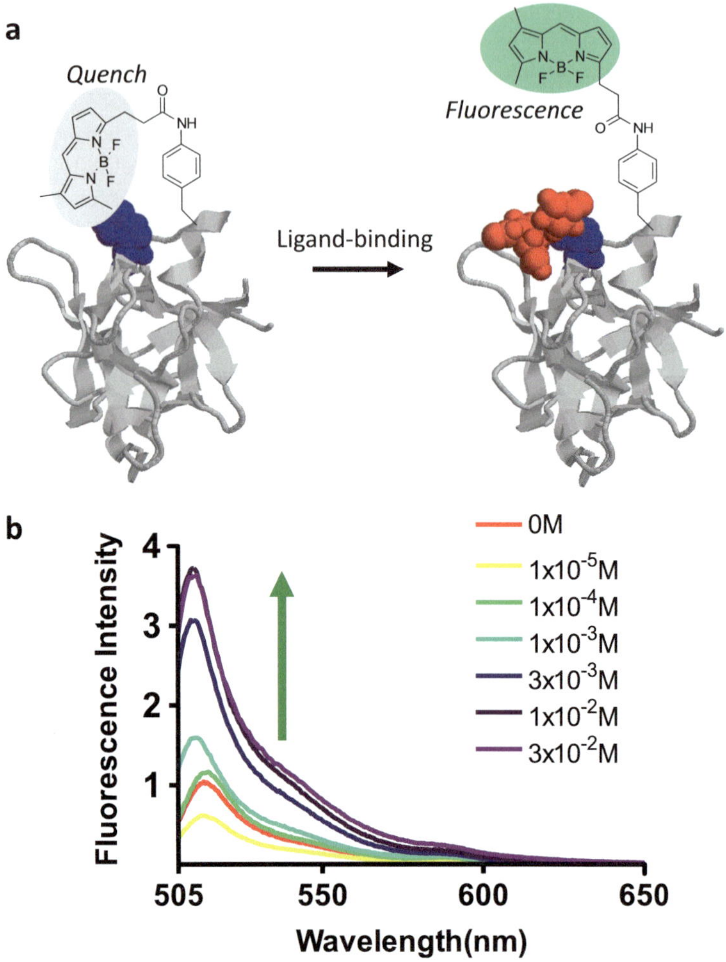

Fig. 5.9 (**a**) Fluorescence detection of the ligand binding of sialic acid-binding lectin containing BODIPYFL. The fluorescence of BODIPYFL is quenched by Trp31 in a ligand-dependent manner. (**b**) Fluorescence spectra of BODIPYFL-containing lectin at Thr38 in the absence and presence of sialyllactose

5.6 Antibody-Based Fluorescent Probes for Antigen Detection

As exemplified above, the site-specific fluorescent labeling and the ligand-dependent fluorescence quenching are highly useful for design and synthesis of protein probes to detect specific ligands. To further generalize the fluorescence quenching detection method, we have applied this strategy to antibodies. Antibodies are custom-made ligand-binding proteins. One class of antibodies, IgG, is

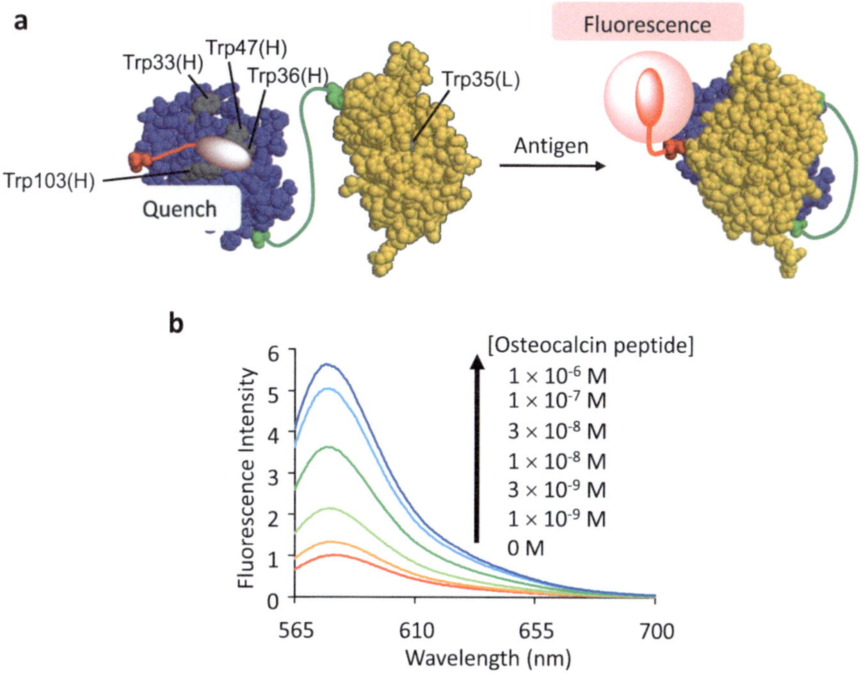

Fig. 5.10 Antibody-based fluorescent probes to detect antigens as fluorescence change. (**a**) Single-chain antibody variable fragment (*scFv*) containing TAMRA-X-aminophenylalanine at the N-terminus shows fluorescence quenching of TAMRA by Trp residues at the interface region of VH and VL domains in the absence of the antigen. The antigen binding to scFv promotes the hydrophobic interaction between VH and VL and eliminates the fluorescence quenching by the Trp residues. (**b**) Fluorescence spectra of the TAMRA-labeled anti-osteocalcin scFv in the absence and presence of the antigen peptide

constituted of two subunits, heavy and light chains. N-terminal variable domains of heavy and light chains (VH and VL domains) associate each other by hydrophobic interaction and form heterodimer to bind specific antigens. The VH and VL association is less efficient in the absence of antigens than that in the presence of antigens, as observed in open-sandwich ELISA method [33]. In open-sandwich ELISA, enzyme-linked VH domain associates with VL domain adsorbed on solid-phase in an antigen-dependent manner. In addition, it is notable that the interface regions in VH and VL domains have several conserved Trp residues for the hydrophobic association of VH and VL. Therefore, there is a possibility that fluorescent-labeled antibodies show antigen-dependent fluorescence quenching by Trp residues if a fluorophore is incorporated at an appropriate site of the antibodies.

We have found that fluorescent-labeled single-chain antibody fragment (scFv) derivatives constituted of VH and VL linked with a flexible peptide linker show antigen-dependent fluorescence quenching (Fig. 5.10a) [34]. Carboxytetramethyl-rhodamine (TAMRA)-linked aminophenylalanine [13] was incorporated into the

N-terminal region of scFv against osteocalcin, a biomarker for bone diseases. The TAMRA-labeled scFv showed significant increase in fluorescence intensity upon the binding of the antigen (Fig. 5.10b). The substitutions of Trp residues by Phe reduced the fluorescence increase compared with wild-type scFv, suggesting that the Trp residues contribute to the antigen-dependent fluorescence quenching.

Because the Trp residues locate at framework region of VH and VL domains and are conserved, the antigen-dependent fluorescence quenching can be applied to other antigen-antibody pairs. In addition to osteocalcin peptide, proteins (lysozyme and serum albumin) and small molecules (bisphenol A, estradiol, and morphine-like drugs) were found to cause the increase in the fluorescence of the corresponding TAMRA-labeled scFvs [34]. Although the extents of the fluorescence increase are different depending on the type of antibodies, the antibody-based fluorescent probes will become a versatile tool for detection of a variety of target molecules.

The fluorescent-labeled antibody-based protein probes can also be derivatized to double-labeled ones for ratiometric detection of antigens. In a similar manner to the case of MBP, double-labeled scFvs were synthesized using two expanded genetic codons. TAMRA-linked aminophenylalanine was introduced at the N-terminal region of scFv as a FRET acceptor in response to the CGGG four-base codon, and carboxyrhodamine dye (CR110)-linked aminophenylalanine (Fig. 5.4a) was introduced at the C-terminus as a FRET donor in response to the amber codon. The N-terminus of VH and the C-terminus of VL in antibody fragments apart from about 35 Å, which is enough short for FRET.

The double-labeled scFv against osteocalcin showed increase in TAMRA fluorescence upon the antigen binding, while CR110 green fluorescence was constant regardless of the presence and absence of the antigen (Fig. 5.11). This result indicates that FRET from CR110 to TAMRA occurs, and subsequently TAMRA is quenched in an antigen-dependent manner. A single-labeled scFv with CR110 at the C-terminus showed no change upon the addition of the antigen. Taken together, the unchanged fluorescence of CR110 in the double-labeled scFv indicates that FRET efficiency is not altered upon the antigen binding.

There are a number of reports on ratiometric FRET protein probes for in vitro and in vivo detection of target molecules including live-cell imaging. Nonetheless, it is not easy to design and synthesize FRET protein probes for the detection of molecules of interest that researchers want to measure. Common strategies are based on FRET change in response to ligand-driven conformational change of proteins; however, most proteins do not always show large conformational change required for FRET change upon the ligand binding. The present antibody-based double-labeled protein probes that show fluorescence ratio change without FRET change are a potential solution to this problem.

While the present method has several advantages as described above, there is a disadvantage that the incorporation of fluorescent nonnatural amino acids requires

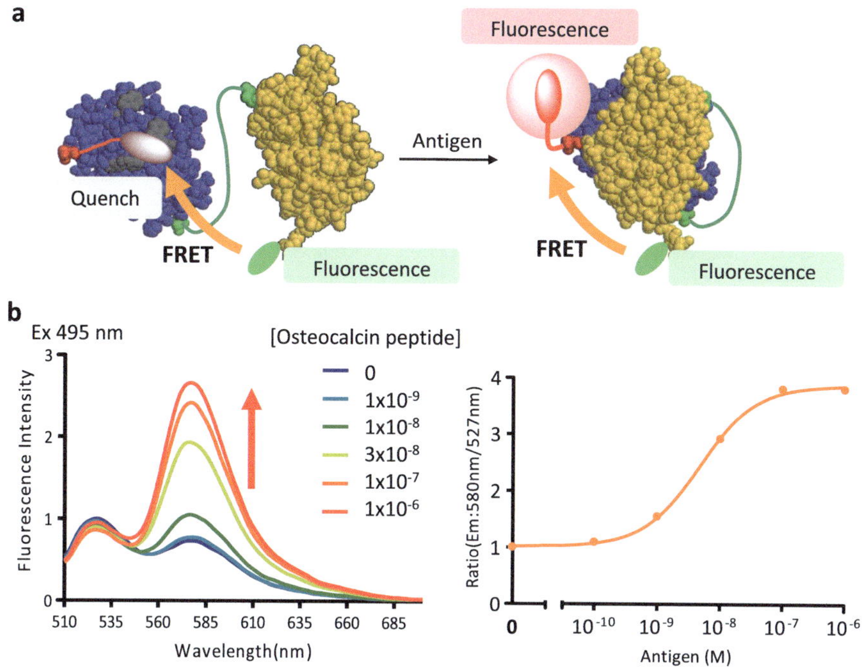

Fig. 5.11 (**a**) Double-labeled scFv shows fluorescence ratio change upon the binding of the antigen based on FRET from CR110 at the C-terminus to TAMRA at the N-terminus and ligand-dependent fluorescence quenching of TAMRA. (**b**) Fluorescence spectra and fluorescence ratio of the double-labeled anti-osteocalcin scFv in the absence and presence of the antigen peptide

cell-free protein synthesis. This limits the use of the labeled proteins in living cells and the large-scale production of labeled proteins required for practical uses. Transfection of labeled proteins to living cells using protein transfection methods is one of the strategies to apply the labeled proteins to in-cell analysis. Alternatively, genetically encoded fluorescent nonnatural amino acids that can be attached to an amber suppressor tRNA by mutant aminoacyl-tRNA synthetase and be incorporated in response to the amber codon in living cells have been developed [35–37]. Although the types of fluorescent nonnatural amino acids are limited at the present time, this method has a potential to apply the site-specific fluorescence labeling technique to a wide variety of basic, diagnostic, and medicinal research fields.

Acknowledgments The author deeply appreciates the contributions of all the coworkers in this research project. This work was partly supported by a Grant-in-Aid for Scientific Research on Innovative Areas (20107005) from the Ministry of Education, Culture, Sports, Science, and Technology, Japan.

References

1. Wang L, Schultz PG (2005) Expanding the genetic code. Angew Chem Int Ed 44:34
2. Budisa N (2004) Prolegomena to future experimental efforts on genetic code engineering by expanding its amino acid repertoire. Angew Chem Int Ed 43:6426
3. Link J, Mock ML, Tirrell DA (2003) Non-canonical amino acids in protein engineering. Curr Opin Biotechnol 14:603
4. Hendrickson TL, de Crécy-Lagard V, Schimmel P (2004) Incorporation of nonnatural amino acids into proteins. Annu Rev Biochem 73:147
5. Zhang WH, Otting G, Jackson CJ (2013) Protein engineering with unnatural amino acids. Curr Opin Struct Biol 23:581
6. Hohsaka T, Sisido M (2002) Incorporation of non-natural amino acids into proteins. Curr Opin Chem Biol 6:809
7. Hohsaka T, Ashizuka Y, Murakami H, Sisido M (1996) Incorporation of nonnatural amino acids into streptavidin through in vitro frame-shift suppression. J Am Chem Soc 118:9778
8. Hohsaka T, Kajihara D, Ashizuka Y, Murakami H, Sisido M (1999) Efficient incorporation of nonnatural amino acids with large aromatic groups into streptavidin in in vitro protein synthesizing systems. J Am Chem Soc 121:34
9. Hohsaka T, Ashizuka Y, Taira H, Murakami H, Sisido M (2001) Incorporation of nonnatural amino acids into proteins by using various four-base codons in an *E. coli* in vitro translation system. Biochemistry 40:11060
10. Robertson SA, Ellman JA, Schultz PG (1991) A general and efficient route for chemical aminoacylation of transfer RNAs. J Am Chem Soc 113:2722
11. Hecht SM, Alford BL, Kuroda Y, Kitano S (1978) "Chemical aminoacylation" of tRNA's. J Biol Chem 253:4517
12. Kajihara D, Abe R, Iijima I, Komiyama C, Sisido M, Hohsaka T (2006) FRET analysis of protein conformational change through position-specific incorporation of fluorescent amino acids. Nat Methods 3:923
13. Abe R, Shiraga K, Ebisu S, Takagi H, Hohsaka T (2010) Incorporation of fluorescent non-natural amino acids into N-terminal tag of proteins in cell-free translation and its dependence on position and neighboring codons. J Biosci Bioeng 110:32
14. Miyawaki A et al (1997) Fluorescent indicators for Ca based on green fluorescent proteins and calmodulin. Nature 388:882
15. Miyawaki A (2003) Visualization of the spatial and temporal dynamics of intracellular signaling. Dev Cell 4:295
16. Hohsaka T, Ashizuka Y, Sasaki H, Murakami H, Sisido M (1999) Incorporation of two different nonnatural amino acids independently into a single protein through extension of the genetic code. J Am Chem Soc 121:12194
17. Taki M, Hohsaka T, Murakami H, Taira K, Sisido M (2002) Position-specific incorporation of a fluorophore-quencher pair into a single streptavidin through orthogonal four-base codon/anticodon pairs. J Am Chem Soc 124:14586
18. Marme N, Knemeyer J-P, Sauer M, Wolfrum J (2003) Inter- and intramolecular fluorescence quenching of organic dyes by tryptophan. Bioconjug Chem 14:1133
19. Iijima I, Hohsaka T (2009) Position-specific incorporation of fluorescent non-natural amino acids into maltosebinding protein for detection of ligand binding by FRET and fluorescence quenching. ChemBioChem 17:999
20. Quiocho FA, Spurlino JC, Rodseth LE (1997) Extensive features of tight oligosaccharide binding revealed in high-resolution structures of the maltodextrin transport/chemosensory receptor. Structure 5:997
21. Li IT, Pham E, Truong K (2006) Protein biosensors based on the principle of fluorescence resonance energy transfer for monitoring cellular dynamics. Biotechnol Lett 28:1971

22. Otsuji T, Okuda-Ashitaka E, Kojima S, Akiyama H, Ito S, Ohmiya Y (2004) Monitoring for dynamic biological processing by intramolecular bioluminescence resonance energy transfer system using secreted luciferase. Anal Biochem 329:230
23. Gammon ST, Villalobos VM, Roshal M, Samrakandi M, Piwnica-Worms D (2009) Rational design of novel redshifted BRET pairs: platforms for real-time single-chain protease biosensors. Biotechnol Prog 25:559
24. Charest PG, Terrillon S, Bouvier M (2005) Monitoring agonist-promoted conformational changes of beta-arrestin in living cells by intramolecular BRET. EMBO Rep 6:334
25. Yamaguchi A, Hohsaka T (2012) Synthesis of novel BRET/FRET protein probes containing light-emitting proteins and fluorescent nonnatural amino acids. Bull Chem Soc Jpn 85:576
26. Verhaegent M, Christopoulos TK (2002) Recombinant Gaussia luciferase. Overexpression, purification, and analytical application of a bioluminescent reporter for DNA hybridization. Anal Chem 74(4378)
27. Gabius H-J, André S, Jimenez-Barbero J, Romeo A, Solís D (2011) From lectin structure to functional glycomics: principles of the sugar code. Trends Biochem Sci 36:298
28. Krishnamoorthy L, Mahal LK (2009) Glycomic analysis: an array of technologies. ACS Chem Biol 4:715
29. Yabe R, Suzuki R, Kuno A, Fujimoto Z, Jigami Y, Hirabayashi J (2007) Tailoring a novel sialic acid-binding lectin from a ricin-B chain-like galactose-binding protein by natural evolution-mimicry. J Biochem 141:389
30. Taira H, Matsushita Y, Kojima K, Shiraga K, Hohsaka T (2008) Comprehensive screening of amber suppressor tRNAs suitable for incorporation of non-natural amino acids in a cell-free translation system. Biochem Biophys Res Commun 374:304
31. Ito Y, Hohsaka T (2013) Incorporation of fluorescent nonnatural amino acid into sialic acid-binding lectin for fluorescence detection of ligand-binding. Bull Chem Soc Jpn 86:729
32. Kuno A, Uchiyama N, Koseki-Kuno S, Ebe Y, Takashima S, Yamada M, Hirabayashi J (2005) Evanescent-field fluorescence-assisted lectin microarray: a new strategy for glycan profiling. Nat Methods 2:851
33. Ueda H et al (1996) Open sandwich ELISA: a novel immunoassay based on the interchain interaction of antibody variable region. Nat Biotechnol 14:1714
34. Abe R, Ohashi H, Iijima I, Ihara M, Takagi H, Hohsaka T, Ueda H (2011) "Quenchbodies": quench-based antibody probes that show antigen-dependent fluorescence. J Am Chem Soc 133:17386
35. Wang J, Xie J, Schultz PG (2006) A genetically encoded fluorescent amino acid. J Am Chem Soc 128:8738
36. Summerer D, Chen S, Wu N, Deiters A, Chin JW, Schultz PG (2006) A genetically encoded fluorescent amino acid. Proc Natl Acad Sci U S A 103:9785
37. Speight LC, Muthusamy AK, Goldberg JM, Warner JB, Wissner RF, Willi TS, Woodman BF, Mehl RA, Petersson EJ (2013) Efficient synthesis and in vivo incorporation of acridonyl-alanine, a fluorescent amino acid for lifetime and Förster resonance energy transfer/luminescence resonance energy transfer studies. J Am Chem Soc 135:18806

Chapter 6
Unfoldomes and Unfoldomics: Introducing Intrinsically Disordered Proteins

Bin Xue and Vladimir N. Uversky

Abstract The last decade and a half has witnessed the fall of one of the major paradigms in structural biology. Contrary to the more than a hundred year-old belief that unique protein function is defined by unique crystal-like protein structure which is encoded in unique amino acid sequence, many biologically active proteins lack stable tertiary and/or secondary structure under physiological conditions in vitro. These intrinsically disordered proteins (IDPs) and IDP regions (IDPRs) are very different from ordered proteins and domains at the variety of levels, starting from the specific and well-recognizable amino acid composition and sequence biases and ending with engagement in biological functions non-accessible to entities with unique and fixed structures. IDPs/IDPRs are highly abundant in nature, and many of them are associated with various human diseases. The functional repertoire of IDPs complements the functions of ordered proteins, with IDPs being often involved in regulation, signaling, and control. Due to their exceptional commonness within the protein universe, combined with a set of specific structural and conformational features, and broad and unique functional repertoire, IDPs and hybrid proteins possessing ordered domains and functional IDPRs clearly comprise unfoldomes within all known proteomes. This chapter provides a brief description of IDPs/IDPRs and shows the place of these exceptionally interesting creatures within the protein kingdom.

B. Xue
Department of Molecular Medicine and USF Health Byrd Alzheimer's Research Institute, Morsani College of Medicine, University of South Florida, 12901 Bruce B. Downs Blvd, MDC07, Tampa, FL 33612, USA

V.N. Uversky (✉)
Department of Molecular Medicine and USF Health Byrd Alzheimer's Research Institute, Morsani College of Medicine, University of South Florida, 12901 Bruce B. Downs Blvd, MDC07, Tampa, FL 33612, USA

Institute for Biological Instrumentation, Russian Academy of Sciences, 142290 Pushchino, Moscow Region, Russia

Biology Department, Faculty of Science, King Abdulaziz University, P.O. Box 80203, Jeddah 21589, Kingdom of Saudi Arabia

Laboratory of Structural Dynamics, Stability and Folding of Proteins, Institute of Cytology, Russian Academy of Sciences, St. Petersburg, Russian Federation
e-mail: vuversky@health.usf.edu

M. Terazima et al. (eds.), *Molecular Science of Fluctuations Toward Biological Functions*, DOI 10.1007/978-4-431-55840-8_6

Keywords Intrinsically disordered protein • Protein structure • Protein function • Protein dynamics • Protein-protein interactions • Posttranslational modification • Alternative splicing • Protein misfolding

6.1 Introducing Unfoldomes and Unfoldomics

Proteins are the major components of the living cell. They play crucial roles in the maintenance of life, and protein dysfunctions may cause development of various pathological conditions. For more than a century, it has been believed that the specific functionality of a given protein is predetermined by its unique 3-D structure encoded by unique amino acid sequence [1, 2]. However, this picture is changing now, and although many proteins are indeed predisposed to fold into unique structures which evolved to possess unique biological functions, some proteins can misfold, and other proteins can stay substantially non-folded. Protein misfolding can occur either spontaneously or due to the mutations and other genetic alterations, problematic processing or posttranslational modifications, or due to exposure to harmful environmental conditions. Such misfolding is now considered as a crucial early step in the development of various protein conformation diseases [3]. As far as non-foldable proteins are concerned, for more than five decades researchers have been discovering individual proteins that possessed no definite ordered 3-D structure but still played important biological roles. The discovery rate for such proteins has been increasing continually and has become especially rapid during the last decade and a half [4]. Such proteins are widely known as intrinsically disordered proteins (IDPs) and are characterized by the lack of a well-defined 3-D structure under physiological conditions. The discovery and characterization of these proteins are becoming one of the fastest growing areas of protein science, and it is recognized now that many such proteins with no unique structure have important biological functions [4–16]. Structural flexibility and plasticity originating from the lack of a definite ordered 3-D structure are believed to represent a major functional advantage for these proteins, enabling them to interact with a broad range of binding partners including other proteins, membranes, nucleic acids, and various small molecules [17–19].

The functions attributed to IDPs were grouped into four broad classes: (1) molecular recognition, (2) molecular assembly, (3) protein modification, and (4) entropic chain activities [5, 6]. IDPs are often involved in regulatory/signaling interactions with multiple partners that require high specificity and low affinity [7, 20]. Some illustrative biological activities of IDPs include regulation of cell division, transcription and translation, signal transduction, protein phosphorylation, storage of small molecules, chaperone action, and regulation of the self-assembly of large multi-protein complexes such as the ribosome [4–11, 13–16, 20–26]. The crucial role of IDPs in signaling is further confirmed by the fact that eukaryotic proteomes, with their extensively developed interaction networks, are highly enriched in IDPs, relative to bacteria and archaea (see below, [27–29]). Application of a data mining tool to over 200,000 proteins from Swiss-Prot database revealed that many protein functions are associated with long disordered regions [13–15]. In fact, of the 711 Swiss-Prot functional keywords that were associated with at least 20 proteins, 262 were found to be strongly positively correlated with long intrinsically disordered regions (IDPRs), whereas 302 were strongly negatively correlated with such regions [13–15]. Therefore, the functional diversity provided by disordered regions complements functions of ordered protein regions.

Although unbound IDPs are disordered in solution, some of them perform their biological functions by binding to their specific partners. This binding often involves a disorder-to-order transition in which IDPs adopt a more structured conformation upon binding to their biological partners [30–37]. In this way, IDPs play diverse roles in regulating the function of their binding partners and in promoting the assembly of supramolecular complexes. Furthermore, because sites within their polypeptide chains are highly accessible, IDPs can undergo extensive posttranslational modifications, such as phosphorylation, acetylation, and/or ubiquitination (sumoylation), etc., allowing for modulation of their biological activity or function. Intriguingly, IDPs were shown to be highly abundant in various diseases, giving rise to the "disorder in disorders" or D^2 concept which generally summarizes work in this area [38].

As the number of IDPs and IDPRs in various proteomes is very large (e.g., for mammals, ~75 % of their signaling proteins are predicted to contain long disordered regions (>30 residues), about half of their total proteins are predicted to contain such long disordered regions, and ~25 % of their proteins are predicted to be fully disordered), and because IDPs and IDPRs have amazing structural variability and possess a very wide variety of functions, the unfoldome and unfoldomics concepts were recently introduced [4, 39, 40]. The use of the suffix "-ome" has a long history, while "-omics" is much more recent. The Oxford English Dictionary (OED) attributes "genome" to Hans Winkler from his 1920 work [41]. While the OED suggests that "genome" arose as a portmanteau of "gene" and "chromosome," this does not seem to be supported by the literature. Instead, Lederberg and McCray suggest that, as a botanist, Winkler must have been familiar with terms such as biome (a biological community), rhizome (a root system), and phyllome (the leaves covering a tree), among others, all of which were in use well before 1920 and all of which signify the collectivity of the units involved [42]. Thus, "ome" implies the complete set of the objects in question, with genome signifying the set of genes of an organism. By changing the "e" in "-ome" to "-ics," the new word is created that indicates the scientific study of the "-ome" in question. For example, officially, the change of "genome" to "genomics" occurred in 1987, when a journal by this name was founded by Victor McKusick and Frank Ruddle [42].

Many additional conversions from -ome to -omics have subsequently occurred, and a large number of "-omes" have been accepted in biology, including but not limited to the following: genome, proteome, interactome, metabolome, transcriptome, diseasome, toxicogenome, nutrigenome, cytome, oncoproteome, epitome, glycome, etc. For a more complete list, the reader is directed to http://omics.org.

Overall, the suffixes -ome and -omics imply a new layer of knowledge, especially when a scientist is dealing with the data produced by the large-scale studies, including the high-throughput experiments and the computational/bioinformatics analyses of the large datasets. The unfoldome and unfoldomics concepts are built on the ideas given above [43]. Unfoldome is attributed to a portion of proteome which includes a set of IDPs (also known as natively unfolded proteins, therefore unfoldome). The term unfoldome is also used to cover segments or regions of proteins that remain unfolded in the functional state. Unfoldomics is the field that focuses on the unfoldome. It considers not only the identities of the set of proteins and protein regions in the unfoldome of a given organism but also their functions, structures, interactions, evolution, etc. [43].

It is clearly recognized now that the disorderedness is linked to the peculiarities of amino acid sequences, as IDPs/IDPRs exhibit low sequence complexity and are generally enriched in polar and charged residues and are depleted of hydrophobic residues (other than proline). These features are consistent with their inability to fold into globular structures and form the basis of computational tools for disorder prediction [8, 10, 44–47]. These same computational tools can also be utilized for the large-scale discovery of IDPs in various proteomes (see below).

Being characterized by specific (and somewhat unique) amino acid sequences, IDPs possess a number of very distinctive structural properties that can be implemented for their discovery. This includes but is not limited to sensitivity to proteolysis [48], aberrant migration during SDS-PAGE [49], insensitivity to denaturing conditions [50], and definitive disorder characteristics visualized by CD spectropolarimetry, NMR spectroscopy, small-angle X-ray scattering, hydrodynamic measurement, and fluorescence, Raman, and infrared spectroscopies [51, 52]. Structurally, intrinsically disordered proteins range from completely unstructured polypeptides to extended partially structured forms to compact disordered ensembles containing substantial secondary structure [4, 8, 9, 23, 53]. Many proteins contain mixtures of ordered and disordered regions. Extended IDPs are known to possess the atypical conformational behavior (such as "turnout" response to acidic pH and high temperature and insensitivity to high concentrations of strong denaturants), which is determined by the peculiarities of their amino acid sequences and the lack of ordered 3-D structure [54]. These unique structural features of extended IDPs and their specific conformational behavior were shown to be useful in elaborating the experimental techniques for the large-scale identification of these important members of the protein kingdom. Three related methods were introduced: a method based on the finding that many proteins that fail to precipitate during perchloric acid or trichloroacetic acid treatment were IDPs [39], a method utilizing the fact that IDPs possessed high resistance toward the aggregation induced by heat treatment [39, 55, 56], and a method based on the heat treatment coupled with a novel 2-D gel methodology to identify IDPs in cell extracts [55]. It is anticipated that these methodologies combined with highly sensitive mass spectrometry-based techniques can be used for the detection and functional characterization of IDPs in various proteomes.

6.2 Peculiarities of the Amino Acid Sequences of IDPs and IDPRs

One of the key arguments about the existence and distinctiveness of IDPs came from various computational analyses. Historically, already at the early stages of the field, simple statistical comparisons of amino acid compositions and sequence complexity indicated that disordered and ordered regions are different to a significant degree [44, 57–59]. These sequence biases were then exploited to predict IDPRs or wholly disordered proteins with relatively high accuracy and to make crucial estimates about the commonness of disordered proteins in the three kingdoms of life [27, 44, 60].

Similar to the "normal" foldable proteins whose correct folding to the unique biologically active conformation is determined by their unique amino acid sequences, the absence of rigid structure in the non-foldable IDPs is also encoded in the specific features of their amino acid sequences. In fact, some of the IDPs have been discovered due to their unusual amino acid sequence compositions, and the absence of regular structure in these proteins has been explained by the specific features of their amino acid sequences including the presence of numerous uncompensated charged groups (often negative) that give rise to high net charges at neutral pH and related extreme pI values for such proteins [61–63] and a low content of hydrophobic amino acid residues [61, 63]. Simple evaluation of these two parameters constitutes a means of the charge-hydropathy (CH) plot, which was built based on the analysis of 275 ordered globular proteins and 91 extended IDPs [44]. From the comparison of the amino acid compositions of protein in these datasets, it has been concluded that the combination of low mean hydrophobicity and relatively high net charge represents an important prerequisite for the absence of compact structure in proteins under physiological conditions [44]. This observation was used to develop a CH plot method to distinguish ordered and disordered proteins based only on their net charges and hydropathy [44]. Since in this plot, IDPs and ordered proteins were separated by a linear boundary, above which a polypeptide chain with a given mean net charge will most probably be unfolded, this CH plot represents a simple and one of the most intuitive predictors of extended intrinsic disorder [44]. Obviously, such highly disordered proteins represent only a small subset of the IDP realm (see below).

More detailed analysis revealed additional information on the compositional difference between ordered protein and IDPs [64, 65]. Here, the IDPs/IDPRs were shown to be significantly depleted in bulky hydrophobic (Ile, Leu, and Val) and aromatic amino acid residues (Trp, Tyr, and Phe), which would normally form the hydrophobic core of a folded globular protein, and also possess low content of Cys and Asn residues. Since these residues (Trp, Tyr, Phe, Ile, Leu, Val, Cys, and Asn) were underrepresented in IDPs/IDPRs, they were proposed to be called order-promoting amino acids. On the other hand, IDPs/IDPRs were shown to be substantially enriched in polar, disorder-promoting, amino acids: Ala, Arg, Gly, Gln, Ser, Glu, and Lys and also in the hydrophobic but structure braking Pro [8, 66, 67].

Later, it has been shown that not only the sequence compositions of ordered and disordered regions were different but also that disordered regions of various lengths were diverse as well. In particular, the comparison of the four groups of protein regions, (a) low B-factor ordered regions, (b) high B-factor ordered regions, (c) short disordered regions, and (d) long disordered regions, revealed that these groups have distinct sequence and physicochemical characteristics [68]. For example, the short disordered and high B-factor regions were shown to be more negatively charged, while long disordered regions were either positively or negatively charged but on average nearly neutral [68].

As the amino acid sequences of the IDPs and IDPRs differ dramatically from those of the ordered proteins and domains, these amino acid sequence differences were used to develop various predictors of intrinsic disorder [69]. Protein disorder is a multi-faced phenomenon; i.e., disordered proteins, being mobile, flexible, and dynamic, might have very different structural features, which range from collapsed molten globule-like conformation to extended coil-like state (see below for more detailed discussion of the structural heterogeneity of IDPs). It has been suggested that just as an ordered protein is comprised of different types of secondary structure (α-helices, β-strands, β-turns, 3_{10}-helices, and others), ID protein can also be made up of distinguishable types of disorder [70]. To check this hypothesis, a partitioning algorithm based on the differential prediction accuracies has been developed [70]. This algorithm used the notion that a specialized predictor built on a given disorder flavor should have significantly higher same-flavor accuracy than other-flavor predictors or than a global predictor applied to the same given flavor. Application of this partitioning algorithm to known disordered proteins identified three distinctive "flavors" of disorder, arbitrarily called V, C, and S [70]. Importantly, the flavor-specific disordered proteins have been shown to be distinguishable not only by their amino acid compositions but also by disordered sequence locations and biological functions. Based on these observations, it was proposed that specific flavor-function relationships do exist, and thus, it is possible (in principle) to identify the functions of disordered regions from their amino acid sequences alone, without any need for specific structural knowledge [70].

Since then, numerous researchers have designed many algorithms to predict disordered proteins utilizing specific biochemical properties and biased amino acid compositions of IDPs. Various prediction ideas and different computing techniques have been utilized. Many of these predictors including PONDR®s [57, 67, 71–74], FoldIndex [75], GlobPlot [76], DisEMBL [77], DISOPRED and DISOPRED2 [28, 78–80], IUPred [81], FoldUnfold [82], RONN [83], DisPSSMP [84], DisPSSMP2 [85], Spritz [86], PrDOS [87], etc. can be accessed via public servers and evaluate intrinsic disorder on a per-residue basis. Since the first predictors were published, more than 50 predictors of disorder have been developed [69]. Several recent reviews summarized the current state of the art in the field of IDP predictions and represent a useful overview of the prediction methods highlighting their advantages and drawbacks [45, 69].

It is important to remember that comparing and combining several predictors on an individual protein of interest or on a protein dataset can provide additional insight regarding the predicted disorder if any exists. For example, the differences in the sensitivity of different predictors to different levels of overall disorderedness were utilized in CDF-CH plot analysis, which allows the ordered and disordered proteins separation in the CH-CDF phase space [88]. Here, CDF stays for the cumulative distribution function. This method is based on the PONDR® VLXT predictor, which predicts the order-disorder class for every residue in a protein [27, 29]. CDF analysis summarizes the per-residue predictions by plotting PONDR® scores against their cumulative frequency, which allows ordered and disordered proteins to be distinguished based on the distribution of prediction scores. As a result, CDF curves for PONDR® VLXT predictions begin at the point with coordinates (0,0) and end at the point with coordinates (1,1) because PONDR® VLXT predictions are defined only in the range [0, 1] with values less than 0.5 indicating a propensity for order and values greater than or equal to 0.5 indicating a propensity for disorder. The optimal boundary that provided the most accurate order-disorder classification can be determined, and the order-disorder classification is based on whether a CDF curve is above or below a majority of boundary points [29]. Since CH plot analysis discriminates protein using only two attributes, mean net charge and mean hydrophobicity, whereas the PONDR-based CDF uses a relatively large feature space, combination of these two methods produces CH-CDF plot, where each spot corresponds to a single protein and its coordinates are calculated as a distance of this protein from the boundary in the corresponding CH plot (Y-coordinate) and an averaged distance of the corresponding CDF curve from the boundary (X-coordinate). Positive and negative Y values correspond to proteins which, according to CH plot analysis, are predicted to be natively unfolded or compact, respectively, whereas positive and negative X values are attributed to proteins that, by the CDF analysis, are predicted to be ordered or intrinsically disordered, respectively. Therefore, this plot has four quadrants: quadrant which contains ordered proteins; quadrant, which contains proteins predicted to be disordered by CDF but compact by CH plot (i.e., proteins with molten globule-like properties or hybrid proteins combining ordered and disordered regions); quadrant that includes proteins predicted to be disordered by both methods; and quadrant including proteins which are predicted to be disordered by CH plot but ordered by the CDF analysis [88].

6.3 Evaluation of Commonness of Disorder in Various Proteomes

The first application of the disorder predictors was the evaluation of the commonness of protein disorder in the Swiss-Prot database [60]. This analysis revealed that 25 % of proteins in Swiss-Prot had predicted ID regions longer than 40 consecutive residues and that at least 11 % of residues in Swiss-Prot were likely to be disordered. Given the existence of a few dozen experimentally characterized disordered regions at the time, this work had significant influence on the recognition of the importance of studying disordered proteins [60].

Next, both PONDR® VLXT [27] and 3 Flavor PONDR® predictors [70] were used to estimate the amount of disorder in various genomes. The predictions counted only disordered regions greater than 40 residues in length, which has a false-positive rate of fewer than 400 residues out of 100,000. The result clearly showed that disorder increases from bacteria to archaea to eukaryotes with over half of the eukaryotic proteins containing predicted disordered regions [27, 70]. One explanation for this trend is a change in the cellular requirements for certain protein functions, particularly cellular signaling. In support of this hypothesis, PONDR analysis of a eukaryotic signal protein database indicates that the majority of known signal transduction proteins are predicted to contain significant regions of disorder [25].

Ward et al. [28] have refined and systematized such an analysis and concluded that the fraction of proteins containing disordered regions of 30 residues or longer (predicted using DISOPRED) were 2 % in archaea, 4 % in bacteria, and 33 % in eukaryotes. In addition, a complete functional analysis of the yeast proteome with respect to the three Gene Ontology (GO) categories was performed. In terms of molecular function, transcription, kinase, nucleic acid, and protein binding activity were the most distinctive signatures of disordered proteins. The most overrepresented GO terms characteristic for the biological process category were transposition, development, morphogenesis, protein phosphorylation, regulation, transcription, and signal transduction. Finally, with respect to cellular component, it appeared that nuclear proteins were significantly enriched in disorder, while terms such as membrane, cytosol, mitochondrion, and cytoplasm were distinctively overrepresented in ordered proteins [28].

The disorderedness of genome-linked proteins VPg from various viruses was evaluated by a combined CH-CDF analysis [89, 90]. The genome-linked protein VPg of potato virus A (PVA; genus *Potyvirus*) has essential functions in all critical steps of PVA infection, i.e., replication, movement, and virulence. The structural analysis of the recombinant PVA VPg revealed that this protein possesses many properties of a native molten globule [89]. In a follow-up study, it has been shown that although VPgs from various viruses (from *Sobemovirus*, *Potyvirus*, and *Caliciviridae* genera) are highly diverse in size, sequence, and function, many of them were predicted to contain long disordered domains [90]. The analysis of VPgs within the CH-CDF phase space revealed that many VPgs are expected to behave as native molten globules [90].

Application of a combined CH-CDF analysis to mice proteins revealed that ~12 % mice proteins are likely to belong to the class of extended IDPs (native coils and native pre-molten globules), whereas ~30 % proteins in mouse genome are potential native molten globules.

Other studies on the fraction of intrinsic disorder in various evolutionary distant species supported the prevalence of disordered proteins in various proteomes and consistently showed that the eukaryotic proteomes had higher fraction of intrinsic disorder than prokaryotic proteomes [28, 91–95]. These observations were

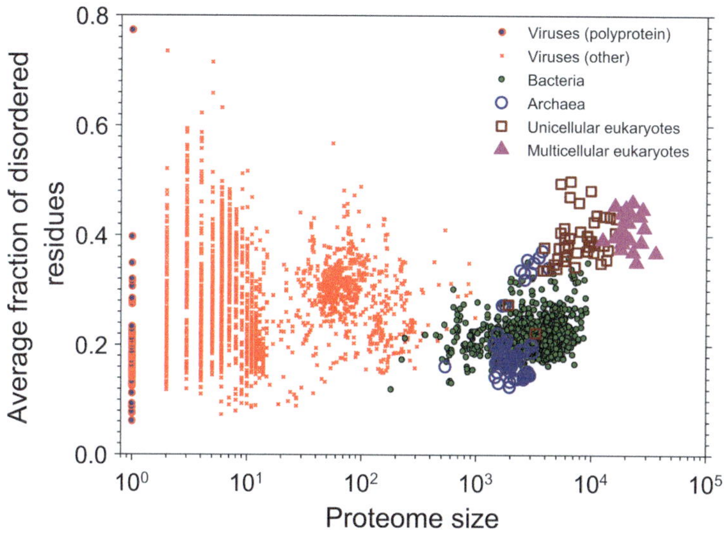

Fig. 6.1 Abundance of intrinsic disorder in various proteomes. Correlation between the intrinsic disorder content and proteome size for 3484 species from viruses, archaea, bacteria, and eukaryotes. Each symbol indicates a species. There are totally six groups of species: viruses expressing one polyprotein precursor (*small red circles* filled with *blue*), other viruses (*small red circles*), bacteria (*small green circles*), archaea (*blue circles*), unicellular eukaryotes (*brown squares*), and multicellular eukaryotes (*pink triangles*). Each viral polyprotein was analyzed as a single polypeptide chain, without parsing it into the individual proteins before predictions. The proteome size is the number of proteins in the proteome of that species and is shown in log base. The average fraction of disordered residues is calculated by averaging the fraction of disordered residues of each sequence over all sequences of that species. Disorder prediction is evaluated by PONDR-VSL2B (Modified from Ref. [98])

explained based on the specific functional repertoire of IDPs/IDPRs, which are often involved in signaling, recognition, and regulation, and there are complex and well-developed regulation networks in eukaryotic and especially in multicellular eukaryotic organisms that might rely on the ability of IDPs/IDPRs to perform the necessary functions [7, 20, 96, 97].

The increased abundance of disorder in higher eukaryotes was repeatedly shown by studies which progressively analyzed more and more completed proteomes, finally culminating with a comprehensive investigation of the disorder distribution in almost 3500 proteomes from viruses and three kingdoms of life [98]. Figure 6.1 represents the results of this analysis by showing the correlation between the intrinsic disorder content and proteome size for 3484 species from viruses, archaea, bacteria, and eukaryotes.

Surprisingly, Fig. 6.1 shows that there is a well-defined gap between the prokaryotes and eukaryotes in the plot of fraction of disordered residues on proteome size, where almost all eukaryotes have 32 % or more disordered residues, whereas the majority of the prokaryotic species have 27 % or fewer disordered residues [98]. Therefore, it looks like the fraction of 30 % disordered residues serves as a boundary between the prokaryotes and eukaryotes and reflects the existence of a complex stepwise correlation between the increase in the organism complexity and the increase in the amount of intrinsic disorder. A gap in the plot of fraction of disordered residues on proteome size parallels a morphological gap between prokaryotic and eukaryotic cells which contain many complex innovations that seemingly arose all at once. In other words, this sharp jump in the disorder content in proteomes associated with the transition from prokaryotic to eukaryotic cells suggests that the increase in the morphological complexity of the cell paralleled the increased usage of intrinsic disorder [98]. The variability of disorder content in unicellular eukaryotes and rather weak correlation between disorder status and organism complexity (measured as the number of different cell types) are likely related to the wide variability of their habitats, with especially high levels of disorder being found in parasitic host-changing protozoa, the environment of which changes dramatically during their life-span [88]. The further support for this hypothesis came from the fact that the intrinsic disorder content in multicellular eukaryotes (which are characterized by more stable and less variable environment of individual cells) was noticeably less variable than that in the unicellular eukaryotes [98].

Concluding, considered above studies clearly showed that IDPs are highly abundant in nature, have numerous crucial functions, and can be reliably identified by various computational means often based solely on the peculiarities of their amino acid sequences.

6.4 Structural Heterogeneity of IDPs

One of the crucial consequences of an extended sequence space and nonhomogeneous distribution of foldability (or the structure-coding potential) within amino acid sequences of IDPs and IDPRs is their astonishing structural heterogeneity. In fact, a typical IDP/IDPR contains a multitude of elements coding for potentially foldable, partially foldable, differently foldable, or not foldable at all protein segments [99, 100]. As a result, different parts of a molecule are ordered (or disordered) to a different degree. This distribution is constantly changing in time where a given segment of a protein molecule has different structures at different time points. As a result, at any given moment, an IDP has a structure which is different from a structure seeing at another moment [99, 100].

Another level of structural heterogeneity is determined by the fact that many proteins are hybrids of ordered and disordered domains and regions, and this mosaic structural organization is crucial for their functions [101]. Also, even when do not possess ordered domains, IDPs are known to have various levels and depth of disorder [20]. Over a few past years, an understanding of the available conformational space of IDPs/IDPRs underwent significant evolution. In fact, for a long time, IDPs were considered mostly "unstructured" or "natively unfolded" polypeptide chains. This was mostly due to the fact that the majority of IDPs analyzed at early

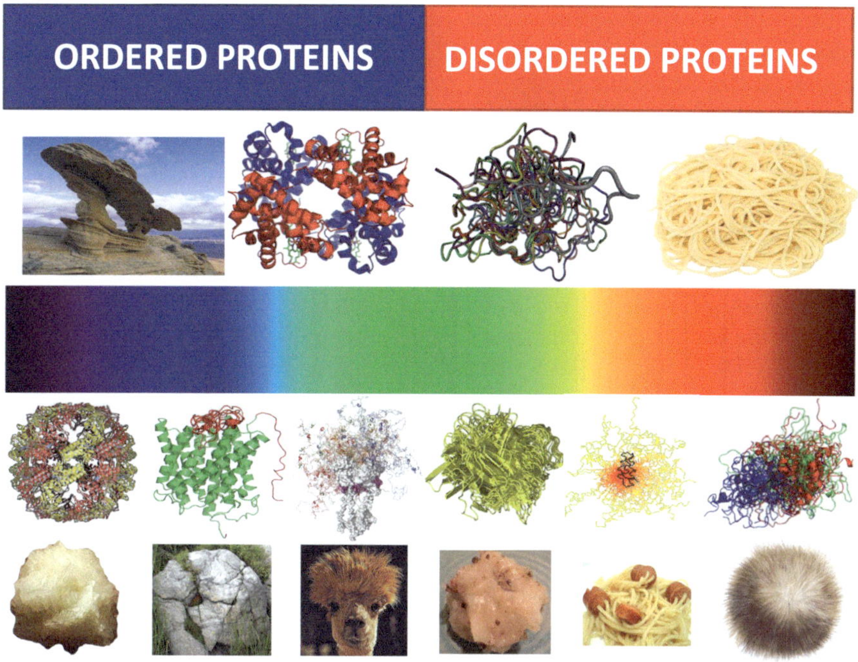

Fig. 6.2 Structural heterogeneity of IDPs/IDPRs. *Top half*: bicolored view of functional proteins which are considered to be either ordered (folded, *blue*) or completely structureless (disordered, *red*). Ordered proteins are taken as rigid rocks, whereas IDPs are considered as completely structureless entities, kind of cooked noodles. *Bottom half*: a continuous emission spectrum representing the fact that functional proteins can extend from fully ordered to completely structureless proteins, with everything in between. Intrinsic disorder can have multiple faces and can affect different levels of protein structural organization, and whole proteins, or various protein regions, can be disordered to a different degree. Some illustrative examples include ordered proteins that are completely devoid of disordered regions (rocklike type), ordered proteins with limited number of disordered regions (grass-on-the-rock type), ordered proteins with significant amount of disordered regions (llama/camel hair type), molten globule-like collapsed IDPs (greasy ball type), pre-molten globule-like extended IDPs (spaghetti-and-sausage type), and unstructured extended IDPs (hair ball type) (Adopted from Ref [100])

stages of the field contained very little ordered structure; i.e., they were really mostly unstructured or unfolded. Finding and characterization of such "structureless" proteins were important to build up a strong case to counterpoint the dominant view represented by the classical sequence-to-structure-to-function paradigm, especially since such fully unstructured, yet functional, proteins clearly represented the other extreme of the protein structure-function spectrum [101]. The top half of Fig. 6.2 illustrates this situation by opposing rocklike ordered proteins and cooked spaghettilike IDPs.

However, already in some early studies, it was indicated that IDPs/IDPRs could be crudely grouped into two major structural classes, proteins with compact and extended disorder [8, 23, 53, 96, 102]. Based on these observations, the protein functionality was ascribed to at least three major protein conformational states, ordered, molten globular, and coil-like [8, 9], indicating that functional IDPs can be less or more compact and possess smaller or larger amount of flexible secondary/tertiary structure [8, 9, 23, 44, 53, 96]. Roughly at the same time, it was emphasized that the extended IDPs (known as natively unfolded proteins) do not represent a uniform entity but contain two broad structural classes, native coils and native pre-molten globules [23].

Currently available data suggest that intrinsic disorder possesses multiple flavors, can have multiple faces, and can affect different levels of protein structural organization, where whole proteins or various protein regions can be disordered to a different degree [99, 100]. This new view of structural space of functional proteins can be visualized to form a continuous spectrum of differently disordered conformations extending from fully ordered to completely structureless proteins, with everything in between (Fig. 6.2, bottom half). Here, functional proteins can be well folded and be completely devoid of disordered regions (rocklike scenario). Other functional proteins may contain limited number of disordered regions (a grass-on-the-rock scenario), or have significant amount of disordered regions (a camel hair scenario), or be molten globule-like (a greasy ball scenario), or behave as pre-molten globules (a spaghetti-and-meatball/sausage scenario), or be mostly unstructured (a hair ball scenario).

Notably, in this representation, there is no boundary between ordered proteins and IDPs, and the structure-disorder space of a protein is considered as a continuum. It is important to remember that even the most ordered proteins do not resemble "solid rocks" and have some degree of flexibility. In fact, a protein molecule is an inherently flexible entity, and the presence of this flexibility (even for the most ordered proteins) is crucial for its biological activity [103]. Also, another important point to remember is that due to their heteropolymeric nature, proteins are never random coils and always have some residual structure [99, 100].

6.5 Discovering and Characterizing the Disease-Related Unfoldomes

Misfolding (the failure of a specific peptide or protein to adopt its functional conformational state) and related dysfunction of many proteins were considered as a major cause for the development of different pathological conditions. Such misfolding and dysfunction can originate from point mutation(s) or result from an exposure to internal or external toxins, impaired posttranslational modifications (phosphorylation, advanced glycation, deamidation, racemization, etc.), an increased probability of degradation, impaired trafficking, lost binding partners, or oxidative damage. All these factors can act independently or in association with one another.

Although the formation of various aggregates represents the most visible consequence of protein misfolding and although these aggregates form the basis for the development of various protein deposition diseases, pathogenesis of many more human diseases does not depend on aggregation being based on protein dysfunction. As many of the proteins associated with the conformational diseases are also involved in recognition, regulation, and cell signaling, it has been hypothesized that many of them are IDPs. In other words, according to the "disorder in disorders" or D^2 concept, IDPs are abundantly involved in the development of the conformational diseases, which therefore may originate from the misidentification, misregulation, and missignaling due to the misfolding of causative IDPs [38].

To support this hypothesis, three approaches were elaborated for estimating the abundance of IDPs in various pathological conditions. The first approach is based on the assembly of specific datasets of proteins associated with a given disease and the computational analysis of these datasets using a number of disorder predictors [25, 38, 104, 105]. In essence, this is an analysis of individual proteins extended to a set of independent proteins. A second approach utilized a network of genetic diseases where the related proteins are interlinked within one disease and between different diseases [40]. A third approach is based on the evaluation of the association of a particular protein function (including the disease-specific functional keywords) with the level of intrinsic disorder in a set of proteins known to carry out this function [13–15]. These three approaches are briefly presented below.

The easiest way to evaluate the abundance of intrinsic disorder in a given disease is based on a simple two-stage protocol, where a set of disease-related proteins is first assembled by searching various databases and then the collected group of proteins is analyzed for intrinsic disorder. The depth of this analysis is based on the breadth of the search for the disease-related proteins and on the number of different computational tools utilized to find disordered proteins/regions [25, 38, 43, 104–106]. Using this approach, it has been shown that many proteins associated with cancer, neurodegenerative diseases, and cardiovascular disease are highly disordered, being depleted in major order-promoting residues (Trp, Phe, Tyr, Ile, and Val) and enriched in some disorder-promoting residues (Arg, Gln, Ser, Pro, and Glu). High level of intrinsic disorder and a substantial number of potential interaction sites were also found using a set of computational tools. Many proteins were predicted to be wholly disordered. Overall, these studies clearly showed that intrinsic disorder is highly prevalent in proteins associated with human diseases, being comparable with that of signaling proteins and significantly exceeding the levels of intrinsic disorder in eukaryotic and in nonhomologous, structured proteins.

Unfoldome of human genetic diseases was assembled via the analysis of a specific network which was built to estimate whether human genetic diseases and the corresponding disease genes are related to each other at a higher level of cellular and organism organization. This network represented a bipartite graph with a network of genetic diseases, the "human disease network," HDN, where two diseases were directly linked if there was a gene that was directly related to both of them, and a network of disease genes, the "disease gene network," DGN, where two genes were directly linked if there was a disease to which they were both directly related [107]. This framework, called the human diseasome, systematically linked the human disease phenome (which includes all the human genetic diseases) with the human disease genome (which contains all the disease-related genes) [107]. The analysis of HDN revealed that of 1284 genetic diseases, 867 had at least one link to other diseases, and 516 diseases formed a giant component, suggesting that the genetic origins of most diseases, to some extent, were shared with other diseases. In the DGN, 1377 of 1777 disease genes were shown to be connected to other disease genes, and 903 genes belonged to a giant cluster HDN. The vast majority of genes associated with genetic diseases was nonessential and showed no tendency to encode hub proteins (i.e., proteins having multiple interactions) [107]. The large-scale analysis of the abundance of intrinsic disorder in transcripts of the various disease-related genes was performed using a set of computational tools which uncovers several important features [40, 43]:

(a) Intrinsic disorder is common in proteins associated with many human genetic diseases.
(b) Different disease classes vary in the IDP contents of their associated proteins.
(c) Molecular recognition features, which are relatively short loosely structured protein regions within mostly disordered sequences and which gain structure upon binding to partners, are common in the diseasome, and their abundance correlates with the intrinsic disorder level.
(d) Some disease classes have a significant fraction of genes affected by alternative splicing, and the alternatively spliced regions in the corresponding proteins are predicted to be highly disordered and in some diseases contain a significant number of molecular recognition features (MoRFs).
(e) Correlations were found among the various diseasome graph-related properties and intrinsic disorder. In agreement with earlier studies, hub proteins were shown to be more disordered.

Another approach is a computational tool elaborated for the evaluation of a correlation between the functional annotations in the UniProt database, and the predicted intrinsic disorder was elaborated [13–15]. The approach is based on the hypothesis that if a function described by a given keyword relies on intrinsic disorder, then the keyword-associated protein would be expected to have a greater level of predicted disorder compared to the protein randomly chosen from the UniProt. To test this hypothesis, functional keywords associated with 20 or more proteins in UniProt were found, and corresponding keyword-associated datasets of proteins were assembled. Next, for each such a keyword-associated set, a length-matching set of random proteins was drawn from the UniProt, and order-disorder predictions were carried out for the keyword-associated sets and for the random sets [13–15]. The application of this tool revealed that out of 710 UniProt keywords, 310 functional keywords were associated with ordered proteins, 238 functional keywords were attributed to disordered proteins, and the remainder 162 keywords yield ambiguity in the likely function-structure associations [13–15]. It has been also shown that keywords describing various diseases were strongly correlated with proteins predicted to be disordered. Contrary to this, no disease-associated proteins were found to be strongly correlated with the absence of disorder [14].

Summarizing, because of the fact that IDPs/IDPRs are highly abundant and play crucial roles in numerous biological processes, it is not too surprising to find that some of them are involved in human diseases, many of which are characterized by the conformational changes, misfolding, and aggregation of an underlying protein. However, there is another side to this coin: protein functionality. In fact, many of the proteins associated with the conformational disorders are also involved in recognition, regulation, and cell signaling. For example, functions ascribed to α-synuclein, a protein involved in several neurodegenerative disorders, include binding fatty acids and metal ions; regulation of certain enzymes, transporters, and neurotransmitter vesicles; and regulation of neuronal survival (reviewed in [108]). Overall, there are about 50 proteins and ligands that interact and/or co-localize with this protein. Furthermore, α-synuclein has amazing structural plasticity and adopts a series of different monomeric, oligomeric, and insoluble conformations (reviewed in [109]). The choice between these conformations is determined by the peculiarities of the protein environment, suggesting that α-synuclein has an exceptional ability to fold in a template-dependent manner. Therefore, the important message here is that the development of the conformational diseases may originate not only from misfolding but also from the misidentification, misregulation, and missignaling of the related proteins.

6.6 IDPs as Novel, Promising, but yet Inconvenient Drug Targets

The possibility of interrupting the action of disease-associated proteins (including through modulation of protein-protein interactions) presents an extremely attractive objective for the development of new drugs. Since many proteins associated with various human diseases are either completely disordered or contain long disordered regions [38, 43] and since some of these disease-related IDPs/IDPRs are involved in recognition, regulation, and signaling, these proteins/regions clearly represent novel potential drug targets [110]. Due to failure to recognize the important role of disorder in protein function, current and evolving methods of drug discovery suffer from an overly rigid view of protein function. In fact, the rational design of enzyme inhibitors depends on the classical view where 3-D structure is an obligatory prerequisite for function. While generally applicable to many enzymatic domains, this view has persisted to influence thinking concerning all protein functions despite numerous examples to the contrary. This is most apparent in the observation that the vast majority of currently available drugs target the active site of enzymes, presumably since these are the only proteins for which the "unique structure-unique function" paradigm is generally applicable.

IDPs often bind their partners with relatively short regions that become ordered upon binding [34, 35, 111]. Targeting disorder-based interactions should enable the development of more effective drug discovery techniques. There are at least two potential approaches for the inhibition of the disorder-based interactions, where small molecule either binds to the binding site of the ordered partner to outcompete the IDPs/IDPRs or interacts directly with the IDP/IDPR. The principles of small molecule binding to IDPRs have not been well studied, but sequence-specific, small molecule binding to short peptides has been observed [112]. An interesting twist here is that small molecules can inhibit disorder-based protein-protein interactions via induction of the dysfunctional ordered structures in targeted IDPR, i.e., via the drug-induced misfolding.

In agreement with these concepts, small molecules "Nutlins" have been discovered that inhibited the p53-Mdm2 interaction by mimicking the inducible α-helix in p53 (residues 13–29) that binds to Mdm2 [113, 114]. Although X-ray crystallographic studies of the p53-Mdm2 complex revealed that the Mdm2 binding region of p53 forms an α-helical structure that binds into a deep groove on the surface of Mdm2 [115] (Fig. 6.3a), NMR studies showed that the unbound N-terminal region of p53 lacks fixed structure, although it does possess an amphipathic helix part of the time [116]. A close examination of the interface between the proteins reveals that Phe_{19}, Trp_{23}, and Leu_{26} of p53 are the major contributors to the interaction, with the side chains of these three amino acids pointing down into a crevice on the Mdm2 surface [113, 114] (Fig. 6.3b). The structure of Nutlin-2 was shown to mimic the crucial residues of p53, with two bromophenyl groups fitting into Mdm2 in the same pockets as Trp_{23} and Leu_{26} and an ethyl-ether side chain filling the spot normally taken by Phe_{19}[117–119] (Fig. 6.3c). Nutlins and related small molecules increased the level of p53 in cancer cell lines. This drastically decreased the viability of these cells, causing most of them to undergo apoptosis. When one of the Nutlins was given orally to mice, a 90 % inhibition of tumor growth compared to the control was induced [113, 117–119].

This successful Nutlin story marks the potential beginning of a new era, the signaling-modulation era, in targeting drugs to protein-protein interactions. Importantly, this druggable p53-Mdm2 interaction involves a disorder-to-order transition. Principles of such transitions are generally understood and therefore can be used to find similar drug targets, which are inducible α-helices [120]. In addition to Nutlins inhibiting p53-Mdm2 interaction, several other small molecules also act by blocking protein-protein interactions [121, 122]. Some of these interactions involve one structured partner and one disordered partner, with disordered segments becoming α-helix upon binding [120]. Therefore, the p53-Mdm2 complex is not a unique exception, and many other disorder-based protein-protein interactions are blocked by a small molecule. All this suggest that there is a cornucopia of new drug targets that would operate by blocking disorder-based protein-protein interactions.

For these p53-Mdm2-type examples, the drug molecules mimic a critical region of the disordered partner (which folds upon binding) and compete with this region for its binding site on the structured partner. These druggable interaction sites operate by the coupled binding and folding mechanism. They are small enough

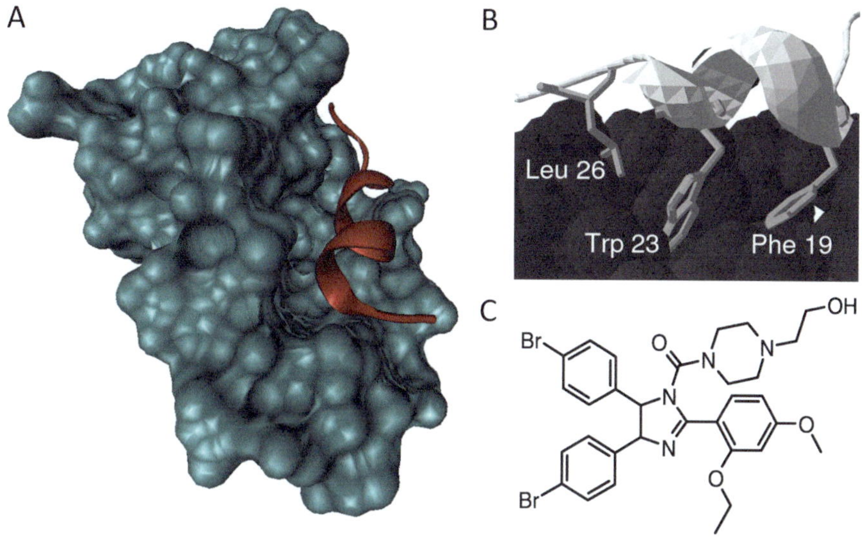

Fig. 6.3 Druggable intrinsic disorder-based interaction between p53 and Mdm2. (**a**) p53 fragment (*red ribbon*) bound to Mdm2 (*light blue surface*) (PDB ID: 1YCR). (**b**) Close-up view of p53 (*light gray ribbon*) bound to Mdm2 (*dark gray surface*). The side chains of p53s crucial residues for the interaction (Phe 19, Trp 23, Leu 26) are shown. (**c**) Structure of Nutlin-2, an inhibitor of the p53-Mdm2 interaction with IC_{50} of 0.14 μM [117] (Figure is adopted from Ref. [126])

and compact enough to be easily mimicked by small molecules [110]. Methods for predicting such binding sites in disordered regions have been developed [123], and the bioinformatics tools to identify which disordered binding regions can be easily mimicked by small molecules have been elaborated [120].

A complementary approach for small molecules to inhibit the disorder-based protein-protein interactions relies on the direct binding of drugs to the IDPs/IDPRs, which is illustrated by the c-Myc-Max story [124]. In order to bind DNA, regulate expression of target genes, and function in most biological contexts, c-Myc transcription factor must dimerize with its obligate heterodimerization partner, Max, which lacks a transactivation segment. Both c-Myc and Max are intrinsically disordered in their monomeric forms (Fig. 6.4a). Upon heterodimerization, they undergo coupled binding and folding of their basic-helix-loop-helix-leucine zipper domains (bHLHZips) (Fig. 6.4a). Since the deregulation of c-Myc is related to many types of cancer, the disruption of the c-Myc-Max dimeric complex is one of the approaches for c-Myc inhibition. Several small molecules were found to inhibit the c-Myc-Max dimer formation [124]. These molecules were shown to bind to one of the three discrete sites within the 85-residue bHLHZip domain of c-Myc, which are composed of short contiguous stretches of amino acids that can selectively and independently bind small molecules [124] (Fig. 6.4b). Inhibitor binding induces only local conformational changes, preserves the overall disorder of c-Myc, and inhibits interaction with Max [124]. Furthermore, binding of inhibitors to c-Myc

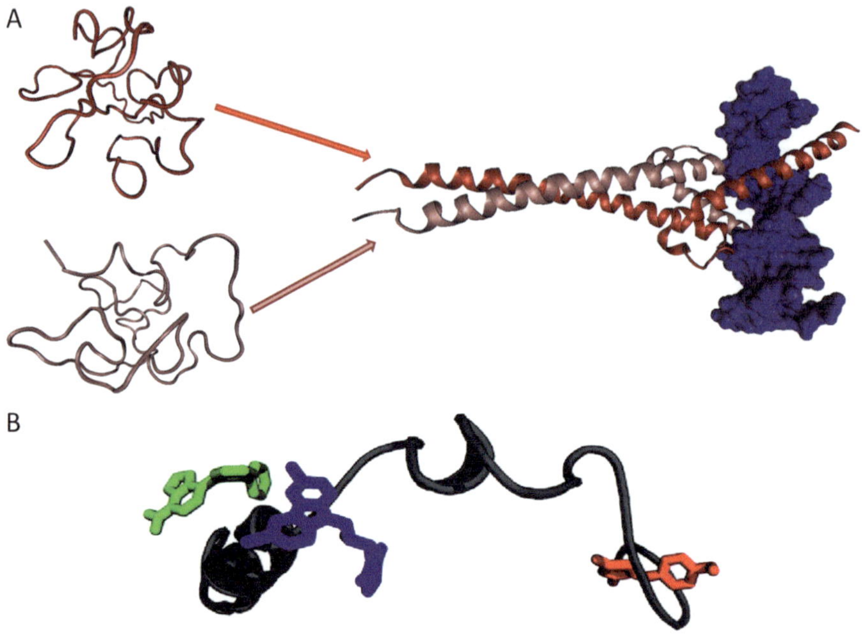

Fig. 6.4 Druggable intrinsic disorder-based interaction between c-Myc and Max. (**a**) Both c-Myc and Max are intrinsically disordered in their unbound states (*left side*) but fold into coiled-coil dimer upon interaction with DNA (*right side*) (PDB ID: 1NKP). c-Myc, Max, and DNA are shown as *red ribbon*, *pink ribbon*, and *blue surface*, respectively. (**b**) c-Myc-Max inhibitors bind to distinct ID regions of c-Myc. These binding sites are composed of short contiguous stretches of amino acids that can selectively and independently bind small molecules. Inhibitor binding induces only local conformational changes, preserves the overall disorder of c-Myc, and inhibits dimerization with Max [124] (Figure is adopted from Ref. [126])

was shown to occur simultaneously and independently on the three independent sites. Based on these observations, it has been concluded that a rational and generic approach to the inhibition of protein-protein interactions involving IDPs may therefore be possible through the targeting of intrinsically disordered sequence [124].

Recently, a functional misfolding concept was introduced to describe a mechanism preventing IDPs from unwanted interactions with nonnative partners [125]. IDPs/IDPRs are characterized by high conformational dynamics and flexibility, the presence of sticky preformed binding elements, and the ability to morph into differently shaped bound configurations. However, detailed analyses of the conformational behavior and fine structure of several IDPs revealed that the preformed binding elements might be involved in a set of nonnative intramolecular interactions. Based on these observations, it was proposed that an intrinsically disordered polypeptide chain in its unbound state can be misfolded to sequester the preformed elements inside the noninteractive or less interactive cage, therefore preventing these elements from the unnecessary and unwanted interactions with nonnative binding partners [125]. It is important to remember, however, that the mentioned functional misfolding is related to the ensemble behavior of transiently populated elements of structure. In other words, it describes the behavior of a globally disordered polypeptide chain containing highly dynamic elements of residual structure, so-called interaction-prone preformed fragments, some of which could potentially be related to protein function [125].

This ability of IDPs/IDPRs to functionally misfold can be used for finding small molecules which would potentially stabilize different members of the functionally misfolded ensemble and therefore prevent the targeted protein from establishing biological interactions [126]. This approach is very different from the above discussed direct targeting of short IDPRs since it is based on a small molecule binding to a highly dynamic surface created via the transient interaction of preformed interaction-prone fragments. In essence, this approach can be considered as an extension of the well-established structure-based rational drug design elaborated for ordered proteins. In fact, if the structure of a member(s) of the functionally misfolded ensemble can be guessed, then this structure can be used to find small molecules that are potentially able to interact with this structure, utilizing tools originally developed for the rational structure-based drug design for ordered proteins [126].

Ideally, a drug that targets a given protein-protein interaction should be tissue specific. Although some proteins are unique for a given tissue, many more proteins have very wide distribution, being present in several tissues and organs. How can one develop tissue-specific drugs targeting such abundant proteins? Often, tissue specificity for many of the abundant proteins is achieved via the alternative splicing of the corresponding pre-mRNAs, which generates two or more protein isoforms from a single gene. Estimates indicate that between 35 % and 60 % of human genes yield protein isoforms by means of alternatively spliced mRNA [127]. The added protein diversity from alternative splicing is thought to be important for tissue-specific signaling and regulatory networks in the multicellular organisms. The regions of alternative splicing in proteins are enriched in intrinsic disorder, and it was proposed that associating alternative splicing with protein disorder enables the time- and tissue-specific modulation of protein function [128]. Since disorder is frequently utilized in protein binding regions, having alternative splicing of pre-mRNA coupled to IDPRs can define tissue-specific signaling and regulatory diversity [128]. These findings open a unique opportunity to develop tissue-specific drugs modulating the function of a given ID protein/region (with a unique profile of disorder distribution) in a target tissue and not affecting the functionality of this same protein (with different disorder distribution profile) in other tissues.

6.7 Concluding Remarks

Intrinsic disorder is highly abundant in nature. According to the genome-based bioinformatics predictions, significant fraction of any given proteome belongs to the class of IDPs. These proteins possess numerous vital functions. Many proteins associated with various human diseases are intrinsically disordered too. High degree of association between protein intrinsic disorder and maladies is due to structural and functional peculiarities of IDPs and IDPRs, which are typically involved in cellular regulation, recognition, and signal transduction. As the number of IDPs is very large and as many of these proteins are interlinked, the concepts of the unfoldome and unfoldomics were introduced. IDPs, especially their extended forms, are characterized by several unique features that can be used for isolation of these proteins from the cell extracts. The corresponding proteomic techniques utilize specific high resistance of IDPs against extreme pH and high temperature, as well as their structural indifference to chemical denaturation. At the computational side, several specific features of the IDP amino acid sequences provide a solid background for the reliable identification of these proteins at the proteome level. These proteomic-scale identification and characterization of IDPs are needed to advance our knowledge in this important field.

Acknowledgments This review would not be possible without many colleagues who, over the years, contributed to the introduction and development of the IDP field. Numerous very helpful and thought-promoting discussions with our colleagues and friends are deeply acknowledged.

References

1. Fischer E (1894) Einfluss der configuration auf die wirkung der enzyme. Ber Dt Chem Ges 27:2985–2993
2. Lemieux UR, Spohr U (1994) How Emil Fischer was led to the lock and key concept for enzyme specificity. Adv Carbohydr Chem Biochem 50:1–20
3. Dobson CM (1999) Protein misfolding, evolution and disease. Trends Biochem Sci 24 (9):329–332
4. Dunker AK, Oldfield CJ, Meng J, Romero P, Yang JY, Chen JW, Vacic V, Obradovic Z, Uversky VN (2008) The unfoldomics decade: an update on intrinsically disordered proteins. BMC Genomics 9(Suppl 2):S1
5. Dunker AK, Brown CJ, Lawson JD, Iakoucheva LM, Obradovic Z (2002) Intrinsic disorder and protein function. Biochemistry 41(21):6573–6582
6. Dunker AK, Brown CJ, Obradovic Z (2002) Identification and functions of usefully disordered proteins. Adv Protein Chem 62:25–49
7. Dunker AK, Cortese MS, Romero P, Iakoucheva LM, Uversky VN (2005) Flexible nets: the roles of intrinsic disorder in protein interaction networks. FEBS J 272(20):5129–5148
8. Dunker AK, Lawson JD, Brown CJ, Williams RM, Romero P, Oh JS, Oldfield CJ, Campen AM, Ratliff CM, Hipps KW, Ausio J, Nissen MS, Reeves R, Kang C, Kissinger CR, Bailey RW, Griswold MD, Chiu W, Garner EC, Obradovic Z (2001) Intrinsically disordered protein. J Mol Graph Model 19(1):26–59

9. Dunker AK, Obradovic Z (2001) The protein trinity – linking function and disorder. Nat Biotechnol 19(9):805–806
10. Radivojac P, Iakoucheva LM, Oldfield CJ, Obradovic Z, Uversky VN, Dunker AK (2007) Intrinsic disorder and functional proteomics. Biophys J 92(5):1439–1456
11. Dunker AK, Silman I, Uversky VN, Sussman JL (2008) Function and structure of inherently disordered proteins. Curr Opin Struct Biol 18(6):756–764
12. Dunker AK, Uversky VN (2008) Signal transduction via unstructured protein conduits. Nat Chem Biol 4(4):229–230
13. Vucetic S, Xie H, Iakoucheva LM, Oldfield CJ, Dunker AK, Obradovic Z, Uversky VN (2007) Functional anthology of intrinsic disorder. 2. Cellular components, domains, technical terms, developmental processes, and coding sequence diversities correlated with long disordered regions. J Proteome Res 6(5):1899–1916
14. Xie H, Vucetic S, Iakoucheva LM, Oldfield CJ, Dunker AK, Obradovic Z, Uversky VN (2007) Functional anthology of intrinsic disorder. 3. Ligands, post-translational modifications, and diseases associated with intrinsically disordered proteins. J Proteome Res 6 (5):1917–1932
15. Xie H, Vucetic S, Iakoucheva LM, Oldfield CJ, Dunker AK, Uversky VN, Obradovic Z (2007) Functional anthology of intrinsic disorder. 1. Biological processes and functions of proteins with long disordered regions. J Proteome Res 6(5):1882–1898
16. Cortese MS, Uversky VN, Keith Dunker A (2008) Intrinsic disorder in scaffold proteins: getting more from less. Prog Biophys Mol Biol 98(1):85–106
17. Russell RB, Gibson TJ (2008) A careful disorderliness in the proteome: sites for interaction and targets for future therapies. FEBS Lett 582(8):1271–1275
18. Oldfield CJ, Meng J, Yang JY, Yang MQ, Uversky VN, Dunker AK (2008) Flexible nets: disorder and induced fit in the associations of p53 and 14-3-3 with their partners. BMC Genomics 9(Suppl 1):S1
19. Tompa P, Csermely P (2004) The role of structural disorder in the function of RNA and protein chaperones. FASEB J 18(11):1169–1175
20. Uversky VN, Oldfield CJ, Dunker AK (2005) Showing your ID: intrinsic disorder as an ID for recognition, regulation and cell signaling. J Mol Recognit 18(5):343–384
21. Wright PE, Dyson HJ (1999) Intrinsically unstructured proteins: re-assessing the protein structure-function paradigm. J Mol Biol 293(2):321–331
22. Tompa P (2002) Intrinsically unstructured proteins. Trends Biochem Sci 27(10):527–533
23. Uversky VN (2002) Natively unfolded proteins: a point where biology waits for physics. Protein Sci 11(4):739–756
24. Dyson HJ, Wright PE (2005) Intrinsically unstructured proteins and their functions. Nat Rev Mol Cell Biol 6(3):197–208
25. Iakoucheva LM, Brown CJ, Lawson JD, Obradovic Z, Dunker AK (2002) Intrinsic disorder in cell-signaling and cancer-associated proteins. J Mol Biol 323(3):573–584
26. Tompa P (2005) The interplay between structure and function in intrinsically unstructured proteins. FEBS Lett 579(15):3346–3354
27. Dunker AK, Obradovic Z, Romero P, Garner EC, Brown CJ (2000) Intrinsic protein disorder in complete genomes. Genome Inform Ser Workshop Genome Inform 11:161–171
28. Ward JJ, Sodhi JS, McGuffin LJ, Buxton BF, Jones DT (2004) Prediction and functional analysis of native disorder in proteins from the three kingdoms of life. J Mol Biol 337 (3):635–645
29. Oldfield CJ, Cheng Y, Cortese MS, Brown CJ, Uversky VN, Dunker AK (2005) Comparing and combining predictors of mostly disordered proteins. Biochemistry 44(6):1989–2000
30. Kriwacki RW, Hengst L, Tennant L, Reed SI, Wright PE (1996) Structural studies of p21Waf1/Cip1/Sdi1 in the free and Cdk2-bound state: conformational disorder mediates binding diversity. Proc Natl Acad Sci U S A 93(21):11504–11509

31. Lacy ER, Filippov I, Lewis WS, Otieno S, Xiao L, Weiss S, Hengst L, Kriwacki RW (2004) p27 binds cyclin-CDK complexes through a sequential mechanism involving binding-induced protein folding. Nat Struct Mol Biol 11(4):358–364
32. Lacy ER, Wang Y, Post J, Nourse A, Webb W, Mapelli M, Musacchio A, Siuzdak G, Kriwacki RW (2005) Molecular basis for the specificity of p27 toward cyclin-dependent kinases that regulate cell division. J Mol Biol 349(4):764–773
33. Dyson HJ, Wright PE (2002) Coupling of folding and binding for unstructured proteins. Curr Opin Struct Biol 12(1):54–60
34. Oldfield CJ, Cheng Y, Cortese MS, Romero P, Uversky VN, Dunker AK (2005) Coupled folding and binding with alpha-helix-forming molecular recognition elements. Biochemistry 44:12454–12470
35. Cheng Y, Oldfield CJ, Meng J, Romero P, Uversky VN, Dunker AK (2007) Mining alpha-helix-forming molecular recognition features with cross species sequence alignments. Biochemistry 46(47):13468–13477
36. Mohan A (2006) MoRFs: a dataset of molecular recognition features. Master of Science, Indiana University, Indianapolis
37. Vacic V, Oldfield CJ, Mohan A, Radivojac P, Cortese MS, Uversky VN, Dunker AK (2007) Characterization of molecular recognition features, MoRFs, and their binding partners. J Proteome Res 6(6):2351–2366
38. Uversky VN, Oldfield CJ, Dunker AK (2008) Intrinsically disordered proteins in human diseases: introducing the D2 concept. Annu Rev Biophys 37:215–246
39. Cortese MS, Baird JP, Uversky VN, Dunker AK (2005) Uncovering the unfoldome: enriching cell extracts for unstructured proteins by acid treatment. J Proteome Res 4(5):1610–1618
40. Midic U, Oldfield CJ, Dunker AK, Obradovic Z, Uversky VN (2009) Protein disorder in the human diseasome: unfoldomics of human genetic diseases. BMC Genomics 10(Suppl 1):S12
41. Winkler H (1920) Verbreitung und Ursache der Parthenogenesis im Pflanzen – und Tierreiche. Verlag Fischer, Jena
42. Lederberg J, McCray AT (2001) 'Ome' sweet 'omics' – a genealogical treasury of words. Scientist 15(7):8
43. Uversky VN, Oldfield CJ, Midic U, Xie H, Xue B, Vucetic S, Iakoucheva LM, Obradovic Z, Dunker AK (2009) Unfoldomics of human diseases: linking protein intrinsic disorder with diseases. BMC Genomics 10(Suppl 1):S7
44. Uversky VN, Gillespie JR, Fink AL (2000) Why are 'natively unfolded' proteins unstructured under physiologic conditions? Proteins 41(3):415–427
45. Ferron F, Longhi S, Canard B, Karlin D (2006) A practical overview of protein disorder prediction methods. Proteins 65(1):1–14
46. Dosztanyi Z, Sandor M, Tompa P, Simon I (2007) Prediction of protein disorder at the domain level. Curr Protein Pept Sci 8(2):161–171
47. Dosztanyi Z, Tompa P (2008) Prediction of protein disorder. Methods Mol Biol 426:103–115
48. Hubbard SJ, Beynon RJ, Thornton JM (1998) Assessment of conformational parameters as predictors of limited proteolytic sites in native protein structures. Protein Eng 11(5):349–359
49. Iakoucheva LM, Kimzey AL, Masselon CD, Smith RD, Dunker AK, Ackerman EJ (2001) Aberrant mobility phenomena of the DNA repair protein XPA. Protein Sci 10(7):1353–1362
50. Reeves R, Nissen MS (1999) Purification and assays for high mobility group HMG-I(Y) protein function. Methods Enzymol 304:155–188
51. Daughdrill GW, Pielak GJ, Uversky VN, Cortese MS, Dunker AK (2005) Natively disordered proteins. In: Buchner J, Kiefhaber T (eds) Handbook of protein folding. Wiley, Weinheim, pp 271–353
52. Receveur-Brechot V, Bourhis JM, Uversky VN, Canard B, Longhi S (2006) Assessing protein disorder and induced folding. Proteins 62(1):24–45
53. Uversky VN (2002) What does it mean to be natively unfolded? Eur J Biochem 269(1):2–12
54. Uversky VN (2009) Intrinsically disordered proteins and their environment: effects of strong denaturants, temperature, ph, counter ions, membranes, binding partners, osmolytes, and macromolecular crowding. Protein J 28:305–325

55. Csizmok V, Szollosi E, Friedrich P, Tompa P (2006) A novel two-dimensional electrophoresis technique for the identification of intrinsically unstructured proteins. Mol Cell Proteomics 5(2):265–273
56. Galea CA, Pagala VR, Obenauer JC, Park CG, Slaughter CA, Kriwacki RW (2006) Proteomic studies of the intrinsically unstructured mammalian proteome. J Proteome Res 5 (10):2839–2848
57. Romero P, Obradovic Z, Kissinger CR, Villafranca JE, Dunker AK (1997) Identifying disordered regions in proteins from amino acid sequences. In: IEEE International Conference Neural Networks, vol 1, pp 90–95
58. Romero P, Obradovic Z, Dunker K (1997) Sequence data analysis for long disordered regions prediction in the calcineurin family. Genome Inform Ser Workshop Genome Inform 8:110–124
59. Xie Q, Arnold GE, Romero P, Obradovic Z, Garner E, Dunker AK (1998) The sequence attribute method for determining relationships between sequence and protein disorder. Genome Inform Ser Workshop Genome Inform 9:193–200
60. Romero P, Obradovic Z, Kissinger CR, Villafranca JE, Garner E, Guilliot S, Dunker AK (1998) Thousands of proteins likely to have long disordered regions. Pac Symp Biocomput:437–448
61. Gast K, Damaschun H, Eckert K, Schulze-Forster K, Maurer HR, Muller-Frohne M, Zirwer D, Czarnecki J, Damaschun G (1995) Prothymosin alpha: a biologically active protein with random coil conformation. Biochemistry 34(40):13211–13218
62. Weinreb PH, Zhen W, Poon AW, Conway KA, Lansbury PT Jr (1996) NACP, a protein implicated in Alzheimer's disease and learning, is natively unfolded. Biochemistry 35 (43):13709–13715
63. Hemmings HC Jr, Nairn AC, Aswad DW, Greengard P (1984) DARPP-32, a dopamine- and adenosine 3':5'-monophosphate-regulated phosphoprotein enriched in dopamine-innervated brain regions. II. Purification and characterization of the phosphoprotein from bovine caudate nucleus. J Neurosci 4(1):99–110
64. Dunker AK, Garner E, Guilliot S, Romero P, Albrecht K, Hart J, Obradovic Z, Kissinger C, Villafranca JE (1998) Protein disorder and the evolution of molecular recognition: theory, predictions and observations. Pac Symp Biocomput:473–484
65. Garner E, Cannon P, Romero P, Obradovic Z, Dunker AK (1998) Predicting disordered regions from amino acid sequence: common themes despite differing structural characterization. Genome Inform Ser Workshop Genome Inform 9:201–213
66. Williams RM, Obradovi Z, Mathura V, Braun W, Garner EC, Young J, Takayama S, Brown CJ, Dunker AK (2001) The protein non-folding problem: amino acid determinants of intrinsic order and disorder. Pac Symp Biocomput:89–100
67. Romero P, Obradovic Z, Li X, Garner EC, Brown CJ, Dunker AK (2001) Sequence complexity of disordered protein. Proteins 42(1):38–48
68. Radivojac P, Obradovic Z, Smith DK, Zhu G, Vucetic S, Brown CJ, Lawson JD, Dunker AK (2004) Protein flexibility and intrinsic disorder. Protein Sci 13(1):71–80
69. He B, Wang K, Liu Y, Xue B, Uversky VN, Dunker AK (2009) Predicting intrinsic disorder in proteins: an overview. Cell Res 19(8):929–949
70. Vucetic S, Brown CJ, Dunker AK, Obradovic Z (2003) Flavors of protein disorder. Proteins 52(4):573–584
71. Li X, Romero P, Rani M, Dunker AK, Obradovic Z (1999) Predicting protein disorder for N-, C-, and internal regions. Genome Inform Ser Workshop Genome Inform 10:30–40
72. Peng K, Vucetic S, Radivojac P, Brown CJ, Dunker AK, Obradovic Z (2005) Optimizing long intrinsic disorder predictors with protein evolutionary information. J Bioinform Comput Biol 3(1):35–60
73. Obradovic Z, Peng K, Vucetic S, Radivojac P, Dunker AK (2005) Exploiting heterogeneous sequence properties improves prediction of disorder. Proteins 61:176–182

74. Peng K, Radivojac P, Vucetic S, Dunker AK, Obradovic Z (2006) Length-dependent prediction of protein intrinsic disorder. BMC Bioinf 7:208
75. Prilusky J, Felder CE, Zeev-Ben-Mordehai T, Rydberg EH, Man O, Beckmann JS, Silman I, Sussman JL (2005) FoldIndex: a simple tool to predict whether a given protein sequence is intrinsically unfolded. Bioinformatics 21(16):3435–3438
76. Linding R, Russell RB, Neduva V, Gibson TJ (2003) GlobPlot: exploring protein sequences for globularity and disorder. Nucleic Acids Res 31(13):3701–3708
77. Linding R, Jensen LJ, Diella F, Bork P, Gibson TJ, Russell RB (2003) Protein disorder prediction: implications for structural proteomics. Structure 11(11):1453–1459
78. Jones DT, Ward JJ (2003) Prediction of disordered regions in proteins from position specific score matrices. Proteins 53(Suppl 6):573–578
79. Ward JJ, McGuffin LJ, Bryson K, Buxton BF, Jones DT (2004) The DISOPRED server for the prediction of protein disorder. Bioinformatics 20(13):2138–2139
80. Bryson K, McGuffin LJ, Marsden RL, Ward JJ, Sodhi JS, Jones DT (2005) Protein structure prediction servers at University College London. Nucleic Acids Res 33(Web Server issue): W36–W38
81. Dosztanyi Z, Csizmok V, Tompa P, Simon I (2005) IUPred: web server for the prediction of intrinsically unstructured regions of proteins based on estimated energy content. Bioinformatics 21(16):3433–3434
82. Galzitskaya OV, Garbuzynskiy SO, Lobanov MY (2006) FoldUnfold: web server for the prediction of disordered regions in protein chain. Bioinformatics 22(23):2948–2949
83. Yang ZR, Thomson R, McNeil P, Esnouf RM (2005) RONN: the bio-basis function neural network technique applied to the detection of natively disordered regions in proteins. Bioinformatics 21(16):3369–3376
84. Su CT, Chen CY, Ou YY (2006) Protein disorder prediction by condensed PSSM considering propensity for order or disorder. BMC Bioinf 7:319
85. Su CT, Chen CY, Hsu CM (2007) iPDA: integrated protein disorder analyzer. Nucleic Acids Res 35(Web Server issue):W465–W472
86. Vullo A, Bortolami O, Pollastri G, Tosatto SC (2006) Spritz: a server for the prediction of intrinsically disordered regions in protein sequences using kernel machines. Nucleic Acids Res 34(Web Server issue):W164–W168
87. Ishida T, Kinoshita K (2007) PrDOS: prediction of disordered protein regions from amino acid sequence. Nucleic Acids Res 35(Web Server issue):W460–W464
88. Mohan A, Sullivan WJ Jr, Radivojac P, Dunker AK, Uversky VN (2008) Intrinsic disorder in pathogenic and non-pathogenic microbes: discovering and analyzing the unfoldomes of early-branching eukaryotes. Mol Biosyst 4(4):328–340
89. Rantalainen KI, Uversky VN, Permi P, Kalkkinen N, Dunker AK, Makinen K (2008) Potato virus a genome-linked protein VPg is an intrinsically disordered molten globule-like protein with a hydrophobic core. Virology 377(2):280–288
90. Hebrard E, Bessin Y, Michon T, Longhi S, Uversky VN, Delalande F, Van Dorsselaer A, Romero P, Walter J, Declerk N, Fargette D (2009) Intrinsic disorder in viral proteins genome-linked: experimental and predictive analyses. Virol J 6:23
91. Feng ZP, Zhang X, Han P, Arora N, Anders RF, Norton RS (2006) Abundance of intrinsically unstructured proteins in P. falciparum and other apicomplexan parasite proteomes. Mol Biochem Parasitol 150(2):256–267
92. Tompa P, Dosztanyi Z, Simon I (2006) Prevalent structural disorder in E. coli and S. cerevisiae proteomes. J Proteome Res 5(8):1996–2000
93. Galea CA, High AA, Obenauer JC, Mishra A, Park CG, Punta M, Schlessinger A, Ma J, Rost B, Slaughter CA, Kriwacki RW (2009) Large-scale analysis of thermostable, mammalian proteins provides insights into the intrinsically disordered proteome. J Proteome Res 8 (1):211–226
94. Xue B, Williams RW, Oldfield CJ, Dunker AK, Uversky VN (2010) Archaic chaos: intrinsically disordered proteins in Archaea. BMC Syst Biol 4(Suppl 1):S1

95. Burra PV, Kalmar L, Tompa P (2010) Reduction in structural disorder and functional complexity in the thermal adaptation of prokaryotes. PLoS One 5(8), e12069
96. Uversky VN, Dunker AK (2010) Understanding protein non-folding. Biochim Biophys Acta 1804(6):1231–1264
97. Uversky VN (2010) The mysterious unfoldome: structureless, underappreciated, yet vital part of any given proteome. J Biomed Biotechnol 2010:568068
98. Xue B, Dunker AK, Uversky VN (2012) Orderly order in protein intrinsic disorder distribution: disorder in 3500 proteomes from viruses and the three domains of life. J Biomol Struct Dyn 30(2):137–149
99. Uversky VN (2013) Unusual biophysics of intrinsically disordered proteins. Biochim Biophys Acta 1834:932–951
100. Uversky VN (2013) A decade and a half of protein intrinsic disorder: biology still waits for physics. Protein Sci 22(6):693–724
101. Dunker AK, Babu MM, Barbar E, Blackledge M, Bondos SE, Dosztányi Z, Dyson HJ, Forman-Kay J, Fuxreiter M, Gsponer J, Han K-H, Jones DT, Longhi S, Metallo SJ, Nishikawa K, Nussinov R, Obradovic Z, Pappu R, Rost B, Selenko P, Subramaniam V, Sussman JL, Tompa P, Uversky VN (2013) What's in a name? Why these proteins are intrinsically disordered. Intrinsically Disord Proteins 1(1): e24157
102. Uversky VN, Gillespie JR, Millett IS, Khodyakova AV, Vasilenko RN, Vasiliev AM, Rodionov IL, Kozlovskaya GD, Dolgikh DA, Fink AL, Doniach S, Permyakov EA, Abramov VM (2000) Zn(2+)-mediated structure formation and compaction of the "natively unfolded" human prothymosin alpha. Biochem Biophys Res Commun 267(2):663–668
103. Ma B, Kumar S, Tsai CJ, Nussinov R (1999) Folding funnels and binding mechanisms. Protein Eng 12(9):713–720
104. Uversky VN, Roman A, Oldfield CJ, Dunker AK (2006) Protein intrinsic disorder and human papillomaviruses: increased amount of disorder in E6 and E7 oncoproteins from high risk HPVs. J Proteome Res 5(8):1829–1842
105. Cheng Y, LeGall T, Oldfield CJ, Dunker AK, Uversky VN (2006) Abundance of intrinsic disorder in protein associated with cardiovascular disease. Biochemistry 45 (35):10448–10460
106. Uversky VN (2009) Intrinsic disorder in proteins associated with neurodegenerative diseases. Front Biosci 14:5188–5238
107. Goh KI, Cusick ME, Valle D, Childs B, Vidal M, Barabasi AL (2007) The human disease network. Proc Natl Acad Sci U S A 104(21):8685–8690
108. Dev KK, Hofele K, Barbieri S, Buchman VL, van der Putten H (2003) Part II: alpha-synuclein and its molecular pathophysiological role in neurodegenerative disease. Neuropharmacology 45(1):14–44
109. Uversky VN (2003) A protein-chameleon: conformational plasticity of alpha-synuclein, a disordered protein involved in neurodegenerative disorders. J Biomol Struct Dyn 21 (2):211–234
110. Dunker AK, Uversky VN (2010) Drugs for 'protein clouds': targeting intrinsically disordered transcription factors. Curr Opin Pharmacol 10(6):782–788
111. Garner E, Romero P, Dunker AK, Brown C, Obradovic Z (1999) Predicting binding regions within disordered proteins. Genome Inform Ser Workshop Genome Inform 10:41–50
112. Chen CT, Wagner H, Still WC (1998) Fluorescent, sequence-selective peptide detection by synthetic small molecules. Science 279(5352):851–853
113. Chene P (2004) Inhibition of the p53-hdm2 interaction with low molecular weight compounds. Cell Cycle 3(4):460–461
114. Chene P (2004) Inhibition of the p53-MDM2 interaction: targeting a protein-protein interface. Mol Cancer Res 2(1):20–28
115. Kussie PH, Gorina S, Marechal V, Elenbaas B, Moreau J, Levine AJ, Pavletich NP (1996) Structure of the MDM2 oncoprotein bound to the p53 tumor suppressor transactivation domain. Science 274(5289):948–953

116. Lee H, Mok KH, Muhandiram R, Park KH, Suk JE, Kim DH, Chang J, Sung YC, Choi KY, Han KH (2000) Local structural elements in the mostly unstructured transcriptional activation domain of human p53. J Biol Chem 275(38):29426–29432
117. Vassilev LT, Vu BT, Graves B, Carvajal D, Podlaski F, Filipovic Z, Kong N, Kammlott U, Lukacs C, Klein C, Fotouhi N, Liu EA (2004) In vivo activation of the p53 pathway by small-molecule antagonists of MDM2. Science 303(5659):844–848
118. Klein C, Vassilev LT (2004) Targeting the p53-MDM2 interaction to treat cancer. Br J Cancer 91(8):1415–1419
119. Vassilev LT (2004) Small-molecule antagonists of p53-MDM2 binding: research tools and potential therapeutics. Cell Cycle 3(4):419–421
120. Cheng Y, LeGall T, Oldfield CJ, Mueller JP, Van YY, Romero P, Cortese MS, Uversky VN, Dunker AK (2006) Rational drug design via intrinsically disordered protein. Trends Biotechnol 24(10):435–442
121. Arkin MR, Wells JA (2004) Small-molecule inhibitors of protein-protein interactions: progressing towards the dream. Nat Rev Drug Discov 3(4):301–317
122. Arkin M (2005) Protein-protein interactions and cancer: small molecules going in for the kill. Curr Opin Chem Biol 9(3):317–324
123. Cochran AG (2000) Antagonists of protein-protein interactions. Chem Biol 7(4):R85–R94
124. Hammoudeh DI, Follis AV, Prochownik EV, Metallo SJ (2009) Multiple independent binding sites for small-molecule inhibitors on the oncoprotein c-Myc. J Am Chem Soc 131 (21):7390–7401
125. Uversky VN (2011) Intrinsically disordered proteins may escape unwanted interactions via functional misfolding. Biochim Biophys Acta-Proteins Proteomics 1814(5):693–712
126. Uversky VN (2012) Intrinsically disordered proteins and novel strategies for drug discovery. Expert Opin Drug Discovery 7(6):475–488
127. Stamm S, Ben-Ari S, Rafalska I, Tang Y, Zhang Z, Toiber D, Thanaraj TA, Soreq H (2005) Function of alternative splicing. Gene 344:1–20
128. Romero PR, Zaidi S, Fang YY, Uversky VN, Radivojac P, Oldfield CJ, Cortese MS, Sickmeier M, LeGall T, Obradovic Z, Dunker AK (2006) Alternative splicing in concert with protein intrinsic disorder enables increased functional diversity in multicellular organisms. Proc Natl Acad Sci U S A 103(22):8390–8395

Chapter 7
Structure, Dynamics, and Function of Staphylococcal Nuclease

Mikio Kataoka

Abstract Protein design principles have been intensively studied using staphylococcal nuclease (SNase) as a model protein. We developed several mutants that take nonnative structures under physiological conditions but still possess enzymatic activity and other mutants that assume stable native structure but are functionally inactive. Using these mutants, we succeeded in decoding the information in the primary sequence. We also describe the nonlocal interactions required for proper folding. Our findings also elucidate the mechanism of ligand-induced folding, which is a good model for coupled folding and binding of intrinsically unfolded proteins.

Keywords Staphylococcal nuclease • Protein folding • Coupling folding and binding • Decoding of information in sequence • Nonlocal interaction • Artificial disulfide bond

7.1 Introduction

Protein folding is a still long-standing problem in protein science that remains important today. Most small proteins can spontaneously fold into their unique tertiary structures under suitable conditions [1, 2]. The protein-folding problem can be considered in terms of the decoding of information contained in the primary structure. Thus, this topic is related to various aspects of studies of protein folding and structure, such as the stability of the native structure, the pathway of folding, the molecular basis of ligand recognition, and the structure–function relationships. We have approached this topic using staphylococcal nuclease (SNase) as a model protein [3–11].

The recent discovery of intrinsically disordered proteins (IDPs) (also known as natively unfolded proteins) opened our eyes to a new aspect of protein folding [12–14]. IDPs assume unfolded or highly disordered structures under physiological conditions, but can still specifically recognize their partner molecules, and fold into the defined tertiary structures upon binding of their partner molecules. Such

M. Kataoka (✉)
Graduate School of Materials Science, Nara Institute of Science and Technology, Ikoma, Japan
e-mail: kataoka@ms.naist.jp

M. Terazima et al. (eds.), *Molecular Science of Fluctuations Toward Biological Functions*, DOI 10.1007/978-4-431-55840-8_7

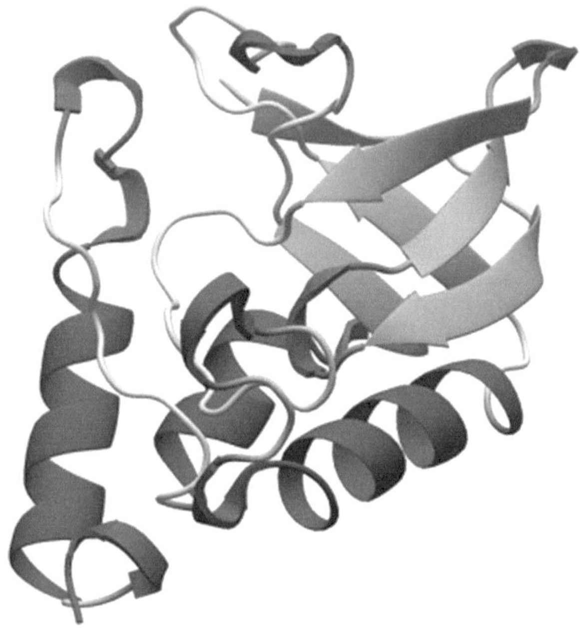

Fig. 7.1 Structure of SNase (1STN) [17]

partner-induced folding, in which binding is closely connected to folding, is called "coupled folding and binding." A better understanding of the mechanism of coupled folding and binding should contribute to our overall understanding of protein folding. We succeeded in creating an interesting category of SNase mutants that are unfolded or highly disordered under physiological conditions but still have substantial enzymatic activity. These mutants can be regarded as model of IDPs.

SNase has been used as a model protein for protein-folding studies since Anfinsen's pioneering work [15]. Multiple high-resolution crystal structures of this protein are available, contributing to a detailed understanding of its catalytic activity [16, 17]. The structure of SNase [17] is shown in Fig. 7.1. SNase is composed of two subdomains, N-terminal and C-terminal, with the substrate-binding site located at the interface between them. The enzyme contains neither disulfide bonds nor prosthetic groups. SNase can catalyze hydrolysis of both DNA and RNA. Our group has been studying protein folding, protein dynamics, and structure–function relationships of SNase. We have intensively applied deletion and insertion as well as the site-directed substitution, which have provided various promising information.

7.2 Information Encoded in the Primary Structure

Systematic deletion studies of SNase revealed that the C-terminal residues play an important role in determining foldability and stability [4, 18]. Figure 7.2 shows CD spectra of various C-terminal deletion mutants. The spectra and enzymatic activities of Δ143–149 and Δ142–149 are identical to those of the wild type [4]. However, the thermal stability of the mutant proteins gradually decreases as a function of the number of deleted residues [4]. These results indicate that the C-terminal residues 142–149 contain no information about structure formation or enzymatic activity but nonetheless contribute to protein stability. By contrast residues 140 (W) and 141 (S) are clearly important for structure formation (Fig. 7.2) [4, 18]. Together, the characteristics of the site-directed substitution and deletion mutants led us to the conclusion that information about structure formation is encoded in the side chain of residue 140 [4, 7]. However, residue 141 is tolerant of substitution and does not need to be S, indicating that the structural information provided by this residue is encoded in the main chain [4]. The NH of S141 forms a hydrogen bond with the main-chain CO group of N138, thereby stabilizing the C-terminal loop structure and fixing W140 in the proper orientation. Further substitution studies revealed that the side chain at position 140 must be an aromatic residue. Among the four aromatic side chains, W is the most suitable at this position [7].

W140 is a key residue for assembly of nearby residues via hydrophobic interactions. According to the crystal structure, W140 is surrounded by G107, K110, E129, A132, K133, and I139. I139 and W140 are part of the C-terminal loop, A132 and K133 are on the C-terminal end of the C-terminal α-helix, and G107 is at the C-terminal end of the second short helix. Thus, W140 assembles three secondary-structure elements of the C-terminal domain of SNase. The residues assembled by W140 form the C-terminal hydrophobic cluster. The cluster is responsible for interactions with the N-terminal domain via hydrophobic or van der Waals interactions [4].

The interacting partners of the C-terminal hydrophobic cluster stabilized by W140 were surveyed by alanine scanning of nearby residues. Site-directed

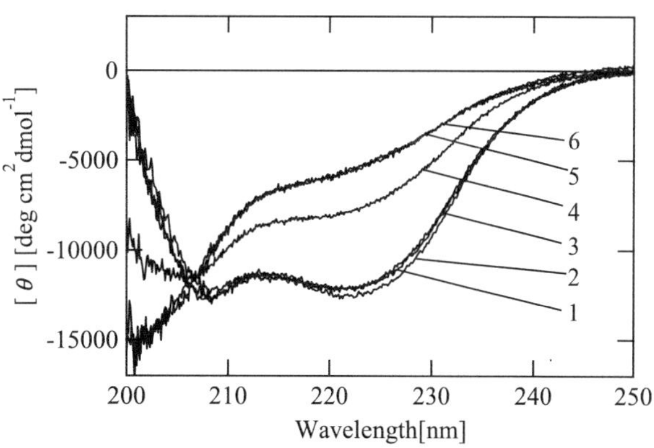

Fig. 7.2 Far-UV spectra of wild-type and C-terminal deletion mutants of SNase. Curves 1–6 are wild type, Δ143–149, Δ142–149, Δ141–149, Δ140–149, and Δ136–149, respectively

mutation studies revealed that L108 is sensitive to alanine substitution and that the preferable residues at the 108th position are L, I, V, W, F, Y, H, and M (M. Kataoka, unpublished results). Both bulkiness and hydrophobicity are the essential properties required at this position. L108 directly contacts Y54, E57, A58, F61, L103, Q106, and N138. Y54, E57, A58, and F61 are located at the central helix. L103 and Q106 are on the second short helix. N138 is located on the C-terminal loop. L108 is located at the N-terminal edge of the loop connected to the second short helix and the C-terminal α-helix. L108 assembles these residues to form another hydrophobic cluster which is the partner of the C-terminal hydrophobic cluster assembled by W140. The role of L108 is quite similar to that of W140. Moreover, Y54 in the L108 cluster and I139 in the W140 cluster interact, and this interaction bridges the C-terminal and the N-terminal subdomains.

Human transcription factor p100 has been described as a protein with SNase-like structure [19], although p100 and SNase share only 15 % amino acid identity [19]. However, L108 and W140 are both conserved, suggesting that these residues are critical for folding of both proteins. On the other hand, residue 141 is not conserved. Thus, these results suggest that a relatively small number of residues are critical for folding. The hydrophobic clusters formed by these critical residues act as building blocks of the tertiary structure. Identification of such critical residues in the N-terminal domain will lead to a more thorough understanding of the folding mechanism of SNase. The idea that a limited number of residues are essential for folding is consistent with the idea that a small number of residues nucleate folding [20].

The deletion of the so-called Ω-loop brings loss of the enzymatic activity of SNase [21]. In order to investigate this further, we examined the characteristics of Δ44–49 SNase. Crystal structure analysis of the mutant reveals that the tertiary fold of the mutant is identical to that of the wild type (M. Kataoka; unpublished result), whereas thermal stability is actually improved by the loop deletion. The mid-temperature of thermal denaturation is 50.7 °C for the wild type but 56.3 °C for the mutant. Therefore, the loop does not contain information essential for structure formation; indeed, on the contrary, the loop destabilizes the overall structure. The mutant can properly bind substrate or inhibitor despite of the loss of enzymatic activity, indicating that it lacks catalytic activity. E43 is involved in the catalytic triad and acts as a general-base catalyst [16]. We found that the programmed motion of the loop controls the orientation of E43 (M. Kataoka, unpublished). The loop dynamics play an essential role in evoking the concerted motion of catalytic residues. The loop sequence is optimized for catalytic activity. All of the site-directed substitutions of the loop examined so far decrease catalytic activity dramatically (M. Kataoka, unpublished). On the other hand, human p100 completely lacks this loop [19], suggesting that it is not required for structure formation and that the functional sites of human p100 are different from those of SNase.

Thus, we succeeded in decoding some of the information encoded in the amino acid sequence of SNase. The decoded information regarding structure formation and enzymatic catalysis is described in Fig. 7.3.

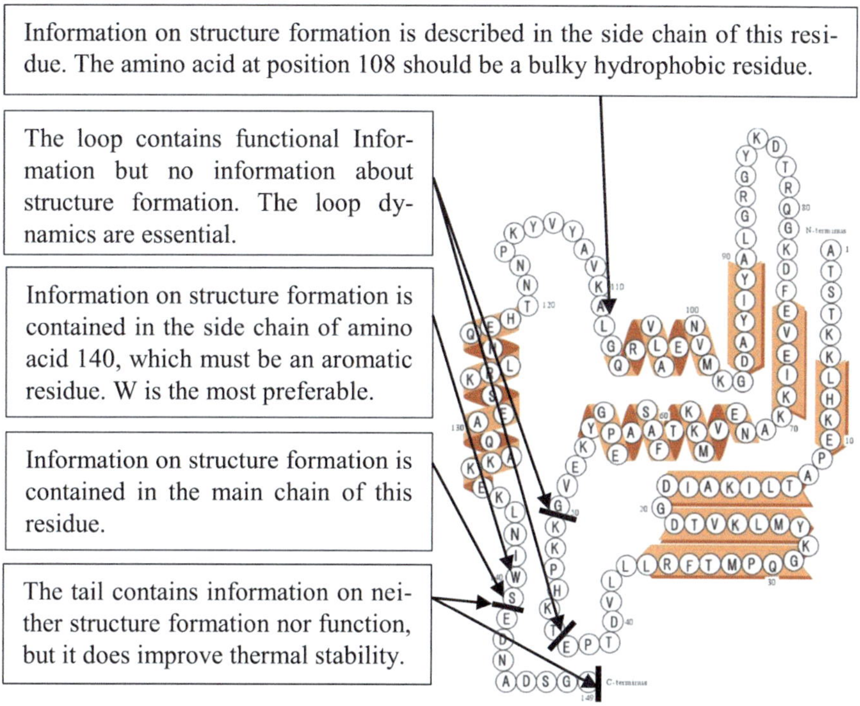

Fig. 7.3 The decoded information described in the amino acid sequence of SNase

7.3 Mechanism of Induced Folding

We generated some SNase mutants that assume nonnative disordered structures under physiological conditions but still possess enzymatic activity. Typical examples include Δ140–149, W140A, and 33A34 [8]. 33A34 is an alanine insertion mutant, in which alanine is inserted between the 33rd and 34th residues. Upon binding of substrate or inhibitor, these mutants adopt a native conformation identical to the wild-type SNase [8]. Ligand-induced folding is similar to the coupled folding and binding of IDP. The mechanism of induced folding is a matter of debate [22]. There are two elementary steps in the process: folding and binding. Depending on the order of two elementary steps, two different mechanisms are considered for the induced folding depending on the order of two elementary steps: (1) $U + S \rightarrow US \rightarrow FS$ (binding before folding) and (2) $U + S \rightarrow F + S \rightarrow FS$ (folding before binding); here U, F, and S denote unfolded structure, folded structure, and substrate (ligand), respectively. It is generally accepted that protein tertiary structure is essential for recognition of ligand (lock-and-key model), i.e., it is most likely that folding occurs before binding. However, an NMR study of an IDP raised the possibility of binding before folding [23].

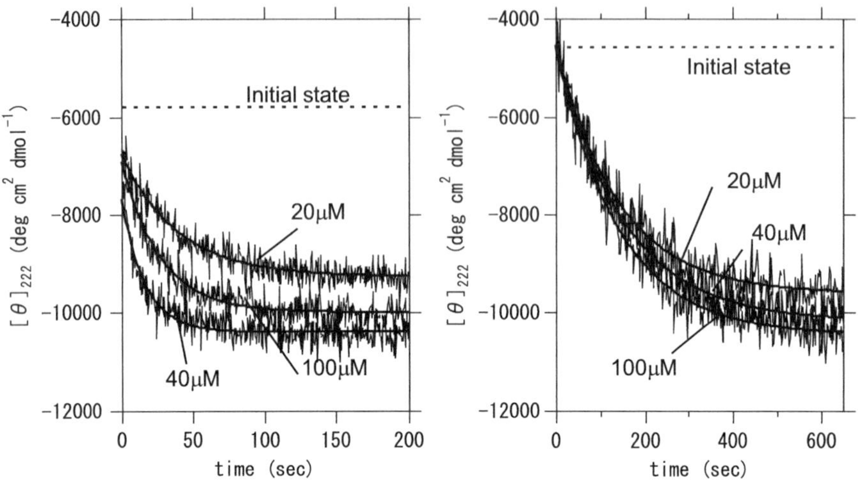

Fig. 7.4 Inhibitor-induced folding reactions of Δ140-149 (*left*) and 33A34 (*right*) monitored by the ellipticity change at 222 nm at three different inhibitor concentrations. The *dashed line* and *solid lines* represent the CD intensity of the initial condition (prAp-free) and the exponential fit of the data, respectively (Modified from Fig. 2 in [8])

Figure 7.4 shows the kinetics of ligand-induced folding of two SNase mutants, at the different concentrations of the inhibitor prAp, obtained by stopped-flow CD measurements. A remarkable difference is observed in the burst phase. The folding curve of Δ140–149 exhibits a clear burst phase whose amplitude depends on the inhibitor concentration. On the other hand, the curves of 33A34 have no burst phase regardless of the inhibitor concentration. The presence of the burst phase suggests that the US complex forms within the dead time of stopped-flow apparatus. Every curve can be fitted with a single exponential. The dependence of the time constant on inhibitor concentration also differs between the two mutants. Analysis of the rate equation led us to conclude that the induced folding of Δ140–149 follows the binding before folding scheme, whereas that of 33A34 follows the folding before binding scheme [8]. In both cases, the folding rate is much slower than the binding rate. The estimated folding rate for the folding before binding scheme is considerably slower than that of the binding before folding scheme. Thus, these findings clearly demonstrate that it is equally possible that either of these two mechanisms could underlie induced folding. The two possibilities are depicted in Fig. 7.5. Note that the folding rate of wild-type SNase never increases as a function of the concentration of inhibitor [24]. Therefore, we need to clarify the reason why the binding before folding mechanism is realized in order to elucidate the molecular mechanism of folding.

Direct proof of the binding before folding mechanism could be provided by structural analysis of the US complex. Because the complex is a transient-state structure, we devised an extended ϕ-value analysis. The inhibitor concentration dependence of the burst phase amplitude gives the equilibrium constant between

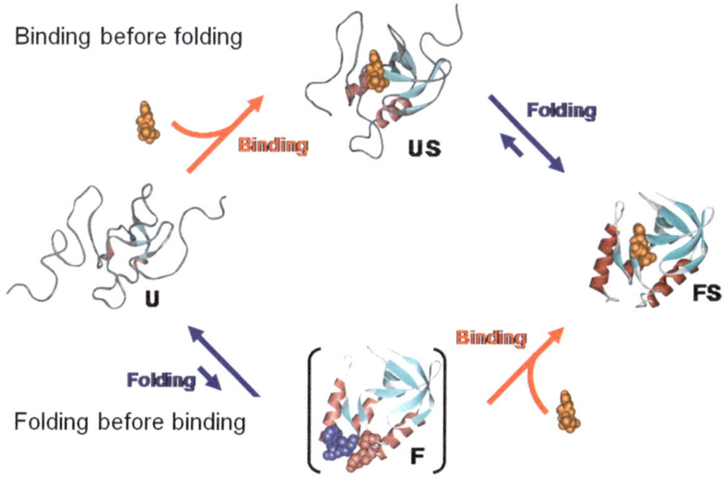

Fig. 7.5 Schematic view of two possible induced-folding mechanisms (Modified from Fig. 4 in [8])

the U and US complexes, whereas that of the final CD value gives the equilibrium constant between U and F. The ratio of the two constants is similar to the ϕ-value used for folding studies. We prepared the alanine substitution mutants of the substrate-binding sites for Δ140–149. Stopped-flow CD experiments of induced folding revealed that R35, R87, and Y115 are N-state-like in the US complex, whereas K84, Y85, and Y113 are U-state-like (M. Kataoka: unpublished results). The results suggest that the structure of the US complex is significantly different from that of the folded native structure. The inhibitor, prAp, initially interacts with R35, R87, and Y115, and then properly bound prAp recruits K84, Y85, and Y115. This recruitment replaces the interaction between the L108 and W140 clusters.

7.4 Role of Nonlocal Interactions in Structure Formation

In the binding before folding scheme, binding is mandatory for folding, and the interactions generated by ligand binding promotes folding. Thus far, all SNase mutants that follow the binding before folding scheme have lacked W140, due to either deletion or substitution. As mentioned above, W140 is a key residue in the assembly of the C-terminal hydrophobic cluster (W140 cluster), and the cluster is responsible for assembling the C-terminal domain with the N-terminal domain. Therefore, the interaction between the C-terminal domain and the N-terminal domain would be lost upon removal of W140. However, as mentioned above, inhibitor binding compensates for the lost interaction. On the other hand, other mutants (such as 33A34) maintain all sets of interactions required for folding. Some

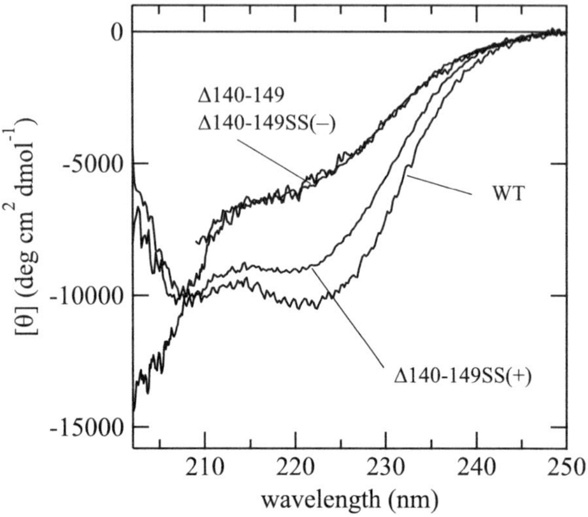

Fig. 7.6 CD spectra of Y54C/I139/C-Δ140–149 in the oxidized form (SS(+)) and the reduced form (SS(−)). CD spectra of wild-type SNase and Δ140–149 are also shown

interactions are only weakened by mutation, resulting in a dramatic deduction in the folding rate. Binding of inhibitor to these mutants shifts the equilibrium toward the N-state.

If these assumptions are reasonable, then the foldability of mutants lacking W140 will be recovered by the introduction of artificial nonlocal interactions that connect the two domains. In order to examine this hypothesis, we prepared double-cysteine mutants (Y54C/I139C) for Δ140–149, W140A, and full-length of SNase [10]. These mutants allow an artificial disulfide bond to be introduced between the C-terminal and the N-terminal subdomains.

Figure 7.6 shows the effect of the artificial disulfide bond on the structure of Δ140–149. The Δ140–149 mutant and the reduced form of Y54C/I139C/Δ140–149 (we call it Δ140–149SS(−)) gave identical CD spectra under physiological conditions. The oxidized form of the double-cysteine mutant (Δ140–149SS(+)) yields a CD spectrum close to that of the wild type [10]. The double-cysteine mutant of full-length SNase yielded a CD spectrum similar to that of Δ140–149SS(+) regardless of the presence or absence of disulfide bond [10]. Therefore, small differences in CD spectra between the wild type and Δ140–149SS(+) or Y54C/I139C suggest that the formation of the W140 cluster is incomplete or slightly deteriorated. The artificial disulfide bond was essential for the folding of Δ140–149, and similar results are also obtained for W140A. From these data, we conclude that the artificial disulfide bond compensates for the interaction lost as a result of removal of W140.

We examined the effect of the artificial disulfide bond on refolding kinetics. The far-UV CD spectra of the acid-unfolded states were identical for the wild type, Y54C/I139C, Δ140–149, and Y54C/I139C/Δ140–149 regardless of the presence or absence of a disulfide bond. However, the molecular sizes were much more compact for Y54C/I139CSS(+) and Δ140–149SS(+) than those without a disulfide bond. The artificial disulfide bond may restrict in conformational space that can be searched. We examined the folding kinetics of the double-cysteine mutants by stopped-flow CD with a pH jump [10].

The refolding kinetics of Y54C/I139CSS(+) are comprised of three phases [10], as in the wild type [25]. The rate constants of the three phases were almost the same as those of the wild type, but the amplitudes were different. These results indicate that the refolding reactions of Y54C/I139CSS(+) and Y54C/I139CSS(−) obey a mechanism similar to that of the wild type within the observed time period. The differences in the amplitudes can be mostly ascribed to the difference in the ellipticities of the native state [10]. The most important finding was that neither Y54C/I139C double mutations nor artificial nonnative disulfide bond formation influenced the folding rates of SNase within the experimental time period. If a nonlocal interaction forms in the wild type during the observed time period, the preexistence of a disulfide bond should affect the folding reaction. Considering the fact that the oxidized form of Y54C/I139C is more compact than the wild type, these results indicate that the intact nonlocal interaction between the N- terminal and C-terminal domains is established during the early stage of the refolding (<22 ms).

The refolding reaction of Δ140–149SS(+) occurs in two phases, in marked contrast to three phases for the wild type and its double-cysteine mutants. The first and the second phases of Δ140–149SS(+) correspond to the first and the third phases of the wild-type and double-cysteine mutant with or without disulfide bond, and their rate constants are almost equivalent. SNase follows a branched folding pathway in which the first and the second phases branch from an early intermediate [25]. We conclude that Δ140–149SS(+) obeys the same pathway, in which the second and the third phases of the wild type are merged [10]. In the case of Δ140–149SS(+), the nonlocal interaction is already formed in the initial state. Nevertheless, the folding pathway and time constant of the first phase are not different from those of the wild type. This strongly supports the idea that the nonlocal interaction between the C-terminal and N-terminal domains is established at a very early stage of wild-type folding. We hypothesize that the nonlocal interaction is generated by the formation of the W140 cluster.

7.5 Conclusion

SNase is quite useful for physicochemical studies of structure formation and enzymatic function. We succeeded in decoding the information encoded in some parts of the primary structure, but much of the whole sequence remains to be decoded. We also described the mechanism of induced folding using SNase mutants that assume disordered structures under physiological conditions but still exhibit enzymatic activity. The results clearly demonstrate that both folding before binding and binding before folding are equally possible. For the realization of the latter scheme, it is necessary to eliminate the essential interactions involved in structure formation and then compensate for the lost interaction by recognition and binding of a target or ligand. Further work should seek to characterize all of the required information and its arrangement or representation in the amino acid sequence. We have proposed that systematic alanine insertion would be extremely useful for the purpose [26]. The ultimate goal of our studies is the realization of rational design of artificial functional proteins.

Acknowledgment The author wishes to sincerely thank Prof. H. Kamikubo, Dr. S. Hirano, Dr. H. Nakagawa, Dr. M. Onitsuka, Dr. S. Kato, and Dr. R. Shiba for their help and discussions during this study and Ms. S. Kayamura, Mr. S. Tadokoro, and Mr. R. Nakagaki for their help in making the measurements. This work was supported by a Grant-in-Aid for Scientific Research to MK (15076208, 20370062). The author is especially grateful for the support from a Grant-in-aid for Scientific Research in Innovative Areas (20107006).

References

1. Anfinsen CB (1973) Principles that govern the folding of protein chains. Science 181:223–230
2. Fersht A (1999) Structure and mechanism in protein science. Freeman, New York
3. Flanagan JM, Kataoka M, Shortle D, Engelman DM (1992) Truncated staphylococcal nuclease is compact but disordered. Proc Natl Acad Sci U S A 89:748–752
4. Hirano S, Mihara K, Yamazaki Y, Kamikubo H, Imamoto Y, Kataoka M (2002) Role of C-terminal region of staphylococcal nuclease for foldability, stability and activity. Proteins 49:255–265
5. Kataoka M, Kamikubo H, Yunoki J, Tokunaga F, Kanaya T, Izumi Y, Shibata K (1999) Low energy dynamics of globular proteins studied by inelastic neutron scattering. J Phys Chem Solid 60:1285–1289
6. Nakagawa H, Kamikubo H, Tsukushi I, Kanaya T, Kataoka M (2004) Protein dynamical heterogeneity derived from neutron incoherent elastic scattering. J Phys Soc Jpn 73:491–495
7. Hirano S, Kamikubo H, Yamazaki Y, Kataoka M (2005) Elucidation of information encoded in tryptophan 140 of staphylococcal nuclease. Proteins 58:271–277
8. Onitsuka M, Kamikubo H, Yamazaki Y, Kataoka M (2008) Mechanism of induced folding: both folding before binding and binding before folding can be realized in staphylococcal nuclease mutants. Proteins 72:837–847
9. Nakagawa H, Joti Y, Kitao A, Kataoka M (2008) Hydration affects both harmonic and anharmonic nature of protein dynamics. Biophys J 95:2916–2923
10. Kato S, Kamikubo H, Hirano S, Yamazaki Y, Kataoka M (2010) Non-local interaction responsible for the tertiary structural formation of Staphylococcal nuclease. Biophys J 98:678–686
11. Nakagawa H, Kataoka M (2010) Percolation of hydration water as a control of protein dynamics. J Phys Soc Jpn 79:083801
12. Wright PE, Dyson HJ (1999) Intrinsically unstructured proteins: re-assessing the protein structure-function paradigm. J Mol Biol 293:32–331
13. Tompa P (2002) Intrinsically unstructured proteins. Trends Biochem Sci 27:527–533
14. Dyson J, Wright P (2005) Intrinsically unstructured proteins and their function. Nat Rev Mol Cell Biol 6:197–208
15. Anfinsen CB (1972) The formation and stabilization of protein structure. Biochem J 128:737–749
16. Cotton FA, Hazen EE, Legg MJ (1979) Staphylococcal nuclease: proposed mechanism of action based on structure of enzyme-thymidine 3′, 5′-bisphosphate-calcium ion complex at 1.5-A resolution. Proc Natl Acad Sci U S A 76:2551–2555

17. Loll PJ, Lattman EE (1989) The crystal structure of the ternary complex of staphylococcal nuclease, Ca^{2+}, and the inhibitor pdTp, refined at 1.65 Å. Proteins 5:183–201
18. Yin J, Jing G (2000) Tryptophan 140 is important, but serine 141 is essential for the formation of the integrated conformation of staphylococcal nuclease. J Biochem 128:113–119
19. Shaw N, Zhao M, Cheng C, Xu H, Saarikettu J, Li Y, Da Y, Yao Z, Silvennoinen O, Yang J, Liu ZJ, Wang BC, Rao Z (2007) The multifunctional human p100 protein 'hoooks' methylated ligands. Nat Struct Mol Biol 14:779–784
20. Fersht AR (2000) Transition-state structure as a unifying basis in protein-folding mechanisms: contact order, chain topology, stability, and the extended nucleus mechanism. Proc Natl Acad Sci U S A 97:1525–1529
21. Poole LB, Loveys DA, Hale SP, Gerlt JA, Stanczyk SM, Bolton PH (1991) Deletion of the Ω-loop in the active site of staphylococcal nuclease. 1. Effect on catalysis and stability. Biochemistry 30:3651–3627
22. Caflish A (2003) Folding for binding or binding for folding. Trends Biotechnol 21:423–425
23. Sugase K, Dyson HJ, Wright PL (2007) Mechanisum of coupled folding and binding of an intrinsically unstructured protein. Nature 447:1021–1025
24. Kalnin NN, Kuwajima K (1995) Kinetic folding and unfolding of staphylococcal nuclease and its six mutants studied by stopped-flow circular dichroism. Proteins 23:163–176
25. Maki K, Ikura T, Hayano T, Takahashi N, Kuwajima K (1999) Effects of proline mutations on the folding of staphylococcal nuclease. Biochemistry 38:2213–2223
26. Shiba R, Umeyama M, Tsukasa S, Kamikubo H, Yamazaki Y, Yamaguchi M, Iwakura M, Kataoka M (2011) Systematic alanine insertion reveals the essential regions that encode structure formation and activity of dihydrofolate reductase. Biophysics 7:1–10

Chapter 8
Theory of Molecular Recognition and Structural Fluctuation of Biomolecules

Fumio Hirata, Norio Yoshida, and Bongsoo Kim

Abstract In this chapter, we present a theoretical treatment on the molecular recognition, one of elementary processes of life phenomena, based on the 3D-RISM/RISM theory. The theory has been applied successfully to a variety of molecular recognition processes. It has proved its ability to discriminate molecules having different charges and structures, which is an essential requirement for the in silico drug design.

The structural fluctuation plays a crucial role in the process where a protein expresses its function. A new theory to characterize the structural fluctuation of protein around its native state, which combines the 3D-RISM/RISM theory with the generalized Langevin theory, is also presented here.

Keywords 3D-RISM theory • Generalized Langevin theory • Molecular recognition • Structural fluctuation

8.1 Introduction

In a living system, a variety of biomolecules including protein, DNA, and sugar are working in order to maintain their life and to inherit it from generation to generation. DNA keeps genetic information which is inherited to the next generation [1]. RNA makes a copy of the information to transfer it to protein. Protein is synthesized according to the information. So synthesized protein serves in our body to maintain our living activity, as a *molecular machine* such as a catalyst (enzyme) of a chemical reaction, an information tranducer between cells (a molecular channel), and molecular carrier (e.g., hemoglobin).

F. Hirata (✉)
Ritsumeikan University, 1-1-1 Nojihigashi, Kusatsu, Shiga 525-8577, Japan
e-mail: hirataf@fc.ritsumei.ac.jp

N. Yoshida
Kyushu University, 744, Motooka, Nishi-ku, Fukuoka 819-0395, Japan

B. Kim
Changwon National University, Changwon 641-773, Korea

M. Terazima et al. (eds.), *Molecular Science of Fluctuations Toward Biological Functions*, DOI 10.1007/978-4-431-55840-8_8

There are two universal molecular processes observed in life phenomena. One of those is "self-organization." The most typical example of the self-organization is cell membrane, which provides a stage for biomolecules to play their roles, is a kind of a molecular assembly, and is build spontaneously from many molecules including phospholipid and cholesterol. The other example is the protein folding. Although protein is a polymer made up from amino acids through polymerization, it forms a specific conformation depending on its amino acid sequence under a proper thermodynamic condition, unlike artificial polymers such as polyethylene. The phenomenon is called "protein folding." It is well documented that the specific conformation is intimately related to its function. As in the two examples, the phenomena in which molecules make an assembly, or a molecules fold into a specific conformation spontaneously, is referred to as "self-organization."

The other physicochemical process universally seen in our body is "molecular recognition." A process in which an enzyme binds substrate molecules in its reaction pocket is a typical example of the molecular recognition. Another example of the molecular recognition is the binding of an antibody to an antigen in an antigen-antibody reaction. A process in which an ion channel accommodates an ion in its pore can also be regarded as a molecular recognition. Namely, "molecular recognition" is an elementary process in life phenomena, through which a biomolecule should always pass, whenever the molecule performs its function.

The two elementary processes in a living system do not work spontaneously in the condition in which the molecules are isolated in vacuum. Let's consider first the case of membrane formation. The situation is illustrated by a cartoon in Fig. 8.1, in which the polar head group and alkyl tail group of phospholipid molecules are depicted, respectively, by a sphere and a rod. If such molecules exist in vacuum, the reaction toward the right direction, or the membrane formation, is quite difficult for two reasons. First, such a configuration in which molecules align closely each other like in (B) is energetically unfavorable, since the polar head groups repel each other due to the electrostatic interaction. The configuration in which molecules are ordered neatly is unstable entropically as well. Then, why do the lipid molecules self-organize? The answer to the question is "water." It is because those molecules are dissolved in water. In this case, water promotes the self-organization by inducing the hydrophobic interactions among the alkyl groups of lipid, and by screening the Coulomb repulsion among the head groups.

In the following, let's think about an enzymatic reaction as an example of the molecular recognition (Fig. 8.2). An enzymatic reaction is characterized essentially with the two equilibrium constants: one that concerns binding of substrate molecules at a reaction pocket, or the molecular recognition. The other equilibrium constant concerns a chemical reaction in the pocket, which is associated with a change in the electronic structure. The theory of reaction rate by Michaelis and Menten takes the two chemical equilibriums into consideration [2].

Now, in general, the reaction pocket of an enzyme binds one or more water molecules and ions sometimes. Thereby, all or some of those water molecules and ions should be disposed from the pocket to bulk solution, in order for a substrate molecule to be accommodated in the active site (Fig. 8.2). On the other hand, a substrate molecule itself is more likely to be hydrated by water molecules.

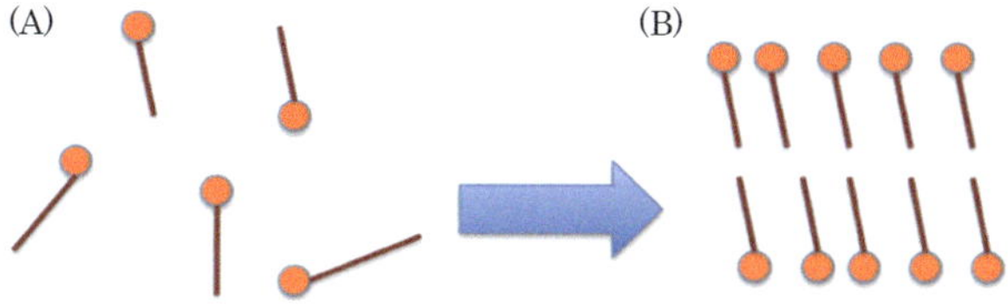

Fig. 8.1 An illustrative example of self-organization (cell membrane formation)

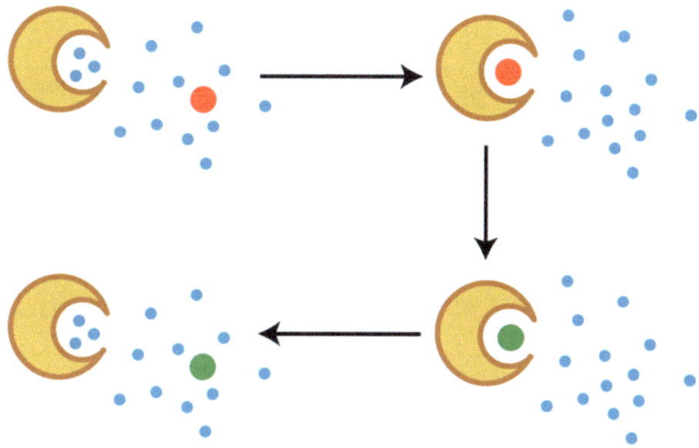

Fig. 8.2 An illustrative example of molecular recognition (enzymatic reaction): *yellow*, protein; *blue*, water; *red*, substrate before reaction; *green*, substrate after reaction

Whether or not a substrate molecule is desolvated upon the molecular recognition is determined by the entire free-energy change concerning the binding process, including the entropy change associated with the desolvation process. It indicates that the molecular recognition process is also governed by water molecules.

It was emphasized so far that water molecules play essential roles in the two physicochemical processes governing life phenomena, "self-organization" and "molecular recognition." But, those are not only the roles water play. In the enzymatic hydrolysis reaction, such as the hydrolysis of ATP, water molecules participate in the reaction as one of the reactants. Ions play crucial roles in many of chemical reactions in our body, but they can exist only in an aqueous environment. A persistent belief that water does not play a crucial role in enzymatic reactions except for the "hydrolysis" is too naive. How many and in what configurations water molecules exist in a reaction pocket are essential factors to determine the reaction field of an enzyme. Any theory of chemical reactions disregarding the factors is senseless. The fact that approximately 70 % of living body is occupied by water possesses a quite essential physicochemical significance.

The molecular recognition process is not only important for our understanding of life phenomena. It is also essential for the practical purpose of developing a new technology for drug discovery. As is mentioned above, protein plays a variety of roles in order to maintain and inherit life. If one of protein functions is lost or is too strong, a living body becomes sick, or dies in the worst case. What we call "drug" is a molecule (many of them are organic compounds) which inhibits or promote activities of protein or DNA. The same statement applies to bacteria and virus. Protein and DNA play a crucial role in maintaining and inheriting their life, and the molecular recognition process works as an elementary process. If we can inhibit the molecular recognition process in those microorganisms, which is vital for their existence and promotion, we can kill the microorganisms. Most of the drug molecules used in the treatment of diseases bind to protein and DNA sel and inhibit the function of the biomolecules. Molecular recognition is nothing but the preferential binding of a drug molecule to the active site of protein and DNA.

8.2 Statistical Mechanics of Molecular Recognition (3D-RISM/RISM)

The most important aspect that distinguishes the processes concerned with biomolecules, protein, and DNA from ordinary chemical processes in solution is "inhomogeneity" of the reaction field in atomic level. For instance, the distribution of solvent, water, and ions, around a hydrophobic residue of protein is entirely different from that around a charged residue. The environment at an active site (or reaction pocket) maybe even more different from that at the surface. Such inhomogeneity of the field in atomic level has been out-of-scope for the statistical mechanics for long time, and it is a target that has been bypassed by the conventional statistical mechanics including the ordinary or the one-dimensional RISM theory [3]. The situation was broken through due to a paper published by Imai et al. in 2005 [4]. They could have probed water molecules bound in an active site of lysozyme by means of the 3D-RISM/RISM theory [3]. The theory has been applied to a variety of the molecular recognition processes concerning functions of biomolecules, and has established its position as the theory of molecular recognition. The present section is devoted to explain the physics behind the 3D-RISM/RISM theory. (The readers are referred to the references [3, 5] for the RISM and 3D-RISM theory.)

Let us consider the average density of solvent molecules at a position around a solute molecule (Fig. 8.3). When the position is far from the solute molecule so as to be regarded as the bulk, the density will be constant which is the same as in the pure solvent. On the other hand, when it is nearby solute, the density will be "perturbed" significantly by the field due to the solute molecule, and be different from that in the bulk, depending on the strength of the perturbation. Physical meaning of the 3D-RISM equation can be readily understood in terms of a *nonlinear* response theory as follows.

Let us denote the average density of solvent atom γ at the bulk, the density nearby solute, and the density response to the perturbation as ρ_γ, $\rho_\gamma(\mathbf{r})$, and $\Delta\rho_\gamma(\mathbf{r})$, respectively. Then, statement above can be expressed as,

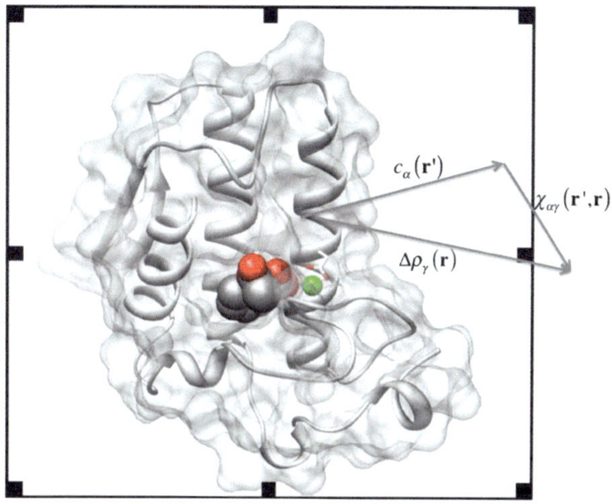

Fig. 8.3 Inhomogeneous density distribution of solvent around a protein

$$\rho_\gamma(\mathbf{r}) = \rho_\gamma + \Delta\rho_\gamma(\mathbf{r}). \tag{8.1}$$

The density response to the perturbation can be expressed in terms of a nonlinear response theory as

$$\Delta\rho_\gamma(\mathbf{r}) = \sum_\alpha \int \chi_{\gamma\alpha}(\mathbf{r},\mathbf{r})\rho_\gamma c_\alpha(\mathbf{r})\mathrm{d}\mathbf{r}, \tag{8.2}$$

where $c_\alpha(\mathbf{r}')$ is a *nonlinear* perturbation due to the solute molecule, and $\chi_{\alpha\gamma}(\mathbf{r},\mathbf{r}')$ is a response function. (The reason why we call it "nonlinear" will be clarified in a moment.) The equation can be viewed as the 3D-RISM equation [3] by identifying $c_a(\mathbf{r})$ as the direct correlation function and $\chi_{\alpha\gamma}(\mathbf{r},\mathbf{r}')$ as the correlation function of density fluctuation in bulk solvent, that is,

$$\rho^2\chi_{\alpha\gamma}(\mathbf{r},\mathbf{r}') = \left\langle \delta\rho_\alpha^{(0)}(\mathbf{r})\delta\rho_\gamma^{(0)}(\mathbf{r}')\right\rangle, \tag{8.3}$$

where $\delta\rho_\alpha^{(0)}(\mathbf{r})$ is the density fluctuation of atom a in the pure liquid defined by $\delta\rho_\alpha^{(0)}(\mathbf{r}) = \rho_\alpha^{(0)}(\mathbf{r}) - \rho_\alpha$.

Several approximate equations have been devised for the direct correlation functions. For example, the hypernetted-chain approximation reads [3],

$$c_\alpha^{\mathrm{uv}}(\mathbf{r}) = \exp\left[-\beta u_\alpha^{\mathrm{uv}}(\mathbf{r}) + h_\alpha^{\mathrm{uv}}(\mathbf{r}) - c_\alpha^{\mathrm{uv}}(\mathbf{r})\right] - 1 - \left[h_\alpha^{\mathrm{uv}}(\mathbf{r}) - c_\alpha^{\mathrm{uv}}(\mathbf{r})\right]. \tag{8.4}$$

In the expression, $u_\alpha(\mathbf{r})$ is the direct interaction potential exerted on the solvent molecules from the solute molecule, and $h_\alpha(\mathbf{r})$ is the density fluctuation in solvent at position $\mathbf{r}$, normalized by the bulk density, namely,

$$h_\alpha(\mathbf{r}) = \Delta\rho_\alpha(\mathbf{r})/\rho. \tag{8.5}$$

The three-dimensional distribution function used in this study is defined from $h_\alpha(\mathbf{r})$ by

$$g_\alpha(\mathbf{r}) = h_\alpha(\mathbf{r}) + 1. \tag{8.6}$$

It is not only the direct interaction $u(\mathbf{r})$ with solute that perturbs the density of solvent at a certain position, but also that from solvent molecules at the other positions, whose density is also perturbed by the existence of the same solute. Such "indirect" perturbations are *renormalized* into the terms including $(h_\alpha(\mathbf{r}) - c_\alpha(\mathbf{r}))$. Such renormalization makes the perturbation highly "nonlinear."

8.3 Molecular Recognition and Drug Design

In the present section, we review some applications of the 3D-RISM/RISM theory to the molecular recognition.

8.3.1 Water Molecules Bound in a Cavity of Lysozyme

The first application of the 3D-RISM/RISM theory to the molecular recognition was made in 2005 by Imai et al., taking water molecules and the hen-egg-white lysozyme as a ligand and a target protein, respectively [4]. They calculated the spatial distribution function g(r) of solvent atoms around and inside the protein based on Eqs. 8.2, 8.3, 8.4, 8.5 and 8.6. Figure 8.4 exhibits the theoretical and experimental results for the distribution of water inside a cavity of the lysozyme, which are surrounded by the residues from Y53 to I58 and from A82 to S91. For simplicity, only the surrounding residues are displayed, except for A82 and L83 which are located in the front side. Figure 8.4a shows the isosurfaces of $g(r) > 8$ for water-oxygen (green) and hydrogen (pink) in the cavity. Four distinct peaks of water oxygen and eight distinct peaks of water hydrogen are found, implying that four water molecules are accommodated in the cavity. From the isosurface plot, we have reconstructed the most probable pose of the water molecules inside the cavity, as shown in Fig. 8.4b. It is interesting to compare the hydration structure obtained by the 3D-RISM theory with crystallographic water sites of X-ray structure [6]. The crystallographic water molecules in the cavity are depicted in Fig. 8.4c, showing four water sites in the cavity, much as the 3D-RISM theory has detected. Moreover, the positions of water molecules obtained by the theory and experiment are almost identical. Thus, we have concluded that the 3D-RISM theory has the ability to probe water molecules recognized by a cavity of protein. Further analysis revealed that the water molecule located at the deepest inside the cavity is stabilized by a

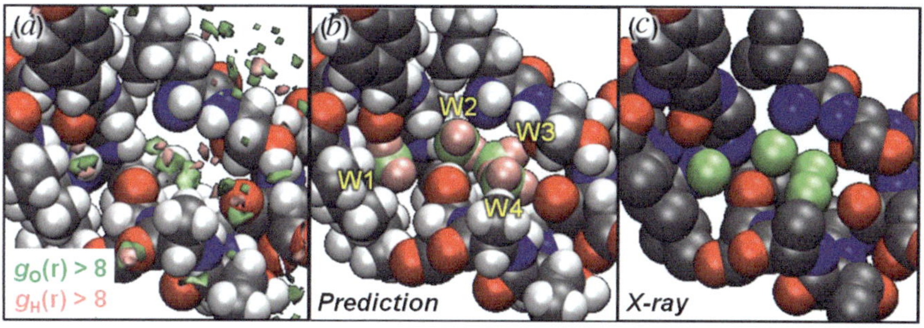

Fig. 8.4 Distribution of water molecules in lysozyme. (**a**) The distribution of water molecules obtained from the 3D-RISM/RISM theory: *green*, oxygen; *purple*, hydrogen. (**b**) Most probable poses of water molecules deduced from the distribution. (**c**) The position of water oxygen determined from the X-ray crystallography

hydrogen-bond with a serine residue, and that the other three water molecules are stabilized by making a hydrogen-bonded train [7].

Before the paper was published, it was widely believed that a statistical mechanics theory cannot give a right solution to a problem in such an inhomogeneous environment as protein cavity. The work by Imai et al. has broken the common understanding in the community. In the same paper, we suggested that the method can be applied for screening of drug compounds by just replacing the solvent (pure water) by aqueous solutions including a drug compound as a component. Should a compound have a high affinity to the active site of protein, one will detect high peak in the spatial distribution (g(r)) of atoms concerning the compounds in the active site. Such a compound can be a great candidate for a drug to the target protein. On the other hand, if one finds small or no peaks in g(r), the compound could not be a good candidate for the drug.

An author of *Chemistry World* published by the Royal Society, however, criticized our suggestion, saying "*The finding is certainly interesting and the method shows some promise* [8]. *However, it is too much to extrapolate from the analysis of just one cavity in one protein and claim that the method is robust and widely applicable*." For the purpose of rebutting against the criticism, we have published many papers including "drug screening" to confirm our statement, selected topics of which are presented below [9–25].

8.3.2 Selective Ion Binding by Protein [9]

Ion binding is essential for a variety of physiological processes. The binding of calcium ions by some proteins triggers the process to induce the muscle contraction and enzymatic reactions [26, 27]. The initial process of the information transmission through the ion channel is the ion binding by channel proteins [28]. The ion binding plays an essential role sometimes to the folding process of a protein by inducing the secondary structure [29]. Such processes are characterized by the

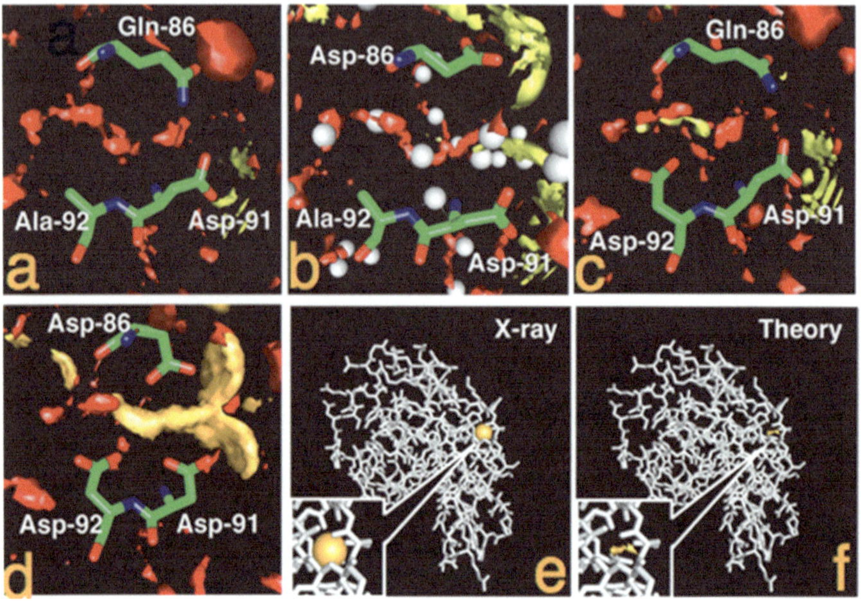

Fig. 8.5 Distribution of water (*red*) and ions (Na^+, *yellow*; K^+, *purple*; Ca^{++}, *orange*) in the active site of the wild type and mutants of human lysozyme: (**a**) wild type, (**b**) Q86D mutant, (**c**) A92D mutant, (**d**) Q86D & A92D mutant, and (**e**, **f**) global view of calcium binding in Q86D & A92D mutant

highly selective ion recognition by proteins. It is of great importance, therefore, for the life science to clarify the origin of ion selectivity in molecular detail.

The 3D-RISM calculation was carried out for aqueous solutions of three different electrolytes $CaCl_2$, NaCl, and KCl, and for four different mutants of the protein: wild type, Q86D, A92D, and Q86D/A92D that have been studied experimentally by Kuroki and Yutani [30].

In the Fig. 8.5, shown are the distributions of water molecules and the cations inside and around the cleft under concern, which consists of amino acid residues from Q86 to A92. The area where the distribution function, g(**r**) is greater than five is painted with a color for each species: oxygen of water, red; Na^+ ion, yellow; Ca^{2+} ion, orange; and K^+ ion, purple. For the wild type of protein in the aqueous solutions of all the electrolytes studied, $CaCl_2$, NaCl, and KCl, there are no distributions (g(**r**) > 5) observed for the ions inside the cleft, as is seen in Fig. 8.5a. The Q86D mutant exhibits essentially the same behavior with that of the wild type, but with the water distribution changed slightly (Fig. 8.5b) (There is a trace of yellow spot that indicates a slight possibility of finding a Na^+ ion in the middle of the binding site, but it would be too small to make a significant contribution to the distribution.) Instead, the distribution corresponding to water oxygen is observed as is shown in the red color in the figure. The distribution covers the region where the crystallographic water molecules have been detected, which are shown with the spheres colored gray. There is a small difference between the theory and the experiment, which is the crystallographic water bound to the backbone of Asp-91. The theory does not reproduce the water molecule by unidentified reasons. Except for the difference, the observation is consistent with the experimental finding, especially, that the protein with the wild type sequence binds neither Na^+ nor Ca^{2+}.

The A92D mutant in the NaCl solution shows conspicuous distribution of a Na^+ ion bound in the recognition site, which is in accord with the experiment (Fig. 8.5c). The Na^+ ion is apparently bound to the carbonyl oxygen atoms of Asp-92, and is distributed around the moieties. There is water distribution observed in the active site, but the shape of the distribution is entirely changed from that in the wild type. The distribution indicates that the Na^+ ion bound in the active site is not naked, but is accompanied by hydrating water molecules. The mutant does not show any indication of binding K^+ ion. (The results are not shown.) This suggests that the A92D mutant discriminates a Na^+ ion from a K^+ ion. The finding demonstrates the capability of the 3D-RISM theory to realize the ion selectivity by protein.

In the lower panels, shown are the distributions of Ca^{2+} ions and of water oxygen in the ion binding site of the holo-Q86D/A92D mutant (Fig. 8.5d). The mutant is known experimentally as a calcium binding protein. The protein, in fact, exhibits a strong calcium binding activity as can be inspected in the figure. The calcium ion is recognized by the carboxyl groups of the three Asp residues, and is distributed around the oxygen atoms. The water distribution at center of the triangle made by the three carbonyl oxygen atoms is reduced dramatically, which indicates that the Ca^{2+} ion is coordinated by the oxygen atoms directly, not with water molecules in between. The Ca^{2+} ion, however, is not entirely naked, because the persistent water distribution is observed at least at two positions where original water molecules were located in the wild type of the protein. Global views of the distribution of the Ca^{2+} ion bound at the active site are shown in Fig. 8.5e and f for the experimental (X-ray) result and for the 3D-RISM/RISM result, respectively, which look almost identical.

8.3.3 Binding of Aspirin to Phospholipase A2 (Application to Intelligent Drug Design) [24]

Here, we present an application of the 3D-RISM/RISM theory to binding of a drug molecule to the active site of a target protein. The target protein is phospholipase A2 [31–34]. The protein is known to catalyze the synthesis of the arachidonic acid which causes pain in our body. Recent experimental studies have disclosed that the protein binds acetylsalicylic acid (or Aspirin), which is popular as a *pain killer*; thereby it is regarded as one of the model compounds in the field of drug design.

We have made an analysis of the binding affinity of the molecule to the protein based on the 3D-RISM/RISM theory, focusing on the effect of position isomer of the ligand. (The binding affinity can be measured in terms of the peak height of the spatial distribution function g(r) of ligand atoms). Ordinary "Aspirin" is a

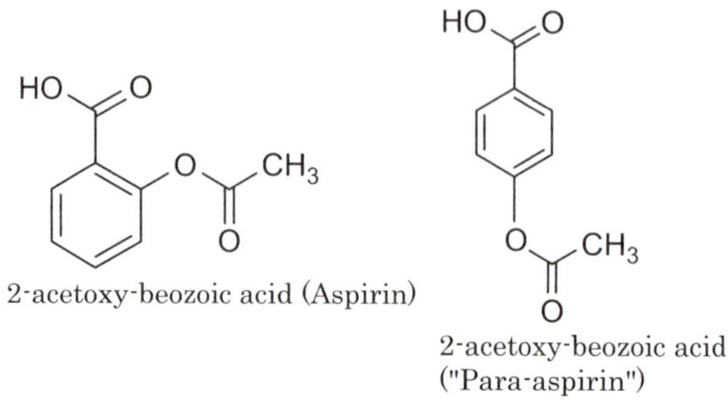

Fig. 8.6 The two positional isomers of acetylsalicylic acid (Aspirin)

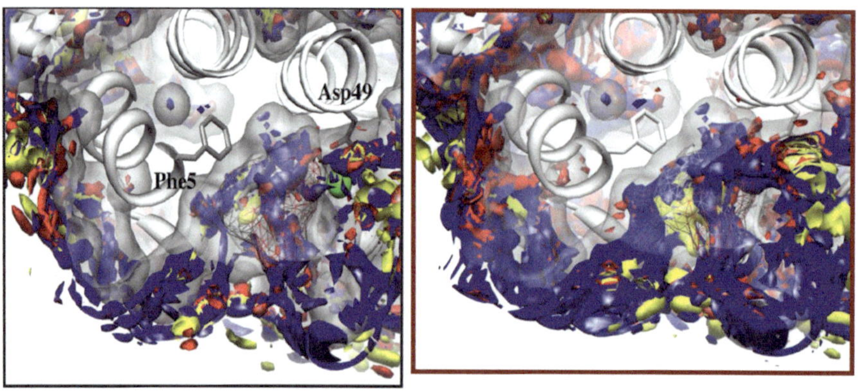

Fig. 8.7 The spatial distribution of Aspirin at the active site of phospholipase A2

compound in which the two functional groups, carboxyl group and acetyl group, are arranged at the ortho-position. A compound in which the two functional groups are located at the para-position can be also synthesized. Let's call them as "ortho-Aspirin" and "para-Aspirin" (Fig. 8.6).

Shown in Fig. 8.7 are the spatial distribution functions of atoms in Aspirin obtained from 3D-RISM/RISM. Both the orth- and para-Aspirin have large distribution around the active site, although the shape of distribution looks different from each other. The results clearly indicate that both the isomers have great affinity to the cavity around the active site of the protein. However, we cannot figure out just from the visual inspection of the distribution how close or different the binding modes are between the two isomers. Therefore, we have made further analysis of the distribution data to determine the most probable binding modes of the two isomers, by minimizing a score function which measures the difference of the trial position and orientation of atoms from the spatial density distribution calculated by 3D-RISM/RISM. The results are shown in Fig. 8.8 along with the binding mode of the ordinary Aspirin, or the ortho-Aspirin, determined from the X-ray crystallography. The results show good agreement between the theoretical prediction and the experiment. On the other hand, the binding-mode for the para-Aspirin is markedly different from that of the X-ray result for the ordinary Aspirin. One might think

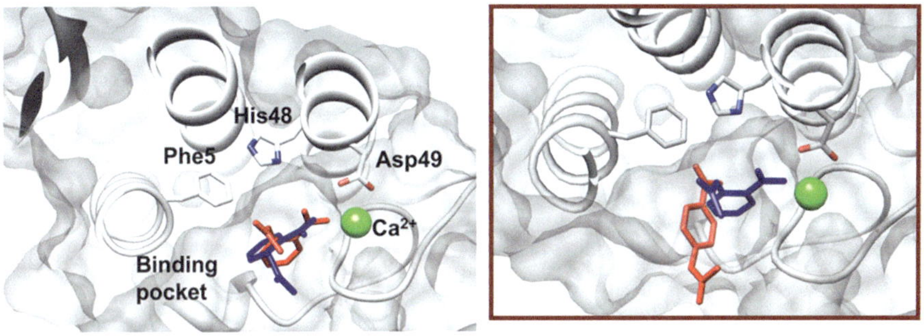

Fig. 8.8 Binding mode (pose) of the orth-Aspirin (*left*) and para-Aspirin (*right*)

the results are trivial, since the structure of the two molecules is different from each other. However, the result proves unequivocally high ability of the theory to discriminate molecular structure in the molecular recognition process. Such high ability to discriminate molecular structure is an essential requirement for a successful method in the in silico screening of drug compounds.

8.4 Structural Fluctuation of Protein and Molecular Recognition

In the preceding sections, we introduced a theoretical method of the molecular recognition based on the 3D-RISM/RISM theory. The method is a powerful tool to describe the affinity of ligands to a protein that has a rigid structure. However, it is well regarded that the structure of protein undergoes spatial and temporal fluctuation around its native state, or an equilibrium conformation, and that the structural fluctuation gives significant influence on the molecular recognition. The so-called *induced fitting* of a substrate to enzyme is a typical example of such phenomena. In the present section, a new statistical mechanics theory [35] to describe the structural fluctuation of protein is briefly sketched.

One of the earliest theoretical efforts to characterize the structural fluctuation of protein was based on the normal mode analysis in vacuum, which dose not involve water [36]. Therefore, the treatment cannot describe a biological fluctuation occurring naturally in living bodies. It is just a "nonbiological" oscillation. There are some phenomenological attempts proposed to take the solvent effect into account (the Poisson-Boltzmann equation and the Langevin theory [37, 38]) in the fluctuation. However, they cannot explain essential factors which give influence upon the structural fluctuation, such as hydration and dehydration upon binding of ligand at an active site of protein, since they disregard the structure of solvent in molecular level. While the molecular dynamics simulation is a quite powerful method to describe the mechanical fluctuation on a potential energy surface, it encounters difficulty for explaining the thermodynamic fluctuation on the nonequilibrium free-energy surface. The difficulty becomes hopeless when the solvent includes ions and other components. It is because the time to sample the nonequilibrium free energy surface of such systems by means of the simulation is so long that a simulation will not converge within the conventional information technologies using the silicon-based electronics.

In the present section, we sketch briefly a new theory that has a great perspective to overcome the difficulties. The theory is formulated by combining the equilibrium (3D-RISM/RISM) and nonequilibrium statistical-mechanics (generalized Langevin equation [39]) theories of liquids [35].

8.4.1 Generalized Langevin Equation of a Protein-Solution System

A generalized Langevin equation (GLE) is derived from the Liouville equation which describes time evolution of physical quantities under concern, or *dynamic variables*, defined in the phase space, via a series of coarse graining or "projection." The equation looks similar to the conventional Langevin equation (LE) constructed heuristically from the Newtonian equation of motion, having the frictional force proportional to velocity, and the random force, both mimicking the force exerted on the particle from solvent [38].

We construct a vector **A**(t) from four physical quantities as dynamic variables: the displacement or fluctuation of atoms in protein from their equilibrium position, $\Delta\mathbf{R}_\alpha(t)$, and their momentum, $\mathbf{P}_\alpha(t)$, the density fluctuation of solvent atoms and their flux.

$$\mathbf{A}(t) = \begin{pmatrix} \Delta\mathbf{R}_\alpha(t) \\ \mathbf{P}_\alpha(t) \\ \delta\rho^a_{\mathbf{k}}(t) \\ \mathbf{J}^a_{\mathbf{k}}(t) \end{pmatrix} \tag{8.7}$$

$\Delta\mathbf{R}_\alpha(t)$: the displacement or fluctuation of atom α in protein from its equilibrium position.
$\mathbf{P}_\alpha(t)$: the momentum of atom α in protein.
$\delta\rho^a_{\mathbf{k}}(t)$: the density fluctuation of solvent atom α, in Fourier space.
$\mathbf{J}^a_{\mathbf{k}}(t)$: the momentum density field of atom α, in Fourier space.

We project all other degrees of freedom in phase space onto the vector A(t) defining the projection operator by

$$P(\cdots)\equiv(\mathbf{A},\cdots)(\mathbf{A},\mathbf{A})^{-1}\mathbf{A}, \tag{8.8}$$

where (**A, A**) represents a scalar product in the Hilbert space, defined by

$$(\mathbf{a},\mathbf{b}) = Z^{-1}\int \mathrm{d}\Gamma \mathbf{a}^{*}\mathbf{b}\exp(-H(\Gamma)/k_B T). \tag{8.9}$$

In Eq. 8.9, H is the Hamiltonian of aqueous solutions including a single protein molecule, defined by the following relations.

$$\begin{aligned} H &= H_0 + H_1 + H_2, \\ H_0 &= \sum_{i=1}^{N}\sum_{a=1}^{n}\left[\frac{\mathbf{p}_i^a\cdot\mathbf{p}_i^a}{2m_a} + \sum_{j\neq i}\sum_{b\neq a} U_0\left(\left|\mathbf{r}_i^a - \mathbf{r}_j^b\right|\right)\right], \\ H_1 &= \sum_{\alpha=1}^{N_u}\left[\frac{\mathbf{p}_\alpha\cdot\mathbf{p}_\alpha}{2M_\alpha} + \sum_{\beta\neq\alpha} U_1\left(\left|\mathbf{R}_\alpha - \mathbf{R}_\beta\right|\right)\right], \\ H_2 &= \sum_{\alpha=1}^{N_u}\sum_{i=1}^{N}\sum_{a=1}^{n} U_{\mathrm{int}}\left(\left|\mathbf{R}_\alpha - \mathbf{r}_i^a\right|\right). \end{aligned} \tag{8.10}$$

where H_0, H_1, and H_2 are the Hamiltonian of solvent, protein, and the interaction between the two, respectively.

By operating the projection operator defined in Eq. 8.9 to the equation of time evolution, or the Liouville equation of the dynamic variables $\mathbf{A}(t)$, or

$$\frac{\mathrm{d}\mathbf{A}(t)}{\mathrm{d}t} = iL\mathbf{A}(t) = \{A, H\}_{PB}, \tag{8.11}$$

we can derive the generalized Langevin equations concerning the fluctuation of atom position in protein, and of solvent density [35].

$$M_\alpha \frac{\mathrm{d}^2\Delta\mathbf{R}_\alpha(t)}{\mathrm{d}t^2} = -k_B T\sum_\beta \left(\mathbf{L}^{-1}\right)_{\alpha\beta}\cdot\Delta\mathbf{R}_\beta(t) - \int_0^t \mathrm{d}s\sum_\beta \Gamma_{\alpha\beta}(t-s)\cdot\frac{\mathbf{P}_\beta(s)}{M_\beta} + W_\alpha(t) \tag{8.12}$$

$$\frac{\mathrm{d}^2\delta\rho_{\mathbf{k}}^a(t)}{\mathrm{d}t^2} = -k^2\sum_{b,c} J_{ac}(k)\chi_{cb}^{-1}(k)\delta\rho_{\mathbf{k}}^b(t) - \frac{i\mathbf{k}}{N}\sum J_{ac}^{-1}(k)\int_0^t \mathrm{d}s M_{\mathbf{k}}^{bc}(t-s)\cdot J_k^b(s) + i\mathbf{k}\cdot\Xi_{\mathbf{k}}^a(t) \tag{8.13}$$

In the equations, $\Gamma_{\alpha\beta}(t)$, $W_\alpha(t)$, $J_{ac}(k)$, $M_{\mathbf{k}}^{bc}(t-s)$, and $\Xi_{\mathbf{k}}^a(t)$ have the following physical meaning.

$\Gamma_{\alpha\beta}(t-s)$: frictional force due to solvent, acting on protein atoms
$W_\alpha(t)$: random force acting on protein atoms
$J_{ac}(k)$: correlation of momentum density between a pair of atoms in solvent
$\chi_{cb}(k)$: density correlation function between a pair of solvent atoms
$M_{\mathbf{k}}^{bc}(t-s)$: frictional force acting on solvent molecules
$\Xi_{\mathbf{k}}^a(t)$: random force acting on solvent atoms

The matrix **L** included in (8.12) is defined by

$$L_{\alpha\beta} = \langle \Delta\mathbf{R}\Delta\mathbf{R}\rangle_{\alpha\beta} \tag{8.14}$$

where $\Delta\mathbf{R}$ is the displacement (fluctuation) of atoms in protein from its equilibrium position. Therefore, the matrix **L** represents a correlation function of structural fluctuation, or the variance-covariance matrix [35].

The structural fluctuation and dynamics of protein in solution can be explored by solving Eqs. 8.12 and 8.13 coupled together. The dynamics has the following features. Firstly, the equation does not include a term related to the force which originates from the first derivative of the free energy surface with respect to the position. The force acting on an atom of protein comes from two terms, one which is proportional to the displacement of the atom from its equilibrium position (the second term in the left hand side), and the other due to the random force (the term in the right hand side). The physical origin why the equation does not include the first derivative of the free energy lies in our treatment based on the generalized Langevin theory. The whole idea of the generalized Langevin theory is to project all the degrees of freedom in the phase space onto few dynamic variables under concern. The projection is carried out using a projection operator, defined by Eqs. 8.8 and 8.9, in terms of an ensemble average of two variables which are fluctuating around an *equilibrium* in the phase space. Obviously, the ensemble average of the displacement of atoms in protein should be zero in equilibrium, as can be readily verified.

$$\langle \Delta\mathbf{R}_\alpha\rangle = \langle \Delta\mathbf{R}_\alpha - \langle \Delta\mathbf{R}_\alpha\rangle\rangle = \langle \Delta\mathbf{R}_\alpha\rangle - \langle \Delta\mathbf{R}_\alpha\rangle = 0 \tag{8.15}$$

Such a force as the first derivative of the free energy, which may cause the complete shift of the equilibrium, is not included in the treatment. The situation is somewhat analogous to the case of a harmonic oscillator, in which an oscillator swings back and forth around a minimum of the harmonic potential. The only force that acts on the system is the restoring force proportional to the displacement from the potential minimum. In our case, too, the only force acting on the protein atoms is the one which restores atom positions from fluctuating to equilibrium ones. However, there is an essential difference in physics between the two systems. The equilibrium position of a harmonic oscillator is the minimum of *mechanical* potential energy, while that of protein in water is the minimum in the *thermodynamic* potential or the free energy, which is concerned not only with energy but also with the entropy both of protein and of water. So, in the case of protein in water, the stochastic character of the dynamics is attributed not only to the random force term, but also to the conformational fluctuation of protein around its equilibrium state, induced by solvent, while the stochastic character is resulted just from the random force term in the case of the coupled harmonic oscillator treated by Wang and Unlenbeck.

The argument above suggests interesting physics implied in Eq. 8.12, and its application to biological functions. If one ignores the friction and random force terms, Eq. 8.12 reads

$$\frac{\mathrm{d}^2 \Delta\mathbf{R}(t)}{\mathrm{d}t^2} = -k_B T \mathbf{L}^{-1} \cdot \Delta\mathbf{R}(t) \tag{8.16}$$

The equation can be viewed as a "harmonic oscillator," "Hessian" of which is $k_B T\mathbf{L}^{-1}$. Considering Eq. 8.14, the "Hessian" matrix is related to the variance-covariance matrix of the positional fluctuation by

$$k_B T\mathbf{L}^{-1} \equiv k_B T \langle \Delta\mathbf{R}\Delta\mathbf{R}\rangle^{-1} \tag{8.17}$$

The observation strongly suggest that the dynamics described by Eq. 8.16 is that of fluctuation around a *minimum* of the *free energy surface* consisting not only of the interactions among atoms in the protein, but of the solvation free energy. In this respect, the configuration corresponding to the free energy minimum is not just one but an ensemble of distinguishable configurations concerning protein and solvent, which can be converted among each other due to the thermal noise. The free energy surface of protein can be given by

$$F(\mathbf{R}) = U(\mathbf{R}) + \Delta\mu(\mathbf{R}) \tag{8.18}$$

where $U(\mathbf{R})$ is the interaction among atoms in a protein, $\Delta\mu(\mathbf{R})$ is the solvation free energy of protein, the conformation of which is denoted by $\mathbf{R}$ [40].

The consideration further suggests a method to evaluate the variance-covariance matrix, which characterizes structural fluctuation of protein, based on the 3D-RISM theory. The variance-covariance matrix is closely related to the Hessian matrix, Eq. 8.17, and the Hessian matrix is the second derivative of the free energy surface, namely,

$$k_B T L_{\alpha\gamma}^{-1} = \frac{\partial^2 F}{\partial R_\alpha \partial R_\gamma} \tag{8.19}$$

Since the free energy $F(\mathbf{R})$ can be obtained by solving the 3D-RISM/RISM equation, Eq. 8.19 provides a way to evaluate the variance-covariance matrix.

The variance-covariance matrix is by itself quite informative for characterizing the structural fluctuation of protein around its native state in atomic detail. As an example, let us consider a hinge-bending motion of protein. In this case, the variance-covariance matrix is expected to have a structure in which a block of elements $\langle \Delta\mathbf{R}_\alpha \Delta\mathbf{R}_\beta \rangle$ for atom pairs, a, b, belonging to the two sides of the hinge-axis have the negative sign, because the direction of the displacements $\Delta\mathbf{R}_\alpha$ and $\Delta\mathbf{R}_\beta$ is opposite.

Usefulness of the variance-covariance matrix is not only to characterize the structural fluctuation around an *equilibrium* state. It can be applied to a relaxation process from an equilibrium conformation to the other induced by small perturbation, such as ligand binding.

Equation 8.19 implies that the free energy of protein at an equilibrium conformation takes the form,

$$F(\{\Delta \mathbf{R}\}) = \frac{1}{2} k_B T \sum_{\alpha,\beta} \Delta \mathbf{R}_\alpha \cdot \left(L^{-1}\right)_{\alpha,\beta} \cdot \Delta \mathbf{R}_\beta \tag{8.20}$$

In the presence of a small perturbation due to, say, ligand binding, the free energy can be changed due to the perturbation as,

$$F(\{\Delta \mathbf{R}\}) = \frac{1}{2} k_B T \sum_{\alpha,\beta} \Delta \mathbf{R}_\alpha \cdot \left(L^{-1}\right)_{\alpha,\beta} \cdot \Delta \mathbf{R}_\beta - \sum_{\alpha} \Delta \mathbf{R}_\alpha \cdot \mathbf{f}_\alpha \tag{8.21}$$

where $\mathbf{f}_\alpha$ is the perturbation on a protein atom. Then, the conformational change due to the perturbation can be determined by the variational principle,

$$\frac{\partial F(\{\Delta \mathbf{R}\})}{\partial \Delta \mathbf{R}_\alpha} = 0 \tag{8.22}$$

With Eq. 8.21, this gives

$$\Delta \mathbf{R}_\alpha = (k_B T)^{-1} \sum \left\langle \Delta \mathbf{R}_\alpha \Delta \mathbf{R}_\beta \right\rangle \cdot \mathbf{f}_\beta \tag{8.23}$$

Therefore, Eq. 8.19 combined with 8.23 provides a theoretical basis for analyzing the conformational relaxation of protein in water due to a perturbation such as ligand binding. Equation 8.23 is first derived by Ikeguchi et al. based on the linear response theory [41].

One possible and illustrative application of the theory is the pH induced gating of an ion channel. Such a channel controls the ion permeation by opening and closing the gate which are composed of amino acid residues, the protonated state of which changes with pH of solution [22]. The perturbation $\mathbf{f}_\beta$ in this case is the electrostatic repulsions among the residues, which close or open the gate depending on the state of protonation. We first calculate the variance-covariance matrix $\langle \Delta \mathbf{R}_\alpha \Delta \mathbf{R}_\beta \rangle$ for the nonperturbed protein or closed state in which no residues are protonated. Then, using Eq. 8.23, we evaluate the structural change of protein due to the perturbation originated from the Coulomb repulsion among charged residues. To find out if the structural change is large enough to accommodate substrate ions inside the pore, we apply the 3D-RISM method to the perturbed structure to calculate the ion distribution inside the channel.

8.5 Concluding Remarks

In this chapter, we presented a theoretical treatment on the molecular recognition, one of the elementary processes of life phenomena, based on the 3D-RISM/RISM theory. For a protein having a rigid structure, the theory has been applied successfully to a variety of molecular recognition processes. It has proved its ability to discriminate molecules having different charges and structures, which is an essential requirement for the in silico drug design. The program has now been available in several application packages in the field of molecular design, including the Molecular Design Frontier (MDF) Suite. The method has been appreciated as a practical tool in the research-and-development site of the intelligent pharmaceutical design, not just as a basic science.

The structural fluctuation plays a crucial role in the process where a protein expresses its function. We presented here a new method to characterize the structural fluctuation of protein around its native state, combining the 3D-RISM/RISM theory with the generalized Langevin theory. The theory can be formulated by defining a projection operator in the phase space, and by projecting or coarse-graining all the degrees of freedom onto the protein structure and solvent density. Due to the projection, the structural fluctuation and its dynamics on its *energy surface* are transformed to those on the *free energy surface*. The variance and covariance matrix characterizing the fluctuation is identified as the second derivative of the free energy surface with respect to the atomic coordinate of the protein. The free energy surface of protein and its first and second derivatives can be calculated by the 3D-RISM/RISM theory. This means that the fluctuation and dynamics of protein can be also characterized by the 3D-RISM/RISM theory. An application of the theory to actual protein is in progress in our group. The theory is expected to be a powerful tool to explore phenomena which are concerned with structural fluctuation of protein, including "induced fitting" and "conformational selection."

References

1. Watson JD et al (eds) (1987) Molecular biology of the gene. The Benjamin/Cummings Publishing, San Francisco
2. Michaelis L, Menten M (1913) Die Kinetik der Invertinwirkung. Biochem Zeitung 49:333
3. Hirata F (ed) (2003) Molecular theory of solvation. Kluwers
4. Imai T, Hiraoka R, Kovalenko A, Hirata F (2005) Water molecules in a protein cavity detected by a statistical-mechanical theory. J Am Chem Soc 127:5334
5. Hirata F (2012) Computational chemistry of macromolecular system. Chapter 7, Kagakudojin, Kyoto
6. Wilson KP, Malcolm BA, Matthews BW (1992) Structural and thermodynamic analysis of compensating mutations within the core of chicken egg white lysozyme. J Biol Chem 267: 10842

7. Imai T, Hiraoka R, Kovalenko A, Hirata F (2007) Locating missing water molecules in protein cavities by the three-dimensional reference interaction site model theory of molecular solvation. Proteins 66:804
8. Evans J (2005) Probing for water in protein cavities. Chem World (On line), October
9. Yoshida N, Phongphanphanee S, Maruyama Y, Imai T, Hirata F (2006) Selective ion-binding by protein probed with the 3D-RISM theory. J Am Chem Soc Commun 128:12042
10. Yoshida N, Phongphanphanee S, Hirata F (2007) Selective ion-binding by protein probed with the statistical mechanical integral equation theory. J Phys Chem B 111:4588
11. Imai T, Ohyama S, Kovalenko A, Hirata F (2007) Theoretical study of the partial molar volume change associated with pressure-induced structural transition of ubiquitin. Protein Sci 16:1927
12. Imai T, Hiraoka R, Seto T, Kovalenko A, Hirata F (2007) Three-dimensional distribution function theory for the prediction of protein-ligand binding sites and affinities: application to the binding of noble gases to hen egg-ehite lysozyme in aqueous solution. J Phys Chem B 111: 11585
13. Phonphanphanee S, Yoshida N, Hirata F (2007) A statistical mechanics study on equilibrium water distribution in aquaporin. Chem Phys Lett 449:196
14. Phongphanphanee S, Yoshida N, Hirata F (2008) On the proton exclusion of aquaporins: a statistical mechanics study. J Am Chem Soc 130:1540
15. Ikuta Y, Karita S, Kitago Y, Watanabe N, Hirata F (2008) Detecting reaction intermediate in enzymatic reaction of Cel44A, GH family 44 endoglucanase with 3D-RISM theory. Chem Phys Lett 465:279
16. Yonetani Y, Maruyama Y, Hirata F, Kono H (2008) Comparison of DNA hydration patterns obtained using two distinct computational methods, molecular dynamics simulation and three-dimensional reference interaction site model theory. J Chem Phys 128:185102
17. Yoshida N, Imai T, Phongphanphanee S, Kovalenko A, Hirata F (2009) Molecular recognition in biomolecules studied by statistical-mechanical integral-equation theory of liquids. J Phys Chem B 113:873
18. Kiyota Y, Hiraoka R, Yoshida N, Maruyama Y, Imai T, Hirata F (2009) Theoretical study of CO escaping pathway in myoglobin with the 3D-RISM theory. J Am Chem Soc (Commun) 131:3852
19. Imai T, Oda K, Kovalenko A, Hirata F, Kidera A (2009) Ligand mapping on protein surfaces by the 3D-RISM theory: toward computational fragment-based drug design. J Am Chem Soc 131:12430
20. Maruyama Y, Yoshida N, Hirata F (2010) Revisiting the salt-induced conformational change of DNA with 3D-RISM theory. J Phys Chem B 114:6464
21. Phongphanphanee S, Yoshida N, Hirata F (2010) Molecular selectivity in aquaporin channels studied by the 3D-RISM theory. J Phys Chem B 114:7967
22. Phongphanphanee S, Rungrotmongkol T, Yoshida N, Hannongbua S, Hirata F (2010) Proton transport through the influenza A M2 channel: 3D-RISM study. J Am Chem Soc 132:9782
23. Maruyama Y, Matsushita T, Ueoka R, Hirata F (2011) Solvent and salt effects on structural stability of human telomere. J Phys Chem B 115:2408
24. Kiyota Y, Yoshida N, Hirata F (2011) A new approach for investigating the molecular recognition of protein: toward structural-based drug design based on the 3D-RISM theory. J Chem Theory Comput 7:3803
25. Sindhikara D, Yoshida N, Hirata F (2012) Placevent: an algorithm for predicting of explicit solvent atom distribution – application to HIV-1 protease and F-ATP synthase. J Comput Chem 33:1536
26. Herzberg O, James MN (1985) Structure of the calcium regulatory muscle protein troponin-C at 2.8 a resolution. Nature 313:653
27. Ikura M, Clore GM, Gronenborn AM, Zhu G, Klee CB, Bax A (1992) Solution structure of a calmodulin-target peptide complex by multidimensional NMR. Science 256:632
28. Hille B (2001) Ionic channels of excitable membranes. Sinauer Associates, Shunderland

29. Tsuda S, Ogura K, Hasegawa Y, Yagi K, Kikuchi K (1990) H NMR study of rabbit skeletal muscle troponin C: Mg2+−induced. Conformational change? Biochemistry 29:495
30. Kuroki R, Yutani K (1998) Structural and thermodynamic responses of mutations at a Ca2+ binding site engineered into human lysozyme. J Biol Chem 273:34310
31. Dennis EA (1994) Diversity of group types, regulation, and function of phospholipase A2. J Biol Chem 269:13057
32. Nicolas JP, Lin Y, Lambeau G, Ghomashchi F, Lazdunski M, Gelb MH (1997) Localization of structural elements of bee venom phospholipase A2 involved in N- type receptor binding and neurotoxicity. J Biol Chem 272:7173
33. Argiolas A, Pisano JJJ (1983) Isolation and charaterization of 2 two new peptides, mastoparan-C and crabrolin, from the venom of the European Hornet, Vespa crabro. Biol Chem 258:13697
34. Chang DT, Oyang Y, Lin J (2005) MEDock: a web server for efficient prediction of ligand binding sites based on a novel optimization algorithm. Nucleic Acids Res 33:W233
35. Kim B, Hirata F (2013) Structural fluctuation of protein in water around its native state: a new statistical mechanics formulation. J Chem Phys 138:054108
36. Cui Q, Bahar I (2006) Normal mode analysis-theory and applications to biological and chemical systems. Chapman & Hall/CRC, New York
37. Ermak DL, McCammon JA (1978) Brownian dynamics with hydrodynamic interactions. J Chem Phys 69:1352
38. Lamm G, Szabo A (1986) Langevin modes of macromolecules. J Chem Phys 85:7334
39. Mori H (1965) Transport, collective motion, and Brownian motion. Prog Theor Phys 33:423
40. Kinoshita M, Okamoto Y, Hirata F, Am J (1998) First-principle determination of peptide conformations in solvents: combination of Monte Carlo simulated annealing and RISM theory. Chem Soc 120(8):1855–1863
41. Ikeguchi M, Ueno J, Sato M, Kidera A (2005) Protein structural change upon ligand binding: linear response theory. Phys Rev Lett 94:078102

Chapter 9
Structural Fluctuations of Proteins in Folding and Ligand Docking Studied by Replica-Exchange Simulations

Yuko Okamoto

Abstract In biomolecular systems with many degrees of freedom such as proteins and nucleic acids, there exists an astronomically large number of local-minimum free energy states. Conventional simulations in the canonical ensemble encounter with great difficulty, because they tend to get trapped in states of these local minima. Enhanced conformational sampling techniques are thus in great demand. A simulation in generalized ensemble performs a random walk in potential energy, volume, and other physical quantities or their corresponding conjugate parameters such as temperature, pressure, etc. and can overcome this difficulty. From only one simulation run, one can obtain canonical ensemble averages of physical quantities as functions of temperature, pressure, etc. by the reweighting techniques. In this chapter, we review uses of the generalized-ensemble algorithms in biomolecular systems. A well-known method, namely, replica-exchange method, is described first. We then present various extensions of the replica-exchange method. The effectiveness of the methods is tested with protein folding and ligand docking simulations.

Keywords Generalized-ensemble algorithm • Replica-exchange method • Replica-exchange umbrella sampling • Reweighting • Molecular simulations • Free energy calculations • Protein folding • Ligand docking

---

Y. Okamoto (✉)
Department of Physics, Graduate School of Science, Nagoya University, Nagoya, Aichi 464-8602, Japan

Structural Biology Research Center, Graduate School of Science, Nagoya University, Nagoya, Aichi 464-8602, Japan

Center for Computational Science, Graduate School of Engineering, Nagoya University, Nagoya, Aichi 464-8603, Japan

Information Technology Center, Nagoya University, Nagoya, Aichi 464-8601, Japan
e-mail: okamoto@phys.nagoya-u.ac.jp

M. Terazima et al. (eds.), *Molecular Science of Fluctuations Toward Biological Functions*, DOI 10.1007/978-4-431-55840-8_9

9.1 Introduction

Many biomolecules are thermally fluctuating within a living cell. Structural fluctuations are often essential for biological functions of biomolecules. However, the amount of fluctuations should be just right for the correct functions. In order to study biomolecules by computer simulations, Monte Carlo (MC) and molecular dynamics (MD) methods in canonical ensemble based on Boltzmann weight factor are usually employed. Because there exist a huge number of local-minimum energy states in complex systems such as biomolecules, conventional MC and MD simulations tend to get trapped in these local-minimum states, giving wrong results. Creating accurate fluctuations of biomolecules is even more difficult. A class of simulation methods, which are referred to as the *generalized-ensemble algorithms*, overcome this difficulty (for reviews see, e.g., Refs. [1–7]). In the generalized-ensemble algorithm, each state is weighted by an artificial, non-Boltzmann probability weight factor so that random walks in potential energy, volume, and other physical quantities or their corresponding conjugate parameters such as temperature, pressure, etc. may be realized. The random walks allow the simulation to escape from any energy barrier and to sample much wider conformational space than by conventional methods. After a single simulation run, we can calculate physical quantities as functions of temperature and other parameters by the reweighting techniques.

One of popular generalized-ensemble algorithms for molecular simulations is the *replica-exchange method* (REM) [8] (the method is also referred to as *parallel tempering* [9]). In this method, a number of noninteracting copies (or replicas) of the original system at different temperatures are simulated independently and exchanged with a specified transition probability. The details of molecular dynamics algorithm for REM, which is referred to as the *replica-exchange molecular dynamics* (REMD), have been worked out in Ref. [10], and this led to a wide application of REMD in the protein and other biomolecular systems.

One is naturally led to a multidimensional (or multivariable) extension of REM, which we refer to as the *multidimensional replica-exchange method* (MREM) [11], which is also referred to as *Hamiltonian replica-exchange method* [12]. A special realization of MREM is *replica-exchange umbrella sampling* (REUS) [11], which combines the conventional umbrella sampling method [13] and REM, and it is particularly useful in free energy calculations for pharmaceutical design. General formulations for multidimensional generalized-ensemble algorithms have also been worked out [14, 15], and special versions for isobaric-isothermal ensemble have been developed [3, 16–18].

In this chapter, we describe the generalized-ensemble algorithms mentioned above. Namely, we review the three methods: REM, MREM, and REUS. Examples of the results in which these methods were applied to the predictions of membrane protein structures and protein-ligand binding structures are then presented.

9.2 Methods

9.2.1 Replica-Exchange Method

Let us consider a system of N atoms of mass m_k ($k = 1,\ldots, N$) with their coordinate vectors and momentum vectors denoted by $q = (\boldsymbol{q}_1,\ldots, \boldsymbol{q}_N)$ and $p = (\boldsymbol{p}_1, \ldots, \boldsymbol{p}_N)$, respectively. The Hamiltonian $H(q,p)$ of the system is the sum of the kinetic energy $K(p)$ and the potential energy $E(q)$:

$$H(q,p) = K(p) + E(q), \tag{9.1}$$

where

$$K(p) = \sum_{k=1}^{N} \frac{\boldsymbol{p}_k{}^2}{2m_k}. \tag{9.2}$$

In the canonical ensemble at temperature T, each state with the Hamiltonian $H(q,p)$ is weighted by the Boltzmann factor:

$$W_{\mathrm{B}}(x;T) = \exp(-\beta H(q,p)), \tag{9.3}$$

where the inverse temperature β is defined by $\beta = 1/k_{\mathrm{B}}T$ (k_{B} is the Boltzmann constant). The average kinetic energy at temperature T is then given by:

$$\langle K(p) \rangle_T = \left\langle \sum_{k=1}^{N} \frac{\boldsymbol{p}_k{}^2}{2m_k} \right\rangle_T = \frac{3}{2} N k_{\mathrm{B}} T. \tag{9.4}$$

Because the coordinates q and momenta p are decoupled in Eq. (9.1), we can suppress the kinetic energy part and can write the Boltzmann factor as

$$W_{\mathrm{B}}(x;T) = W_{\mathrm{B}}(E;T) = \exp(-\beta E). \tag{9.5}$$

The canonical probability distribution of potential energy $P_{\mathrm{NVT}}(E;T)$ is then given by the product of the density of states $n(E)$ and the Boltzmann weight factor $W_{\mathrm{B}}(E;T)$:

$$P_{\mathrm{NVT}}(E;T) \propto n(E) W_{\mathrm{B}}(E;T). \tag{9.6}$$

The conventional MC or MD simulations at constant temperature are expected to yield $P_{\mathrm{NVT}}(E;T)$. An MC simulation based on the Metropolis algorithm [19] is performed with the following transition probability from a state x of potential energy E to a state x' of potential energy E':

$$w(x \to x') = \min\left(1, \frac{W_B(E';T)}{W_B(E;T)}\right) = \min(1, \exp(-\beta \Delta E)), \tag{9.7}$$

where

$$\Delta E = E' - E. \tag{9.8}$$

However, in practice, it is very difficult to obtain accurate canonical distributions of complex systems at low temperatures by conventional MC or MD simulation methods. This is because simulations at low temperatures tend to get trapped in one or a few of local-minimum energy states. This difficulty is overcome by, for instance, the generalized-ensemble algorithms, which greatly enhance conformational sampling.

The *replica-exchange method* (REM) is one of effective generalized-ensemble algorithms. The system for REM consists of M *noninteracting* copies (or replicas) of the original system in the canonical ensemble at M different temperatures $T_m (m = 1, \ldots, M)$. We arrange the replicas so that there is always exactly one replica at each temperature. Then there exists a one-to-one correspondence between replicas and temperatures; the label $i(= 1, \ldots, M)$ for replicas is a permutation of the label $m(= 1, \ldots, M)$ for temperatures, and vice versa:

$$\begin{cases} i = i(m) \equiv f(m), \\ m = m(i) \equiv f^{-1}(i), \end{cases} \tag{9.9}$$

where $f(m)$ is a permutation function of m and $f^{-1}(i)$ is its inverse.

Let $X = \left\{x_1^{[i(1)]}, \ldots, x_M^{[i(M)]}\right\} = \left\{x_{m(1)}^{[1]}, \ldots, x_{m(M)}^{[M]}\right\}$ stand for a "state" in this generalized ensemble. Each "substate" $x_m^{[i]}$ is specified by the coordinates $q^{[i]}$ and momenta $p^{[i]}$ of N atoms in replica i at temperature T_m:

$$x_m^{[i]} \equiv \left(q^{[i]}, p^{[i]}\right)_m. \tag{9.10}$$

Because the replicas are noninteracting, the weight factor for the state X in this generalized ensemble is given by the product of Boltzmann factors for each replica (or at each temperature):

$$\begin{aligned} W_{\text{REM}}(X) &= \prod_{i=1}^{M} \exp\left\{-\beta_{m(i)} H\left(q^{[i]}, p^{[i]}\right)\right\} \\ &= \prod_{m=1}^{M} \exp\left\{-\beta_m H\left(q^{[i(m)]}, p^{[i(m)]}\right)\right\}, \end{aligned} \tag{9.11}$$

where $i(m)$ and $m(i)$ are the permutation functions in Eq. (9.9).

We now consider exchanging a pair of replicas in this ensemble. Suppose we exchange replicas i and j which are at temperatures T_m and T_n, respectively:

$$X = \left\{\ldots, x_m^{[i]}, \ldots, x_n^{[j]}, \ldots\right\} \to X' = \left\{\ldots, x_m^{[j]\prime}, \ldots, x_n^{[i]\prime}, \ldots\right\}. \tag{9.12}$$

The exchange of replicas can be written in more detail as

$$\begin{cases} x_m^{[i]} \equiv \left(q^{[i]}, p^{[i]}\right)_m \to x_m^{[j]\prime} \equiv \left(q^{[j]}, p^{[j]\prime}\right)_m, \\ x_n^{[j]} \equiv \left(q^{[j]}, p^{[j]}\right)_n \to x_n^{[i]\prime} \equiv \left(q^{[i]}, p^{[i]\prime}\right)_n, \end{cases} \tag{9.13}$$

where the definitions for $p^{[i]\prime}$ and $p^{[j]\prime}$ will be given below.

In the original implementation of REM [8], Monte Carlo algorithm was used, and only the coordinates q (and the potential energy function $E(q)$) had to be taken into account. In molecular dynamics algorithm, on the other hand, we also have to deal with the momenta p. We proposed the following momentum assignment in Eq. (9.13) [10]:

$$\begin{cases} p^{[i]\prime} \equiv \sqrt{\dfrac{T_n}{T_m}} p^{[i]}, \\ p^{[j]\prime} \equiv \sqrt{\dfrac{T_m}{T_n}} p^{[j]}, \end{cases} \tag{9.14}$$

which we believe is the simplest and the most natural. This assignment means that we just rescale uniformly the velocities of all the atoms in the replicas by the square root of the ratio of the two temperatures so that the temperature condition in Eq. (9.4) may be satisfied immediately after replica exchange is accepted. We remark that similar momentum rescaling formulae for various constant temperature algorithms have been worked out in Ref. [20].

The transition probability of this replica-exchange process is given by the usual Metropolis criterion:

$$w(X \to X') \equiv w\left(x_m^{[i]} \middle| x_n^{[j]}\right) = \min\left(1, \frac{W_{\text{REM}}(X')}{W_{\text{REM}}(X)}\right) = \min(1, \exp(-\Delta)), \tag{9.15}$$

where in the second expression (i.e., $w\left(x_m^{[i]} \middle| x_n^{[j]}\right)$), we explicitly wrote the pair of replicas (and temperatures) to be exchanged. From Eqs. (9.1), (9.2), (9.11), and (9.14), we have

$$\Delta = \beta_m\left(E\left(q^{[j]}\right) - E\left(q^{[i]}\right)\right) - \beta_n\left(E\left(q^{[j]}\right) - E\left(q^{[i]}\right)\right) \tag{9.16}$$

$$= (\beta_m - \beta_n)\left(E\left(q^{[j]}\right) - E\left(q^{[i]}\right)\right). \tag{9.17}$$

Note that after introducing the momentum rescaling in Eq. (9.14), we have the same Metropolis criterion for replica exchanges, i.e., Eqs. (9.15) and (9.17), for both MC and MD versions.

Without loss of generality we can assume that $T_1 < T_2 < \ldots < T_M$. The lowest temperature T_1 should be sufficiently low so that the simulation can explore the experimentally relevant temperature region, and the highest temperature T_M should be sufficiently high so that no trapping in an energy-local-minimum state occurs. An REM simulation is then realized by alternately performing the following two steps:

1. Each replica in canonical ensemble of the fixed temperature is simulated *simultaneously* and *independently* for a certain MC or MD steps.
2. A pair of replicas at neighboring temperatures, say, $x_m^{[i]}$ and $x_{m+1}^{[j]}$, are exchanged with the probability $w\left(x_m^{[i]}|x_{m+1}^{[j]}\right)$ in Eq. (9.15).

A random walk in "temperature space" is realized for each replica, which in turn induces a random walk in potential energy space. This alleviates the problem of getting trapped in states of energy local minima.

After a long production run of a replica-exchange simulation, the canonical expectation value of a physical quantity A at temperature $T_m(m = 1, \ldots, M)$ can be calculated by the usual arithmetic mean:

$$\langle A \rangle_{T_m} = \frac{1}{n_m} \sum_{k=1}^{n_m} A(x_m(k)), \tag{9.18}$$

where $x_m(k)(k = 1, \ldots, n_m)$ are the configurations obtained at temperature T_m and n_m is the total number of measurements made at $T = T_m$. The expectation value at any intermediate temperature T $(=1/k_B\beta)$ can also be obtained as follows:

$$\langle A \rangle_T = \frac{\sum_E A(E) P_{\rm NVT}(E;T)}{\sum_E P_{\rm NVT}(E;T)} = \frac{\sum_E A(E) n(E) \exp(-\beta E)}{\sum_E n(E) \exp(-\beta E)}. \tag{9.19}$$

The density of states $n(E)$ in Eq. (9.19) is given by the multiple-histogram reweighting techniques or the weighted histogram analysis method (WHAM) [21, 22] (see also Ref. [23]) as follows. Let $N_m(E)$ and n_m be, respectively, the potential energy histogram and the total number of samples obtained at temperature $T_m = 1/k_B\beta_m(m = 1, \ldots, M)$. The best estimate of the density of states is then given by:

$$n(E) = \frac{\sum_{m=1}^{M} N_m(E)}{\sum_{m=1}^{M} n_m \exp(f_m - \beta_m E)}, \tag{9.20}$$

where we have for each $m(= 1, \ldots, M)$

$$\exp(-f_m) = \sum_E n(E)\exp(-\beta_m E). \tag{9.21}$$

Note that Eqs. (9.20) and (9.21) are solved self-consistently by iteration [21, 22] to obtain the density of states $n(E)$ and the dimensionless Helmholtz free energy f_m. Namely, we can set all the $f_m (m = 1, \ldots, M)$ to, e.g., zero initially. We then use Eq. (9.20) to obtain $n(E)$, which is substituted into Eq. (9.21) to obtain next values of f_m, and so on.

Moreover, the ensemble averages of any physical quantity A (including those that cannot be expressed as functions of potential energy) at any temperature $T (=1/k_B\beta)$ can now be obtained from the "trajectory" of configurations of the production run. Namely, we first obtain $f_m (m = 1, \ldots, M)$ by solving Eqs. (9.20) and (9.21) self-consistently, and then we have [23]

$$\langle A \rangle_T = \frac{\sum_{m=1}^{M} \sum_{k=1}^{n_m} A(x_m(k)) \frac{1}{\sum_{l=1}^{M} n_l \exp(f_l - \beta_l E(x_m(k)))} \exp(-\beta E(x_m(k)))}{\sum_{m=1}^{M} \sum_{k=1}^{n_m} \frac{1}{\sum_{l=1}^{M} n_l \exp(f_l - \beta_l E(x_m(k)))} \exp(-\beta E(x_m(k)))}, \tag{9.22}$$

where $x_m(k) (k = 1, \ldots, n_m)$ are the configurations obtained at temperature T_m.

9.2.2 *Multidimensional Replica-Exchange Method*

We now present the multidimensional/multivariable extension of the replica-exchange method, which we refer to as the *multidimensional replica-exchange method* (MREM) [11, 14]. Let us consider a generalized potential energy function $E_{\boldsymbol{\lambda}}(x)$, which depends on L parameters $\boldsymbol{\lambda} = \left(\lambda^{(1)}, \ldots, \lambda^{(L)}\right)$, of a system in state x. Although $E_{\boldsymbol{\lambda}}(x)$ can be any function of $\boldsymbol{\lambda}$, we consider the following specific generalized potential energy function for simplicity:

$$E_{\lambda}(x) = E_0(x) + \sum_{l=1}^{L} \lambda^{(l)} V_l(x). \tag{9.23}$$

Here, there are $L+1$ energy terms, $E_0(x)$ and $V_l(x) (l = 1, \ldots, L)$, and $\lambda^{(l)}$ are the corresponding coupling constants for $V_l(x)$ (we collectively write $\boldsymbol{\lambda} = \left(\lambda^{(1)}, \ldots, \lambda^{(L)}\right)$).

The crucial observation that led to MREM is that as long as we have M *noninteracting* replicas of the original system, the Hamiltonian $H(q,p)$ of the system does not have to be identical among the replicas, and it can depend on a parameter with different parameter values for different replicas [11]. The system for MREM consists of M noninteracting replicas of the original system in the "canonical ensemble" with $M(= M_0 \times M_1 \times \ldots \times M_L)$ different parameter sets $\boldsymbol{\Lambda}_m (m = 1, \ldots, M)$, where $\boldsymbol{\Lambda}_m \equiv (T_{m_0}, \boldsymbol{\lambda}_m) \equiv \left(T_{m_0}, \lambda^{(1)}_{m_1}, \ldots, \lambda^{(L)}_{m_L}\right)$ with $m_0 = 1, \ldots, M_0, m_l = 1, \ldots, M_l (l = 1, \ldots, L)$. Because the replicas are noninteracting, the weight factor is given by the product of Boltzmann-like factors for each replica:

$$W_{\mathrm{MREM}} \equiv \prod_{m_0=1}^{M_0} \prod_{m_1=1}^{M_1} \cdots \prod_{m_L=1}^{M_L} \exp\left(-\beta_{m_0} E_{\boldsymbol{\lambda}_m}\right). \tag{9.24}$$

Without loss of generality, we can order the parameters so that $T_1 < T_2 < \ldots < T_{M_0}$ and $\lambda^{(l)}_1 < \lambda^{(l)}_2 < \ldots < \lambda^{(l)}_{M_l}$ (for each $l = 1, \ldots, L$). An MREM simulation is realized by alternately performing the following two steps:

1. For each replica, a "canonical" MC or MD simulation at the fixed parameter set is carried out *simultaneously* and *independently* for certain steps.
2. We exchange a pair of replicas i and j which are at the parameter sets $\boldsymbol{\Lambda}_m$ and $\boldsymbol{\Lambda}_{m+1}$, respectively. The transition probability for this replica-exchange process is given by

$$w(\boldsymbol{\Lambda}_m \leftrightarrow \boldsymbol{\Lambda}_{m+1}) = \min(1, \exp(-\Delta)), \tag{9.25}$$

where we have

$$\Delta = \left(\beta_{m_0} - \beta_{m_0+1}\right)\left(E_{\boldsymbol{\lambda}_m}\left(q^{[j]}\right) - E_{\boldsymbol{\lambda}_m}\left(q^{[i]}\right)\right), \tag{9.26}$$

for T-exchange and

$$\Delta = \beta_{m_0}\left[\left(E_{\boldsymbol{\lambda}_{m_l}}\left(q^{[j]}\right) - E_{\boldsymbol{\lambda}_{m_l}}\left(q^{[i]}\right)\right) - \left(E_{\boldsymbol{\lambda}_{m_l+1}}\left(q^{[j]}\right) - E_{\boldsymbol{\lambda}_{m_l+1}}\left(q^{[i]}\right)\right)\right], \tag{9.27}$$

for $\lambda^{(l)}$ exchange (for one of $l = 1, \ldots, L$). Here, $q^{[i]}$ and $q^{[j]}$ stand for coordinate vectors for replicas i and j, respectively, before the replica exchange.

Suppose we have made a single run of a short MREM simulation with $M(= M_0 \times M_1 \times \ldots \times M_L)$ replicas that correspond to M different parameter sets $\boldsymbol{\Lambda}_m(m = 1, \ldots, M)$, let $N_{m_0,m_1,\ldots,m_L}(E_0, V_1, \ldots, V_L)$ and $n_{m_0,m_1,\ldots,m_L}$ be, respectively, the $(L+1)$ dimensional potential energy histogram and the total number of samples obtained for the m-th parameter set $\boldsymbol{\Lambda}_m = \left(T_{m_0}, \lambda^{(1)}_{m_1}, \ldots, \lambda^{(L)}_{m_L}\right)$. The generalized WHAM equations are then given by [14]

$$n(E_0, V_1, \ldots, V_L) = \frac{\sum\limits_{m_0, m_1, \ldots, m_L} N_{m_0,m_1,\ldots,m_L}(E_0, V_1, \ldots, V_L)}{\sum\limits_{m_0, m_1, \ldots, m_L} n_{m_0,m_1,\ldots,m_L} \exp\left(f_{m_0,m_1,\ldots,m_L} - \beta_{m_0} E_{\boldsymbol{\lambda}_m}\right)}, \tag{9.28}$$

and

$$\exp\left(-f_{m_0,m_1,\ldots,m_L}\right) = \sum_{E_0, V_1, \ldots, V_L} n(E_0, V_1, \ldots, V_L) \exp\left(-\beta_{m_0} E_{\boldsymbol{\lambda}_m}\right). \tag{9.29}$$

The density of states $n(E_0, V_1, \ldots, V_L)$ and the dimensionless free energy $f_{m_0,m_1,\ldots,m_L}$ are obtained by solving Eqs. (9.28) and (9.29) self-consistently by iteration.

We now present the equations for calculating ensemble averages of physical quantities with any temperature T and any parameter $\boldsymbol{\lambda}$ values. After a long production run of MREM simulations, the canonical expectation value of a physical quantity A with the parameter values $\boldsymbol{\Lambda}_m(m = 1, \ldots, M)$, where $\boldsymbol{\Lambda}_m \equiv (T_{m_0}, \boldsymbol{\lambda}_m) \equiv \left(T_{m_0}, \lambda^{(1)}_{m_1}, \ldots, \lambda^{(L)}_{m_L}\right)$ with $m_0 = 1, \ldots, M_0, m_l = 1, \ldots, M_l (l = 1, \ldots, L)$, and $M(= M_0 \times M_1 \times \ldots \times M_L)$, can be calculated by the usual arithmetic mean:

$$\langle A \rangle_{T_{m_0}, \boldsymbol{\lambda}_m} = \frac{1}{n_m} \sum_{k=1}^{n_m} A(x_m(k)), \tag{9.30}$$

where $x_m(k)(k = 1, \ldots, n_m)$ are the configurations obtained with the parameter values $\boldsymbol{\Lambda}_m(m = 1, \ldots, M)$ and $n_m(= n_{m_0,m_1,\ldots,m_L})$ is the total number of measurements made with these parameter values. The expectation values of A at any intermediate T $(=1/k_{\rm B}\beta)$ and any $\boldsymbol{\lambda}$ can also be obtained from

$$\langle A \rangle_{T,\boldsymbol{\lambda}} = \frac{\sum\limits_{E_0, V_1, \ldots, V_L} A(E_0, V_1, \ldots, V_L) n(E_0, V_1, \ldots, V_L) \exp(-\beta E_{\boldsymbol{\lambda}})}{\sum\limits_{E_0, V_1, \ldots, V_L} n(E_0, V_1, \ldots, V_L) \exp(-\beta E_{\boldsymbol{\lambda}})}, \tag{9.31}$$

where the density of states $n(E_0, V_1, \ldots, V_L)$ is obtained from the WHAM equations in Eqs. (9.28) and (9.29).

Moreover, the ensemble average of the physical quantity A (including those that cannot be expressed as functions of E_0 and $V_l(l = 1, \ldots, L)$) can be obtained from the "trajectory" of configurations of the production run [23]. Namely, we first obtain $f_{m_0, m_1, \ldots, m_L}$ for each $m_0 = 1, \ldots, M_0, m_l = 1, \ldots, M_l(l = 1, \ldots, L)$ by solving Eqs. (9.28) and (9.29) self-consistently, and then we have [14]

$$\langle A \rangle_{T,\boldsymbol{\lambda}} = \frac{\displaystyle\sum_{m_0=1}^{M_0} \cdots \sum_{m_L=1}^{M_L} \sum_{x_m} A(x_m) \frac{\exp(-\beta E_{\boldsymbol{\lambda}}(x_m))}{\displaystyle\sum_{m_0=1}^{M_0} \cdots \sum_{m_L=1}^{M_L} n_{n_0, \ldots, n_L} \exp\left(f_{n_0, \ldots, n_L} - \beta_{n_0} E_{\boldsymbol{\lambda}_n}(x_m)\right)}}{\displaystyle\sum_{m_0=1}^{M_0} \cdots \sum_{m_L=1}^{M_L} \sum_{x_m} \frac{\exp(-\beta E_{\boldsymbol{\lambda}}(x_m))}{\displaystyle\sum_{m_0=1}^{M_0} \cdots \sum_{m_L=1}^{M_L} n_{n_0, \ldots, n_L} \exp\left(f_{n_0, \ldots, n_L} - \beta_{n_0} E_{\boldsymbol{\lambda}_n}(x_m)\right)}}, \tag{9.32}$$

where x_m are the configurations obtained at $\boldsymbol{\Lambda}_m \equiv (T_{m_0}, \boldsymbol{\lambda}_m) \equiv \left(T_{m_0}, \lambda_{m_1}^{(1)}, \ldots, \lambda_{m_L}^{(L)}\right)$. Here, the trajectories x_m are stored for each $\boldsymbol{\Lambda}_m$ separately.

We now describe a free energy calculation method based on MREM, which we refer to as *replica-exchange umbrella sampling* (REUS) [11]. In Eq. (9.23), we consider that $E_0(q)$ is the original unbiased potential and that $V_l(q)(l = 1, \ldots, L)$ are the biasing (umbrella) potentials with $\lambda^{(l)}$ being the corresponding coupling constants $\left(\boldsymbol{\lambda} = \left(\lambda^{(1)}, \ldots, \lambda^{(L)}\right)\right)$. Introducing a "reaction coordinate" ξ, the umbrella potentials are usually written as harmonic restraints:

$$V_l(q) = k_l(\xi(q) - d_l)^2, (l = 1, \ldots, L), \tag{9.33}$$

where d_l are the midpoints and k_l are the strengths of the restraining potentials. We prepare M replicas with M different values of the parameters $\boldsymbol{\Lambda}_m = (T_{m_0}, \boldsymbol{\lambda}_m)$, and the replica-exchange simulation is performed. Because the umbrella potentials $V_l(q)$ in Eq. (9.33) are all functions of the reaction coordinate ξ only, we can take the histogram $N_m(E_0, \xi)$ instead of $N_m(E_0, V_1, \ldots, V_L)$. The WHAM equations can then be written:

$$P_{T,\boldsymbol{\lambda}}(E_0, \xi) = \left[\frac{\displaystyle\sum_{m=1}^{M} N_m(E_0, \xi)}{\displaystyle\sum_{m=1}^{M} n_m \exp\left(f_m - \beta_{m_0} E_{\boldsymbol{\lambda}_m}\right)}\right] \exp(-\beta E_{\boldsymbol{\lambda}}), \tag{9.34}$$

and

$$\exp(-f_m) = \sum_{E_0,\xi} P_{T_{m_0},\lambda_m}(E_0,\xi). \tag{9.35}$$

The expectation value of a physical quantity A is now given by

$$<A>_{T,\lambda} = \frac{\sum_{E_0,\xi} A(E_0,\xi) P_{T,\lambda}(E_0,\xi)}{\sum_{E_0,\xi} P_{T,\lambda}(E_0,\xi)}. \tag{9.36}$$

The potential of mean force (PMF), or free energy as a function of the reaction coordinate, of the original, unbiased system at temperature T is given by

$$W_{T,\lambda=\{0\}}(\xi) = -k_{\mathrm{B}}T\ln\left[\sum_{E_0} P_{T,\lambda=\{0\}}(E_0,\xi)\right], \tag{9.37}$$

where $\{0\} = (0,\ldots,0)$.

9.3 Results

We now present some examples of the simulation results by REM and its extensions described in the previous section.

We first give the results of membrane protein folding and structure predictions. We included only transmembrane helices in the replica-exchange MC simulations with constraints that the helices stay within the membrane regions [24–26]. Helix backbones were treated as rigid bodies, but side chains were flexible. The MC moves were rigid translation of helices, rigid rotation of helices, and torsion angle rotations in side chains of helices.

Recently, we also allowed deformations of helices [27], because many transmembrane helices are distorted in the native structures. We applied the replica-exchange MC method to a helix dimer system of glycophorin A. The number of amino acid residues in the helix is 18, and the sequence is identical and TLIIFGVMAGVIGTILLI. Twenty replicas were used with the following temperatures: 300, 333, 371, 413, 460, 512, 571, 635, 707, 767, 857, 956, 1047, 1110, 1170, 1235, 1300, 1370, 1435, and 1500 K. The total number of MC steps was 100,000,000. For the initial states, ideal helices were placed at the center of the membrane and perpendicularly to the membrane boundary surfaces, and the centers of two helices were placed 20 Å apart.

We first examine how the replica-exchange simulation performed. Figure 9.1a shows the time series of the replica index at the lowest temperature of 300 K. We see that the minimum temperature was taken by different replicas many times during the replica-exchange simulation, and we observe a random walk in the

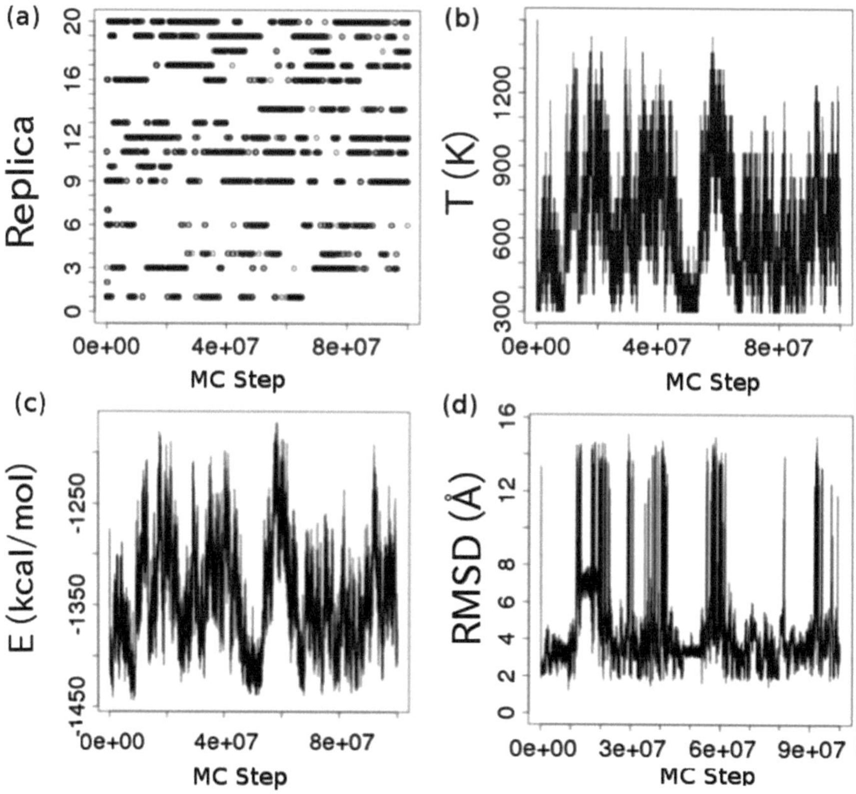

Fig. 9.1 Time series of various quantities for the replica-exchange simulation of glycophorin A. (**a**) Time series of replica index at 300 K. (**b**) Time series of temperature change for Replica 6. (**c**) Time series of total potential energy change for Replica 6. (**d**) Time series of the RMSD (in Å) with respect to all the C^{α} atoms from the PDB structure (PDB code: 1AFO) for Replica 6 (Reprinted from Ref. [27] with kind permission of the Physical Society of Japan (2015))

replica space. The complementary picture is the temperature exchange for each replica. Figure 9.1b shows the time series of temperatures for one of the replicas (Replica 6). We see that Replica 6 took various temperature values during the replica-exchange simulation. We observe random walks in the temperature space between the lowest and highest temperatures. Other replicas behaved similarly. Figure 9.1c shows the corresponding time series of the total potential energy for Replica 6. We see a strong correlation between the time series of temperatures (Fig. 9.1b) and that of potential energy (Fig. 9.1c), as expected. We next examine how widely the conformational space was sampled during the replica-exchange simulation. We plot the time series of the root-mean-square deviation (RMSD) of all the atoms from the experimental structure of Protein Data Bank (PDB) (PDB code: 1AFO) for Replica 6 in Fig. 9.1d. When the temperature becomes high, the RMSD takes large values, and when the temperature becomes low, the RMSD takes

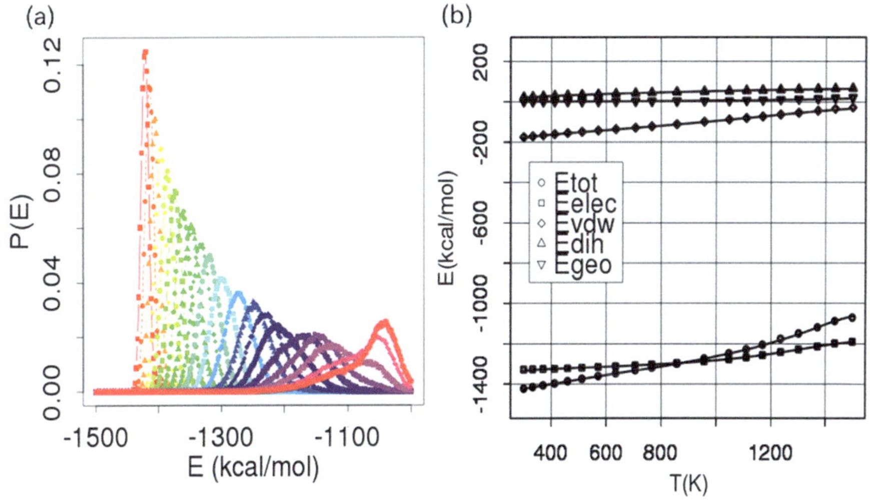

Fig. 9.2 (**a**) Canonical probability distributions of total potential energy at each temperature from the replica-exchange simulation of glycophorin A. The distributions correspond to the following temperatures (from *left* to *right*): 300, 333, 371, 413, 460, 512, 571, 635, 707, 767, 857, 956, 1047, 1110, 1170, 1235, 1300, 1370, 1435, and 1500 K. (**b**) Averages of total potential energy E_{tot} of glycophorin A and its component terms: electrostatic energy E_{elec}, Lennard-Jones energy E_{vdw}, dihedral energy E_{dih}, and constraint energy E_{geo} as functions of temperature (Reprinted from Ref. [27] with kind permission of the Physical Society of Japan (2015))

small values. By comparing Fig. 9.1b and d, we see that there is a positive correlation between the temperature and RMSD. The fact that the RMSDs at high temperatures are large implies that our simulations did not get trapped in local-minimum potential energy states. These results confirm that the replica-exchange simulation was properly performed.

Figure 9.2a shows the canonical probability distributions of the potential energy obtained from the replica-exchange simulation at 20 temperatures. We see that the distributions have sufficient overlaps between the neighboring temperature pairs. This ensures that the number of replicas was sufficient. In Fig. 9.2b, the average potential energy and its components, namely, the electrostatic energy E_{elec}, Lennard-Jones energy E_{vdw}, torsion energy E_{dih}, and constraint energy E_{geo}, are shown as functions of temperature, which were calculated by using Eq. (9.22). Because the helices are generally far apart from each other at high temperatures, the energy components, especially E_{elec} and E_{vdw}, are higher at high temperatures. At low temperatures, on the other hand, the side-chain packing among helices is expected. We see that as the temperature becomes lower, E_{vdw}, E_{dih}, and E_{elec} decrease almost linearly down to about 1200 K, and as a result, E_{tot} is also almost linearly decreasing down to about 1200 K. On the other hand, when the temperature becomes <1200 K, E_{vdw} contributes more to the decrease in E_{tot}. This is reasonable because E_{vdw} decreases as a result of side-chain packing and the stability of the conformation increases.

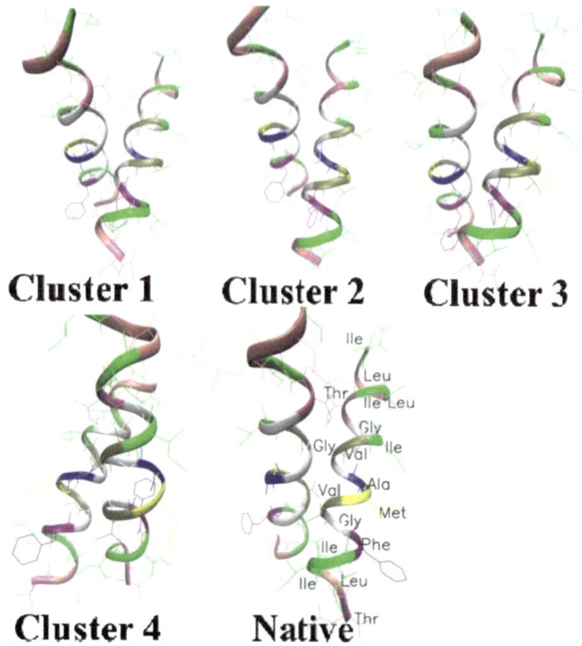

Fig. 9.3 Typical structures of glycophorin A in each cluster of similar structures selected as the local-minimum free energy state. The RMSD from the native conformation (PDB code: 1AFO) with respect to all the C^{α} atoms is 3.1, 3.4, 3.3, and 6.5 Å for Cluster 1, Cluster 2, Cluster 3, and Cluster 4, respectively (Reprinted from Ref. [27] with kind permission of the Physical Society of Japan (2015))

We next examine the typical local-minimum free energy state structures in each cluster of similar structures. Representative structures were selected in the highest density regions within the clusters. In Fig. 9.3, the representative structure of each cluster and the solution NMR structure (PDB code: 1AFO) are shown. We confirm that the structure of Cluster 1 is the closest to the experimental one. The structures of Clusters 2 and 3 have different packing in GXXXG motifs compared with the native and Cluster 1 structures. Note that each helix in all these structures has a similar structure, which is close to an ideal helix structure. This means that glycophorin A has basically ideal helix structures as local-minimum free energy states in this simulation, although we allowed the helix to be distorted or bent during the simulation.

The generalized-ensemble algorithms are particularly useful for drug design. We next examine how our docking method based on REUS performs in the prediction of protein-ligand binding structures [28, 29]. The reaction coordinate ξ is the distance between the ligand and the binding site in the protein.

We selected five protein-ligand complexes from PDB: 1T4E, 1SNC, 1OF6, 1ROB, and 2JJC, which correspond to the protein and ligand pairs, MDM2 and benzodiazepine, staphylococcal nuclease and PD*TP, aldolase and tyrosine, ribonuclease A and 2′-2CMP, and heat shock protein and pyrimidine-2-amine, respectively.

Figure 9.4 shows the structures of the five ligand molecules complexed to proteins, as registered in the PDB. We immersed each protein-ligand complex in a water box with at least 10.0 Å solvation from the protein surface. The total

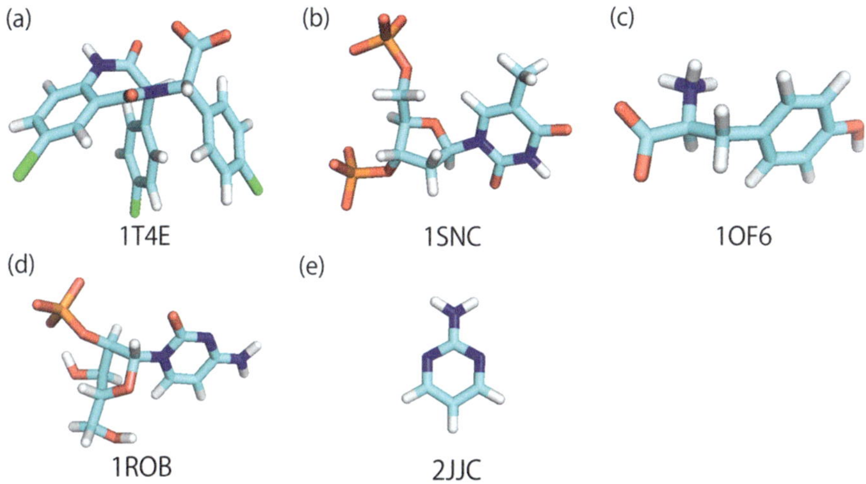

Fig. 9.4 Ligand molecule conformations complexed to proteins (from PDB). PDB IDs are (**a**) 1T4E, (**b**) 1SNC, (**c**) 1OF6, (**d**) 1ROB, and (**e**) 2JJC (Reprinted from Ref. [28] with kind permission of Wiley (2011))

number of water molecules was 5971 for the 1T4E system, 7561 for the 1SNC system, 14,670 for the 1OF6 system, 7093 for the 1ROB system, and 8812 for the 2JJC system. The following 24 umbrella potentials were prepared: $k_m(\xi(q) - d_m)^2$, where the midpoint values were $d_m = 5.0, 5.5, 6.0, 6.5, 7.0, 7.5, 8.0, 8.5, 9.0, 9.5, 10.0, 10.5, 11.0, 12.0, 13.0, 14.0, 15.0, 16.0, 17.5, 19.0, 20.5, 22.0, 23.5$, and 25.0 Å, and the strengths of the restraining potentials were $k_m = 1.0\text{kcal}/\left(\text{mol Å}^2\right)$ for $d_m \leq 13.0$Å and $k_m = 0.5\text{kcal}/\left(\text{mol Å}^2\right)$ for $d_m > 13.0$Å.

Figure 9.5 shows typical snapshots from the umbrella sampling simulations with the largest midpoint value of 25 Å (22 Å only for 1OF6). We see that ligand molecules are apart from proteins and are completely outside the protein pockets. The production runs were performed for 110–200 ns and the last 100 ns data was used for the analyses.

Figure 9.6 shows the time series of various quantities that we obtained from the REUS simulations of 1T4E as a representative system. Figure 9.6a shows the time series of the reaction coordinate (the distance of the ligand from the pocket) through umbrella potential exchange for some of the replicas. We observe that the reaction coordinate goes up and down between the lowest value and the highest one and that sufficient sampling has been achieved. Other replicas behaved similarly.

Figure 9.6b shows the time series of the distances of the center positions of umbrella potentials corresponding to Fig. 9.6a. When the center position of umbrella position is high, the reaction coordinate is high, and when the former is low, so is the latter. There is a strong correlation between Figs. 9.6a, b, as is to be expected.

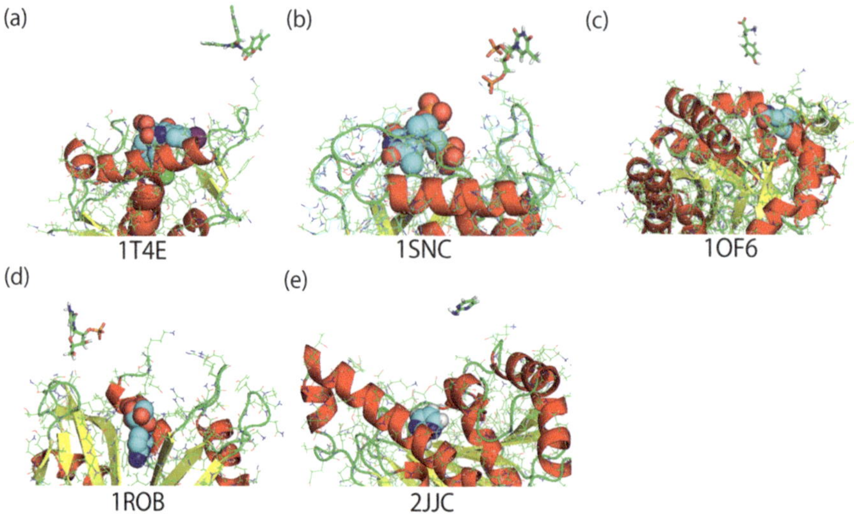

Fig. 9.5 Snapshots from the umbrella sampling simulations with the largest midpoint value of 25 Å (22 Å only for 1OF6). PDB IDs are shown below each figure. The space-filled molecules, which do not exist in these simulations, actually show the correct ligand binding positions (from PDB) as references. Water molecules are not shown to see the protein-ligand conformations clearly (Reprinted from Ref. [28] with kind permission of Wiley (2011))

Figure 9.6c shows the time series of RMSD of the ligand from PDB structure. We see that the RMSD goes up and down between the lowest and highest values found and that wide conformational space along the reaction coordinate has indeed been sampled during the simulations. Comparing Fig. 9.6a and c, we see that Replica 1 (black) has the distance close to the correct binding distance in the beginning, but it has actually a very different conformation and has a large RMSD from the experimental binding structure in PDB. However, we observe that the correct binding structures were found independently by different replicas, while the REUS simulation proceeded.

Figure 9.6d is the time series of replica exchange. The time series of replica number at an umbrella potential with the midpoint of 6.0 Å is shown. We see that many replicas experienced the umbrella potential with the midpoint value of 6.0 Å. These figures show that the replica-exchange simulations have been properly performed. Thus far, we examined how well the REUS simulation worked in the case of 1T4E system. We also confirmed that REUS simulations have been performed properly in all other systems.

Figure 9.7 shows the PMF for the five systems that we examined in this study. We collected about 50,000 ligand structures with the global-minimum PMF inside the pocket for each system. The global-minimum PMF inside the pocket was also the global-minimum PMF over all distance range for (a) 1T4E, (b) 1SNC, (c) 1OF6, and (d) 1ROB systems. On the other hand, we found that the structures with the global-minimum PMF for (e) 2JJC system were not inside the pockets. Therefore,

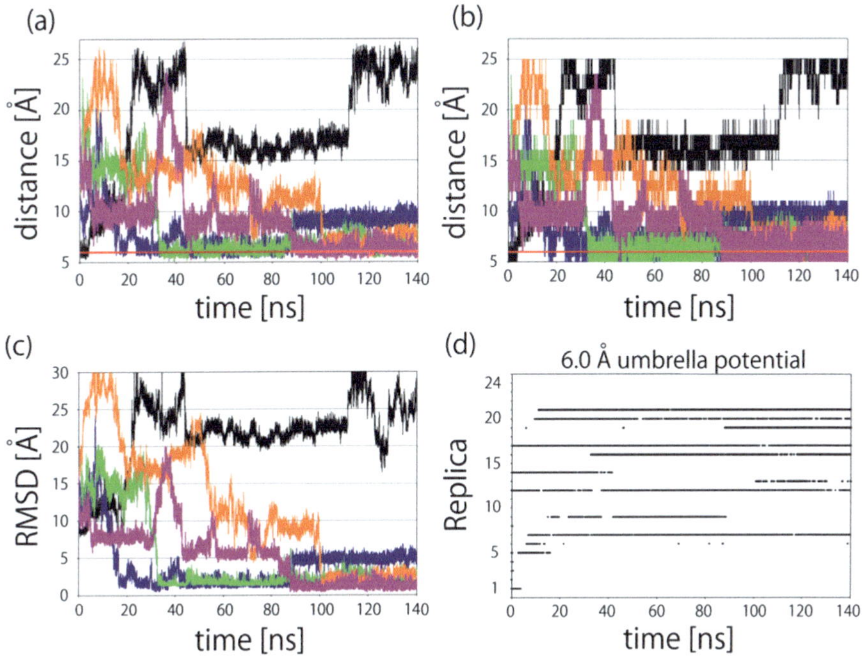

Fig. 9.6 Time series of (**a**) the reaction coordinate ξ (distance from the protein pocket), (**b**) the center position of umbrella potential from the protein pocket, (**c**) RMSD (in Å) from the correct binding mode, and (**d**) replica exchange for the third umbrella potential (kcal/(mol Å2) and Å), obtained from the REUS simulation of 1T4E system. Time series of Replicas 1 (*black*), 9 (*blue*), 13 (*orange*), 16 (*green*), and 19 (*pink*) are shown for (**a**), (**b**), and (**c**) as representatives. The *bold red lines* at 6.0 Å in (**a**) and (**b**) show the correct binding distance from PDB (Reprinted from Ref. [28] with kind permission of Wiley (2011))

we collected the structures with the first local-minimum PMF, which has the lowest PMF inside the protein pocket.

Figure 9.8 shows the comparisons of the predicted binding modes (green) with the experimental ones (blue) from PDB. We see that the predicted binding modes are in excellent agreement with the experimental binding modes especially for (a) 1T4E and (b) 1SNC, while we obtained essentially the correct binding modes also for (c) 1OF6, (d) 1ROB, and (e) 2JJC, considering the fluctuations of both protein and ligand structures. Note that docking software GOLD made incorrect predictions for (b) 1SNC and (d) 1ROB and these two systems are categorized as "significant error" in the original paper [30].

To investigate the need for protein flexibility for the protein-ligand association from a different point of view, we have performed the comparative REUS simulations with the restraints on the heavy atoms in the protein for one of the systems (PDB ID: 1OVE) [29]. All of the heavy atoms were constrained harmonically with a bond constant of 1.0 kcal/(mol $Å^2$) to the reference structure after the equilibration run, whereas the heavy atoms near the pocket were not constrained in the original

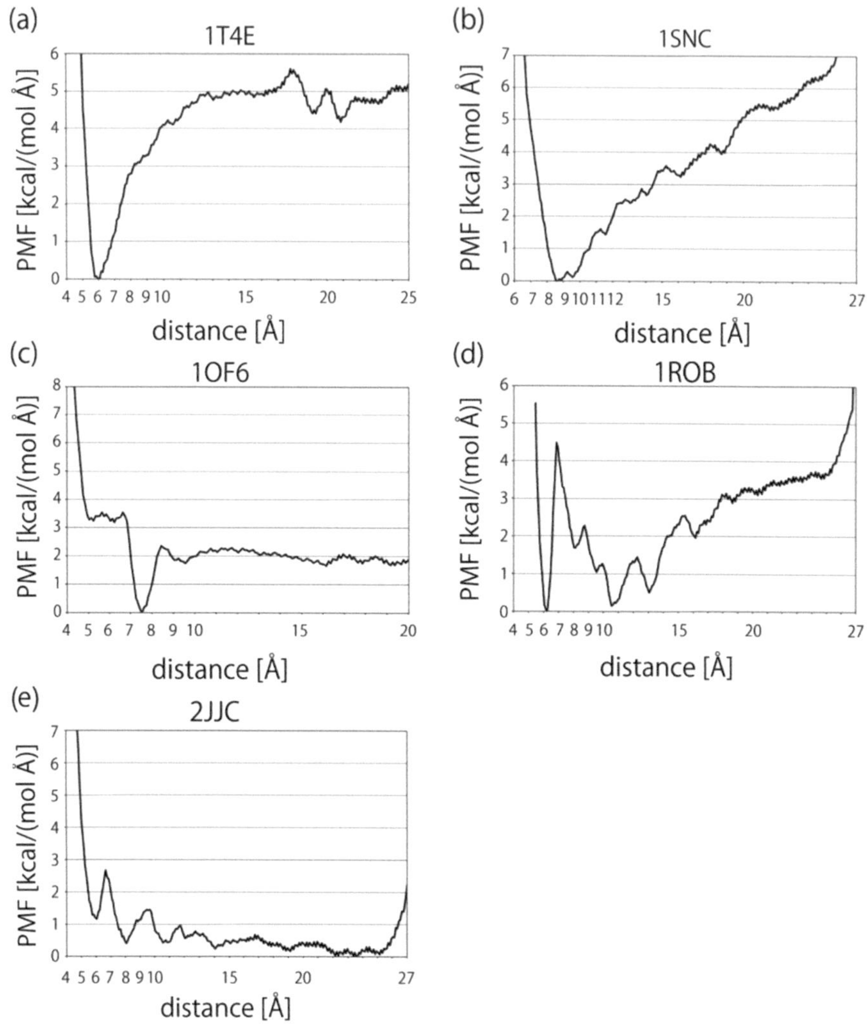

Fig. 9.7 Potential of mean force profiles along the reaction coordinate ξ (distance from the protein pocket) for all the five systems (**a**) 1T4E, (**b**) 1SNC, (**c**) 1OF6, (**d**) 1ROB, and (**e**) 2JJC (Reprinted from Ref. [28] with kind permission of Wiley (2011))

conditions. The pocket shape of the reference structure should be suitable for the ligand because the reference structure was prepared by the equilibration run with the protein-ligand complex correctly binding to each other.

Figure 9.9 shows the time series of the reaction coordinate ξ and RMSD that we obtained from the REUS simulations of 1OVE with and without restraints on the heavy atoms in the protein. Figures 9.9a, b are the time series of the reaction coordinate ξ (the distance of the ligand from the pocket) at the umbrella potential with the midpoint of 8.0 Å experienced by different replicas (a) with the original

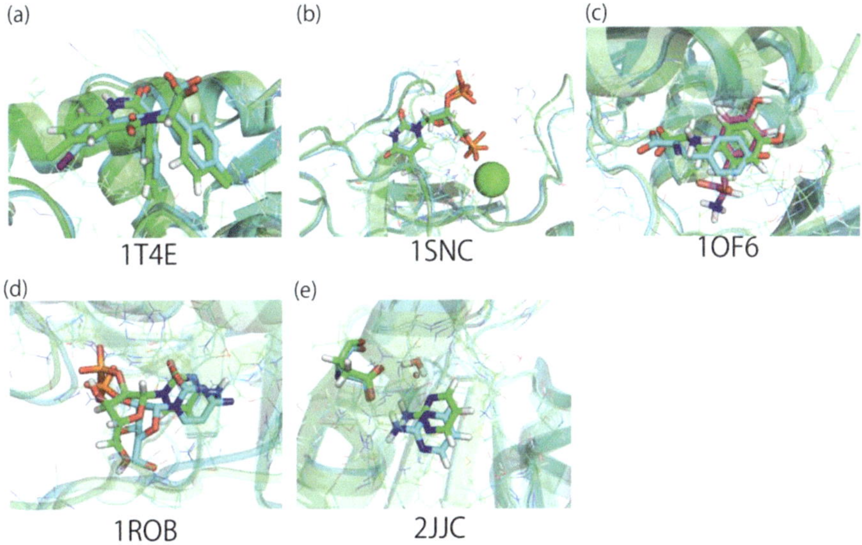

Fig. 9.8 Comparisons of the predicted binding modes by the REUS simulations (*green*) with the experimental ligand binding modes from PDB (*blue*). In (**c**) the REUS simulation predicted two different binding modes (*green* and *pink*), which were detected as equally stable ones on the free energy landscape (Reprinted from Ref. [28] with kind permission of Wiley (2011))

conditions (without restraints near the pocket) and (b) with the restraints on all the heavy atoms of the protein, respectively. We observe that both simulations properly sampled the structures with the distance specified by the umbrella potential. However, it is shown that the structures sampled by these two REUS simulations were very different. Figures 9.9c, d shows the time series of RMSD of a ligand molecule from the correct binding structure from PDB corresponding to Figs. 9.9a, b, respectively. We see that the simulation under the original conditions in Fig. 9.9c sampled the structures close to the experimental one correctly. On the other hand, the simulation under the restraint condition on all the heavy atoms in Fig. 9.9d did not find the correct binding structure in PDB. This is a reasonable result when we compare the PDB structure with structures from typical snapshots of the simulation corresponding to Fig. 9.9d.

Figure 9.10 compares the PDB structure of 1OVE and a typical snapshot from the REUS simulation with the restraints on all the heavy atoms of the protein, which could not find the correct binding structure. We see that the ligand in PDB is surrounded by the protein atoms and the gateway seems to be too narrow if the heavy atoms in the protein are fixed. Therefore, the ligand could not approach the correct binding region by the REUS simulation with the restraints in Fig. 9.10b. However, when the protein atoms around the pocket region were flexible, the ligand successfully found the correct binding structure by the REUS simulation (see Fig. 9.10a). These results imply the importance of the protein flexibility for a ligand binding to a protein.

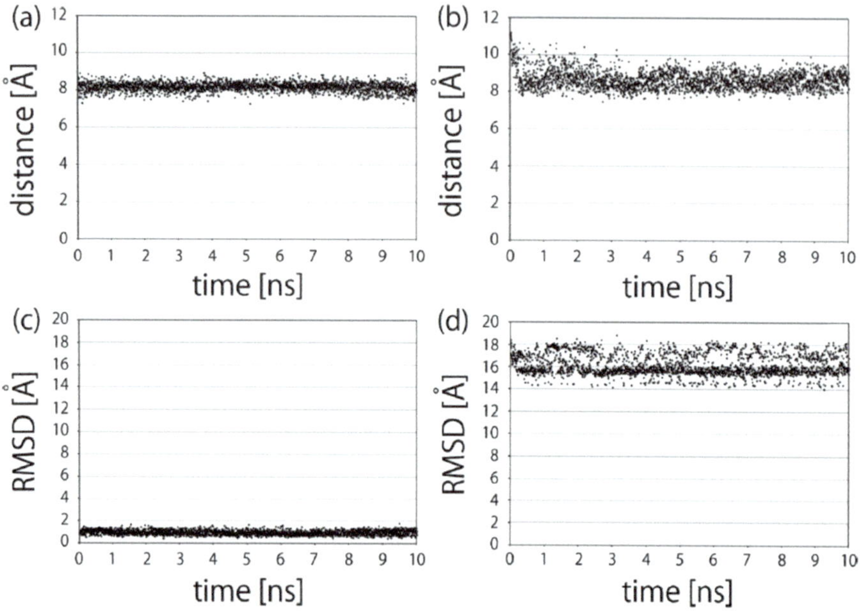

Fig. 9.9 Time series of the reaction coordinate ξ (the distances from the protein pocket) at the umbrella potential with the midpoint of 8.0 Å obtained from the REUS simulation of the 1OVE system (**a**) without any restraints on all the atoms within 12 Å around a ligand and (**b**) with the restraints ($k = 1.0$ kcal/(mol Å^2)) on all the heavy atoms of a protein. (**c**) and (**d**) are the time series of RMSD of a ligand molecule from the correct binding mode, corresponding to (**a**) and (**b**), respectively. The correct binding distance from PDB is 8.3 Å (Reprinted from Ref. [29] with kind permission of the American Chemical Society (2013))

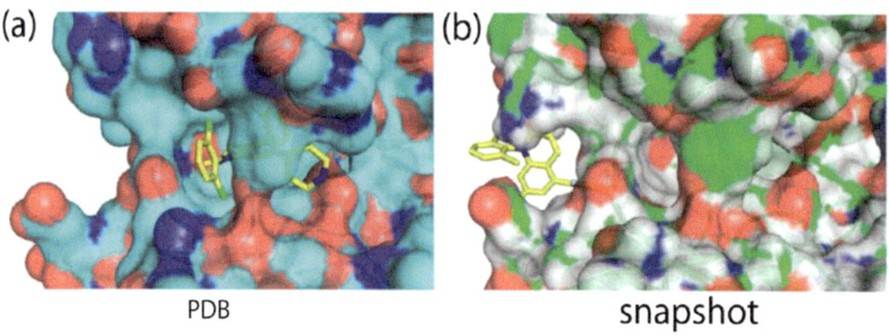

Fig. 9.10 (**a**) PDB structure of 1OVE and (**b**) a snapshot from the REUS simulation with the restraints on all the heavy atoms of the protein for the replica with the umbrella potential of the midpoint of 8.0 Å at 10 ns (Reprinted from Ref. [29] with kind permission of the American Chemical Society (2013))

9.4 Conclusions

In this chapter, we introduced a powerful generalized-ensemble algorithm, replica-exchange method (REM). We then described an extension of REM, multidimensional replica-exchange method (MREM). We also presented a special example of MREM, replica-exchange umbrella sampling (REUS). They can greatly enhance conformational sampling and structural fluctuations of biomolecular systems, allowing accurate free energy calculations. We then presented the results of REM MC simulations for membrane protein folding and those of REUS simulations for ligand docking to target proteins. We have obtained very promising results. Namely, we have succeeded in predicting the transmembrane helix structures of a membrane protein and the ligand binding structures both in excellent agreement with the experimental data from PDB. The latter is especially useful for drug design. Further improvement of the REUS-based docking method is now in progress. For example, a two-dimensional REM by the introduction of additional dimension to REUS has been developed [31, 32] and seems very promising.

Acknowledgments The author thanks his co-workers for useful discussions. In particular, he is grateful to Drs. A. Kitao, H. Kokubo, A. Mitsutake, Y. Sugita, T. Tanaka, and R. Urano for collaborations that led to the results presented in the present chapter. This work was supported, in part, by the Grant-in-Aid for Scientific Research on Innovative Areas ("Fluctuations and Biological Functions") from MEXT, Japan.

References

1. Hansmann UHE, Okamoto Y (1999) New Monte Carlo algorithms for protein folding. Curr Opin Struct Biol 9:177–183
2. Mitsutake A, Sugita Y, Okamoto Y (2001) Generalized-ensemble algorithms for molecular simulations of biopolymers. Biopolymers 60:96–123
3. Sugita Y, Okamoto Y (2002) Free-energy calculations in protein folding by generalized-ensemble algorithms. In: Schlick T, Gan HH (eds) Lecture notes in computational science and engineering. Springer, Berlin, pp 304–332, e-print: cond-mat/0102296
4. Okumura H, Itoh SG, Okamoto Y (2012) Generalized-ensemble algorithms for simulations of complex molecular systems. In: Leszezynski J, Shukla MK (eds) Practical aspects of computational chemistry II. Springer, Dordrecht, pp 69–101
5. Mitsutake A, Mori Y, Okamoto Y (2012) Enhanced sampling algorithms. In: Monticelli L, Salonen E (eds) Biomolecular simulations: methods and protocols. Humana Press, New York, pp 153–195
6. Okamoto Y, Kokubo H, Tanaka T (2013) Ligand docking simulations by generalized-ensemble algorithms. In: Karabencheva-Christova T (ed) Advances in protein chemistry and structural biology, vol 92. Academic, Burlington, pp 63–91
7. Yoda T, Sugita Y, Okamoto Y (2014) Protein folding simulations by generalized-ensemble algorithms. In: Han K-L, Zhang X, Yang M-J (eds) Protein conformational dynamics, advances in experimental medicine and biology, vol 805. Springer, Berlin, pp 1–27
8. Hukushima K, Nemoto K (1996) Exchange Monte Carlo method and application to spin glass simulations. J Phys Soc Jpn 65:1604–1608
9. Marinari E, Parisi G, Ruiz-Lorenzo JJ (1997) Numerical simulations of spin glass systems. In: Young AP (ed) Spin glasses and random fields. World Scientific, Singapore, pp 59–98

10. Sugita Y, Okamoto Y (1999) Replica-exchange molecular dynamics method for protein folding. Chem Phys Lett 314:141–151
11. Sugita Y, Kitao A, Okamoto Y (2000) Multidimensional replica-exchange method for free-energy calculations. J Chem Phys 113:6042–6051
12. Fukunishi F, Watanabe O, Takada S (2002) On the Hamiltonian replica exchange method for efficient sampling of biomolecular systems: application to protein structure prediction. J Chem Phys 116:9058–9067
13. Torrie GM, Valleau JP (1997) Nonphysical sampling distributions in Monte Carlo free-energy estimation: umbrella sampling. J Comput Phys 23:187–199
14. Mitsutake A, Okamoto Y (2009) Multidimensional generalized-ensemble algorithms for complex systems. J Chem Phys 130:214105 (14 pages)
15. Mitsutake A (2009) Simulated-tempering replica-exchange method for the multidimensional version. J Chem Phys 131:094105 (15 pages)
16. Okabe T, Kawata M, Okamoto Y, Mikami M (2001) Replica-exchange Monte Carlo method for the isobaric-isothermal ensemble. Chem Phys Lett 335:435–439
17. Okumura H, Okamoto Y (2006) Multibaric-multithermal ensemble molecular dynamics simulations. J Comput Chem 27:379–395
18. Mori Y, Okamoto Y (2010) Generalized-ensemble algorithms for the isobaric-isothermal ensemble. J Phys Soc Jpn 79:074003 (5 pages)
19. Metropolis N, Rosenbluth AW, Rosenbluth MN, Teller AH, Teller E (1953) Equation of state calculations by fast computing machines. J Chem Phys 21:1087–1092
20. Mori Y, Okamoto Y (2010) Replica-exchange molecular dynamics simulations for various constant temperature algorithms. J Phys Soc Jpn 79:074001 (8 pages)
21. Ferrenberg AM, Swendsen RH (1989) Optimized Monte Carlo data analysis. Phys Rev Lett 63:1195–1198
22. Kumar S, Bouzida D, Swendsen RH, Kollman PA, Rosenberg JM (1992) The weighted histogram analysis method for free-energy calculations on biomolecules. 1. The method. J Comput Chem 13:1011–1021
23. Mitsutake A, Sugita Y, Okamoto Y (2003) Replica-exchange multicanonical and multicanonical replica-exchange Monte Carlo simulations of peptides I. Formulation and benchmark test. J Chem Phys 118:6664–6675
24. Kokubo H, Okamoto Y (2004) Prediction of transmembrane helix configurations by replica-exchange simulations. Chem Phys Lett 383:397–402
25. Kokubo H, Okamoto Y (2004) Prediction of membrane protein structures by replica-exchange Monte Carlo simulations: case of two helices. J Chem Phys 120:10837–10847
26. Kokubo H, Okamoto Y (2009) Analysis of helix-helix interactions of bacteriorhodopsin by replica-exchange simulations. Biophys J 96:765–776
27. Urano R, Kokubo H, Okamoto Y (2015) Predictions of tertiary structures of α-helical membrane proteins by replica-exchange method with consideration of helix deformations. J Phys Soc Jpn 84:084802 (12 pages)
28. Kokubo H, Tanaka T, Okamoto Y (2011) *Ab initio* prediction of protein-ligand binding structures by replica-exchange umbrella sampling simulations. J Comput Chem 32:2810–2821
29. Kokubo H, Tanaka T, Okamoto Y (2013) Prediction of protein-ligand binding structures by replica-exchange umbrella sampling simulations: application to kinase systems. J Chem Theory Comput 9:4660–4671
30. Jones G, Willett P, Glen RC, Leach AR, Taylor R (1997) Development and validation of a genetic algorithm for flexible docking. J Mol Biol 267:727–748
31. Kokubo H, Tanaka T, Okamoto Y (2013) Two-dimensional replica-exchange method for predicting protein-ligand binding structures. J Comput Chem 34:2601–2614
32. Okamoto Y, Kokubo H, Tanaka T (2014) Prediction of ligand binding affinity by the combination of replica-exchange method and double-decoupling method. J Chem Theory Comput 10:3563–3569

Chapter 10
Spatiotemporal Fluctuations of Protein Folding in Living Cells

Minghao Guo and Martin Gruebele

Abstract Protein folding landscapes and protein-protein interaction landscapes are subject to modulation by many factors inside living cells: crowding, electrostatics, hydrophobic interactions, and even hydrodynamic phenomena. The resulting spatiotemporal fluctuations in protein folding rates, protein stability, protein structure, protein function, and protein interactions are the subject of a recent generation of experiments, as well as in-cell modeling. We discuss some specific examples of how the cell modulates energy landscapes and leads to fluctuations that may, in some cases, simply be a consequence of statistical physics, but, in other cases, may be exploited by the cell to improve its fitness.

Keywords Crowding • Fluorescence microscopy • FReI • In-cell folding • Quinary interaction • Temperature jump

10.1 Introduction

Proteins are constantly subject to a fluctuating environment. Some of these fluctuations are internal to the protein itself: these macromolecules are large enough to produce their own environment. For example, self-friction contributes to folding kinetics [1, 2], and fluctuating loops can regulate enzyme activity [3, 4]. Other fluctuations result from the environment. For example, solvation water dynamics are affected by the protein's presence [5–7] and in turn affect protein dynamics [8], and inside cells, proteins interact with a myriad of macromolecules and molecular machinery.

The fluctuations of the protein energy landscape inside cells are the topic of this chapter. They occur both on different spatial (length) scales [9] and time scales

M. Guo
Department of Physics, University of Illinois, Urbana-Champaign, IL 61801, USA

M. Gruebele (✉)
Department of Physics, University of Illinois, Urbana-Champaign, IL 61801, USA

Department of Chemistry and Center for Biophysics and Computational Biology, University of Illinois, Urbana-Champaign, IL 61801, USA
e-mail: mgruebel@illinois.edu

M. Terazima et al. (eds.), *Molecular Science of Fluctuations Toward Biological Functions*, DOI 10.1007/978-4-431-55840-8_10

[10], resulting in processes such as anomalous protein diffusion in the cell [11] or variations of protein stability in different parts of the cytoplasm or different organelles in the cell [12]. Some of these fluctuations may simply be the unavoidable physics of small numbers – a protein in a cell is typically surrounded by only a dozen or so macromolecules – but some of them may have evolved to optimize the cell for survival [13]. In effect, the crowded interior of the cell may give living systems another handle for controlling their protein machinery, a level of control exerted close to the subtle thermal fluctuations on the order of k_BT. At that level, hydrodynamics [14], transiently associated complexes (quinary interactions) [15], water-mediated interactions [16], and ultrastructure of the cell [17, 18] play a role in modulating the energy landscape of the cell. In most cases, we do not expect such modulation to fundamentally alter the energy landscape of proteins studied in vitro, but to tweak it in advantageous ways. In some cases (e.g., disordered polypeptides on the verge of folding), small modulations may be amplified drastically [19]. Indeed, in complex interaction networks, sensitivities can arise that amplify fluctuations even in the limit of some of the constituents occurring in large numbers [20].

A number of experimental techniques have been developed during the past decade to investigate fluctuations of protein dynamics inside cells. Translational dynamics (e.g., anomalous diffusion) has been studied by tracer particle or labeling techniques [21, 22]. Stability of proteins has been altered by additives tolerated by live cells [23]. NMR techniques now can look at overexpressed proteins inside cells [24], and mass spectrometry techniques can analyze average stability of proteins inside cells [25]. For a look at fluctuations, fluorescence techniques are currently the most promising, owing to their spatial resolution (ca. 300 nm at the diffraction limit, potentially 10s of nm in super-resolution, which is just becoming possible in live cells [26, 27]). We developed Fast Relaxation Imaging (FReI) as a simple and robust microscopy technique to look at fluctuations of protein stability, protein folding kinetics, protein-protein interactions, and protein function inside living cells. The idea is as follows: FRET-labeled proteins are introduced in a cell through injection, or plasmid expression, or intrinsic expression. The proteins are imaged inside the cell with millisecond (and potentially even faster) time resolution. An infrared laser either subjects the cell to a simple temperature jump [28] or modulates the cell temperature with other waveforms [29, 30]. The response of the FRET-labeled protein (or FRET-labeled pair or triplet of proteins) is then imaged throughout the cell as a function of time, recording responses such as protein unfolding, protein refolding, or transient protein-protein interactions.

The following discussion is based on a lecture given at the International Symposium on Molecular Science of Fluctuations towards Biological Functions and discusses some of our work in the area of spatiotemporal fluctuations of protein stability and folding kinetics inside living cells.

10.2 Protein Structure and Thermodynamics: Consequences of Environmental Modulation

Inside living cells exists a crowded environment, with macromolecule concentration up to 400 mg/ml [31]. The size exclusion effect of crowders has a large impact on protein stability. It limits the space a protein can occupy, thus distorting the folding energy landscape in favor of more compact structures. With noninteracting crowders, the unfolded state is entropically disfavored because it occupies more space, stabilizing the protein. There are also interacting "attractive" crowders, which enthalpically stabilize the unfolded state, bringing down the melting temperature of a protein [32, 33].

In vitro, the uncharged carbohydrate Ficoll is often used in crowding experiment for the comparison of the folding thermodynamics and kinetics with protein in aqueous buffer [34]. This 70 kDa inert macromolecule provides a simple homogeneous crowding environment. The crowding effect alters the protein folding energy landscape by a small amount, but significant enough to alter protein populations substantially. For example, the stability of the native state of CRABP1 is changed in Ficoll by up to 5 kJ/mol [35]. Due to the exponential sensitivity of the population to free energy (the Boltzmann factor $\exp[-\Delta G/k_BT]$), this small modulation is enough to alter the population distribution by sevenfold at constant temperature. Extensive studies have shown that protein folding energy landscape is modulated on the order of a few k_BT in the presence of crowders [36–38].

Proteins can also be destabilized by crowding or in vivo [39, 40]. For example, proteins crowded by other proteins in vitro are in some cases stabilized [41] and in other cases destabilized [39]. This shows that hydrophobic or electrostatic interactions with crowders can create "stickiness" in the protein's environment, thus stabilizing less compact unfolded states of the protein. In cells, a rich mixture of environments with different hydrophobic and electrostatic properties exists in different organelles and even in ultrastructure within the cytoplasm.

We have used the 415 residue two-domain enzyme PGK (phosphoglycerate kinase) extensively to study crowding-induced averages and fluctuations of protein structure and stability inside cells. Even the structure of PGK (specifically, the relative orientation and contact among the two domains) can be modulated by a crowding environment [34]. FRET-labeled PGK shows larger fluorescence energy transfer in a Ficoll-crowded environment and in living cells compared to aqueous solution. Figure 10.1a shows that the donor/acceptor fluorescence ratio (D/A) drops to half of the original value as the Ficoll concentration increases from 0 to 300 mg/ml. This experimental observation was confirmed by coarse-grained molecular dynamic simulation. The structures found in these simulations reveal that the folded state of PGK in Ficoll populates a more compact structure than the crystal structure. The protein population undergoes a shift from the native crystal structure (C) with two well-separated domains, through collapsed crystal structure (CC), and finally to a spherical compact state (Sph) as it unfolds in a crowded environment. The radii of gyration of the CC and Sph states are ca. 2 Å smaller than in the C state, and such

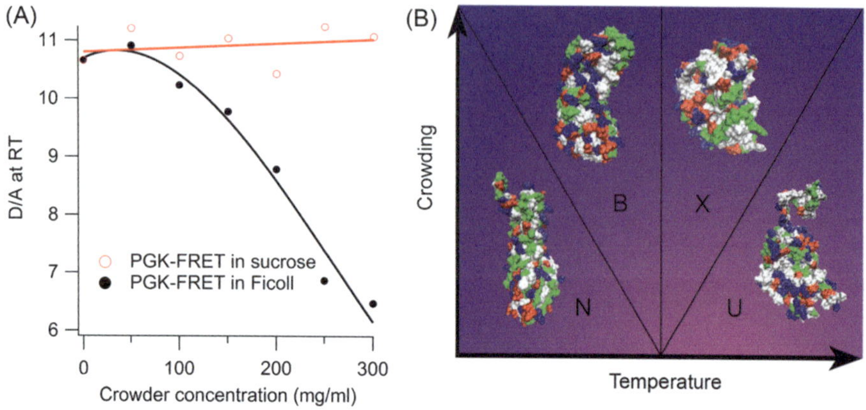

Fig. 10.1 Protein structural change in response to crowding. (**a**) FRET efficiency of PGK-FRET in terms of the donor acceptor ratio D/A in sucrose (*open circles*) and Ficoll (*dots*). Sucrose elicits no change in compactness, but Ficoll crowder leads to a much smaller FRET label separation [34]. (**b**) The four computed conformations of VlsE in different crowding and denaturant conditions [31]. The bean-like shape occurs in crowded environment. It unfolds to X and U states in sequence

compactness is favored in a crowded environment. The stability of folded state increases in the presence of Ficoll as a result of confinement: the melting temperature of PGK in 200 mg/ml Ficoll increases by 2 °C. The drop in D/A and gain in stability of PGK in Ficoll are consistent with the in vivo result measured for PGK [34]. The comparison implies that the influence of the cytoplasm on PGK folding can be described by the macromolecular crowding effect without large enthalpic contributions that stabilize the unfolded state.

The story is entirely different for the protein VlsE (variable major protein-like sequence expressed). *Borrelia burgdorferi* VlsE is an extracellular protein that assists in the pathogenicity of the Lyme disease agent. VlsE has multiple variable surfaces to prevent antibody recognition in the infected host [42]. It is an elongated, football-shaped protein. Experiments have shown that VlsE is sensitive to a simple crowded environment. In the presence of Ficoll, more helical contents of VlsE are observed in the far UV CD spectrum, as compared to aqueous solution. The native state (crystal structure) in absence of crowders has only 52 % helicity, while it rises to 80 % in the folded state in 400 mg/ml Ficoll 70 solution [43]. MD simulation suggests that VlsE achieves a more compact bean-shaped (B) structure rich in nonnative helical contents in crowded environment (Fig. 10.1b). This B state has a bigger N-to-C terminal distance than native state, even though it is more structured and compact [31]. This increase of the N-to-C distance agrees with the in-cell experiment of VlsE FRET labeled at the N and C termini, which shows that the donor/acceptor ratio D/A of the folded state at room temperature increases from ~2 in buffer to ~3–4 in cells [44].

However, the Ficoll and in-cell experiments part ways when it comes to VlsE stability. In the simple in vitro crowded environment, VlsE is stabilized just like

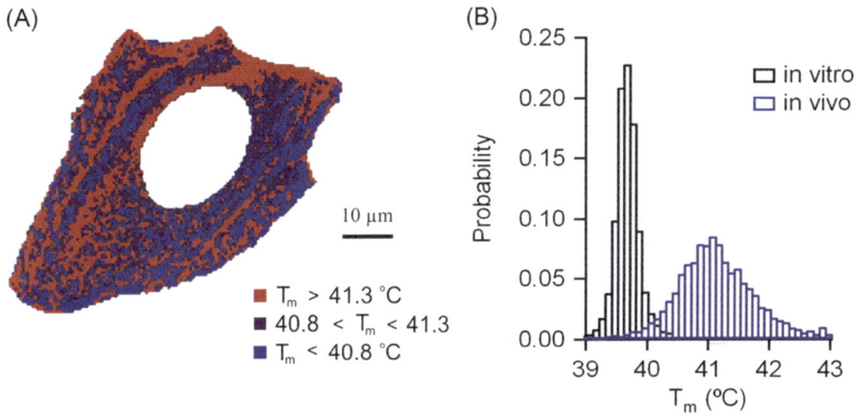

Fig. 10.2 (**a**) The spatial distribution of melting cooperativity of PGK-FRET measured in the cytoplasm of a live U2OS cell (the nuclear area has been excluded from the analysis). (**b**) A histogram of the distribution. The melting temperature ranges from 39 °C to 43 °C in vivo, much wider distributed than in vitro

PGK. The result can be rationalized mainly by an excluded volume effect. Yet in vivo, in U2OS human bone carcinoma cells, VlsE has a melting temperature 3 °C lower than in vitro [44]. This interesting result demonstrates that chemical interactions play a role and can overcome purely non-interaction confinement effects [32, 33]. If so, the interactions must be mainly hydrophobic, as the isoelectric points of VlsE and PGK are very similar, so at pH ~7.2 inside a cell, electrostatics of the two proteins should not be too different. In addition to chemical interactions with crowders inside the cell, a more complex volume exclusion mechanism is also possible. For example, the states N (native in aqueous solution) and B (the bean-shaped state stabilized by crowding) may have free energies that tune at different rates upon crowding. While the X state is destabilized by crowding relative to the B state, the B state may not be stabilized enough to make the overall effect in the cell a stabilizing one. It will be very interesting to investigate chemical (mainly enthalpic) versus volume exclusion/crowding (mainly entropic) effects in the complex environment inside cells.

A subcellular resolution study of PGK folding in cells shows how PGK stability fluctuates among different compartments in the cell and even within the cytoplasm. Figure 10.2 is a pixel-resolution map (ca. 0.5 μm) and histogram of the melting temperature of PGK inside U2OS cells [9]. There is a nonuniform pattern of melting temperature distributions inside the cytoplasm. This 2 °C wide distribution corresponds to about 4 kJ/mol stability fluctuations of the protein in various parts of the cytoplasm. In some stabilized areas, there is only 1/3 unfolded population, whereas in some other areas, there is 2/3 unfolded population. This map clearly shows protein stability strongly depends on the local environment. Comparison with intracellular membrane location shows that the patterns are not co-localized with membranes; they may be associated with the cytoskeleton. For a protein like PGK, the fluctuations are relatively small near the cell's natural temperature of

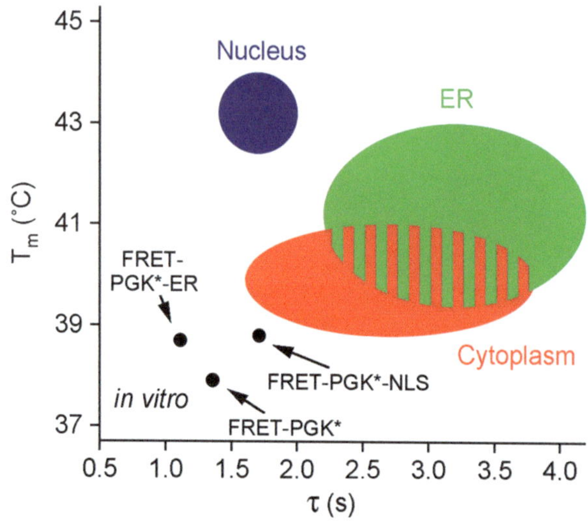

Fig. 10.3 The distribution of melting temperature and folding rates of PGK-FRET in cytoplasm (*red*), nucleus (*blue*), and endoplasmic reticulum (*green*) compared with in vitro (*black*) [12]. The PGK constructs are all stabilized and have slower folding rates with a wider rate distribution in vivo

37 °C. However, an intrinsically disordered signaling protein with a sharp folding-unfolding transition dependent on the presence of a binding partner might show very large fluctuations of unfolded-unbound versus folded-bound population inside a cell, a possible location-dependent regulation mechanism.

Protein stability can be modulated also by organelles within the cell. Dhar et al. compared the FRET-labeled PGK in different organelles of mammalian cells [12]. It was shown that PGK in the endoplasmic reticulum is more compact than in the cytoplasm. Nuclear-localized PGK is even more compact. Their stabilities vary in the same order: least stable in cytoplasm, median in the ER, and most stable in nucleus (Fig. 10.3). The dramatic stabilization in the nucleus compared to ER and cytoplasm must be due to more than the crowded environment (nuclear crowding is similar to cytoplasmic crowding), perhaps histone-bound nucleic acids provide a sufficiently different chemical environment than cytoplasm. In recent experiments, Wirth et al. [10] showed that PGK co-localized with chromosomal material during mitosis also is more stable than in the rest of the cytoplasm, lending support to this idea.

10.3 Protein Folding, Binding, and Biological Function: Environmental Modulation of the Energy Landscape

Structure and stability are not the only protein properties that fluctuate in the intracellular environment. Protein folding kinetics is also highly sensitive to the free energy landscape, which in turn is modulated by a protein's surroundings. For example, the crowding effect that destabilizes the unfolded state lowers the activation energy at the same time [45]. That would speed up protein refolding. At the

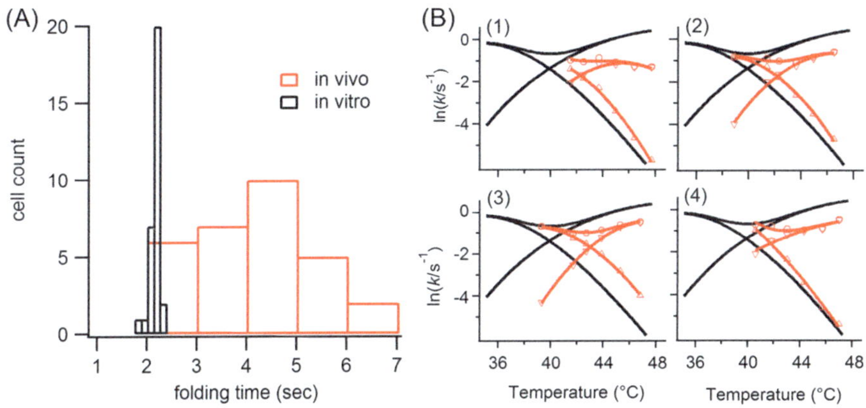

Fig. 10.4 Comparison of folding rates of PGK in cells. (**a**) Distribution of folding times In vivo and in vitro, all measured at 43 °C. In vivo folding is slowed down with a bigger standard deviation [46]. (**b**) Folding rates of PGK as functions of temperature in six cells are shown in different colors. The in vitro data is shown in the black curve [29]

same time, crowding eventually (at very high concentrations) increases the local viscosity and decreases the local diffusion coefficient. That would slow down protein folding. Viscosity inside cells is a complex phenomenon because of the different length scales of transport involved. For example, protein rotation and folding are very localized phenomena (length scale <10 nm), and depend on local viscosity, which may be relatively similar compared to the bulk in interstitial spaces between the crowders. At the other extreme, long-range diffusional transport may encounter jammed environments with very high effective viscosity, slowing down the dynamics.

Thus, folding rates inside cells could go up or down, depending on whether confinement or viscosity dominates [36]. For PGK folding in living cells, the increased local viscosity plays an important role. In Ficoll, PGK folding actually speeds up relative to aqueous buffer at concentrations up to 100 mg/ml [34]. The average in vivo folding rate of PGK is slowed down by a factor of 2 compared to aqueous buffer. At the same time, the fluctuations of in vivo folding rates are large. Figure 10.4a compares the distribution of the folding rates in vitro and in different cells from a U2OS cell population. The in vitro data shows negligible variation, which mainly arises from the experiment error, while the in vivo folding rates vary from 1/2 to 1/6 s^{-1}. It is estimated that the cell-to-cell variation of the activation energy of folding is only about 0.6 kJ/mol. The estimate is based on measuring fluctuations of the protein stability (see previous section) and combining this with an average Phi-value of 0.3 (The Phi value [47] is a measure of how much the transition state free energy fluctuates compared to the folding free energy; it lies between 0.3 and 0.5 for most proteins on average, i.e., a stress that destabilizes a protein by 1 kJ/mol destabilizes the transition state on average by 0.3–0.5 kJ/mol). The total folding rate varies by a factor of 3, more than can be accounted for by 0.6 kJ/mol, so the majority of the fluctuations are still attributed to local viscosity fluctuations inside the cell [48].

Figure 10.4b illustrates how the folding rate depends on temperature and how it is modulated from cell to cell (with similar fluctuations of the rate occurring also within cells). The red curves (in-cell) are similar in appearance to the black curves (in vitro). Thus, the fundamental folding mechanism is not altered in vivo. They do however differ both systematically and fluctuate from cell to cell. Some of the systematic variations have already been discussed: the k_f (folding) and k_u (unfolding) rate constants intersect at higher temperature and at a lower rate than in vitro. It means that the protein is stabilized inside the cell and that the observed relaxation rate near the protein melting temperature is slower in cells than in vitro. The following can be observed additionally in the kinetics results: the folding rate is always faster in cell than in vitro, whereas the unfolding rate is always slower. The former is in agreement with the crowding scenario, in which the unfolded state is destabilized. On top of these averages, different cells show different "tilts" of the rate curves in Fig. 10.4b. For example, cells 1 and 4 are more tilted clockwise. This could be due to a weakening of the crowding effect at higher temperature, such that the black in-cell curves approach the red in vitro curves. Clearly this does not happen in all cells, so the crowding effect behaves differently for different cells within a population (and also differently in different parts of the cytoplasm).

The different behavior of in vivo folding rates can be observed also in organelle-localized folding experiment. Even though nuclear-localized PGK is stabilized by over 4 °C relative to aqueous buffer, the folding rate of is almost same as in vitro. On the other hand, PGK in ER and in cytoplasm are less stabilized (only 2 °C, see Fig. 10.3), but folding is slowed down twofold compared to in vitro [12]. PGK folding in the nucleus is more similar to folding in 200 mg/ml Ficoll [34] than it is to folding in the ER: the protein is stabilized by a few degrees, rates are not changed by much, and the folding mechanism remains similar (The evidence for the latter is that the stretched exponential constant β ranges from 0.5 to 0.6 both in vitro and in the nucleus in the fitting formula $\exp[-(t/\tau)^{\beta}]$ used to fit kinetic relaxation after a T-jump; in the ER, β is closer to 1, and PGK folds more like a two-state folder with simple exponential kinetics).

A temporal correlation study of the folding rate in cytoplasm allows us to learn for how long the microenvironment in live cells persists and how they rearrange with time. The PGK translational diffusion time is faster than folding time on a 2 μm spatial scale. The fluctuations of folding rates and mechanism are therefore washed out below the 2 μm length scale due to rapid exchange of protein by diffusion. But we can examine the granular and filamentous cytoplasmic microenvironment on a larger spatial scale that persists much longer than the folding time scale. A set of T-jumps performed in a time series to study the time correlation of the folding rate distribution pattern in the cytoplasm is shown in Fig. 10.5. This correlation function decays in about 70 s, which is the typical time scale that a protein molecule travels across the whole cell, for example, due to cytoplasmic convection in addition to diffusion.

The diffusion of proteins inside cells is strongly dependent on the folding state and not necessarily for the obvious reason of different sizes of the two states [11]. When FRET-labeled PGK is subject to unfolding in the cell, its diffusion

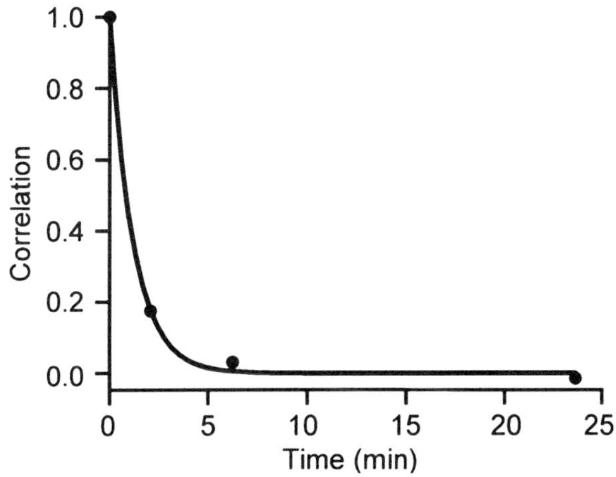

Fig. 10.5 Temporal correlation of the folding rate distribution in live cell

coefficient decreases much more than what is expected just based on size. The exposed hydrophobic surface area interacts with a "sticky" cytoplasm to slow down diffusion of the protein much more than expected based on the hydrodynamic radius of the unfolded state, which is only modestly larger than that of the folded state. Thus, interactions with biomolecular surfaces and chaperonins in the cytoplasm can be as important as excluded volume effects. This has also been observed in structural in-cell NMR experiments [49], where "quinary interactions" (a term created by McConkey to describe loose transient interactions in the cell and discussed in detail in Ref. [13]) can overcome crowding.

10.4 Protein Binding and Function Fluctuate Within the Cell

Binding is another important effect that modulates protein folding free energy landscapes in living cells. We discussed above that enthalpically driven "sticking" of unfolded protein to crowders can destabilize a protein. Chaperones may act in a similar way by binding to hydrophobic-exposed patches on partly or completely unfolded proteins, storing denatured proteins until they can refold or assisting the unfolding of proteins so that refolding may be attempted [50]. Ultimately, unfolding can be a precursor to proper refolding in vivo. For example, unfolding-refolding experiments done with PGK in live cells on a 10 s time scale show that refolding is more reversible in the cell than in vitro. This could be due to the general properties of the cytoplasm or it could be due to binding and "storage" of unfolded protein by chaperones, or both. Future experiments to look at co-localization of unfolded proteins and chaperones inside cells will show whether chaperones can contribute to protein maintenance even on such short-time scales.

Intrinsically disordered polypeptides (IDPs) are another example of proteins whose functions can be modulated by crowding and other weak interactions in the cell [19]. IDPs do not maintain folded structure in vitro, due to an unfavorable charge to hydrophobicity ratio [51]. In cellular environment, some can fold into stable structures because crowding and electrostatics overcome the unfavorable charge/hydrophobicity ratio. NCBD and ACTR are two IDPs that synergistically fold in multiple stages of binding and folding in a highly cooperative way [48, 52, 53]. A Go model simulation by Ganguly et al. reveals that these two proteins fold only when they bind to each other. In the absence of intermolecular interaction, these two proteins can only form a metastable mini folding core of NCBD helices $\alpha 1$ and $\alpha 2$. At this first stage of folding, the conformational fluctuations are very large because of very small free energy barriers of a few kcal/mol or less. Thus, the helices are flexible enough to sample a large configurational space. Once the mini folding core is formed, binding of NCBD with ACTR and subsequent folding are enabled. The bind process is fast, limited only by diffusional encounter [54]. This fast binding-folding cooperative behavior is important for IDPs functioning in the cells without being identified and degraded by protein degradation machinery.

Protein conformational fluctuations can be highly correlated with biological function and are also modulated within living cells. The conformational fluctuation of CypA can be coarse-grained into a three-state process that involves a free state and two enzyme-substrate complexes [55]. The time scale of interconversion between these states is very fast – sub-ms. Conformational fluctuations of CypA are observed even without substrate bound. The conformation fluctuations are global, rather than being limited to the substrate-binding site [56]. An interesting question for future study arises in connection with the "conformational selection" and "induced fit" scenarios of substrate processing. In conformational selection, the protein already has a significant probability of occupying structures required for tight substrate binding. In induced fit, it is the actual binding of the substrates that lowers the free energy of the proper conformation sufficiently to be populated. The energy landscape modulation on the order of k_BT that occurs in the cell may be sufficient to favor one over the other. For example, states that are not populated in vitro may actually be conformationally selected in the cell because they are stabilized in the cytoplasm.

The crowded cytoplasm has significant effects on enzyme turnover rates. Enzyme activity can vary over a wide range. The Michaelis constant K_m in crowded matrices differs from 10 times larger to 3 times smaller relative to that in buffer without crowding agents, depending on the nature of crowders and enzymes themselves [57–59]. The concentrations of enzymes in the cell are kept at values near Km by crowders, allowing efficient function despite any fluctuation of the substrate concentration. PGK is an example of an enzyme whose activity is dramatically increased in crowded Ficoll solution due to accelerated substrate binding [34] (At fivefold-higher enzyme concentrations, enzyme-product dissociation is not affected by crowding [57]). FRET measurements and computer simulations of PGK structure provide some evidence that upon crowding in Ficoll, the two halves of the active site (the Mg-ATP and the 3PGA-binding site) come closer together. The crowded environment favors a compact structure that may be enzymatically more active. It is calculated that the distance between the two halves of the active site is reduced by 24–42 % in the presence of crowders compared to crowder-free aqueous solution [34]. In this context, crowding may also improve multienzyme processing of substrates. There are some evidences [60] that diffusion coefficients of metabolic enzymes inside cells are reduced, possibly because multiple such enzymes loosely cluster inside cells, thereby decreasing the diffusion time of substrates from one enzyme to the next. Such weak associations (termed "quinary structure" [15]) promoted by crowding and chemical interactions within cells may confer additional fitness to cells, although they are difficult to detect by conventional pull-down assays. Such weak and transient protein-protein association can only be quantified by the next generation of in vivo experiments.

10.5 Outlook

Spatiotemporal fluctuations of proteins occur at many levels inside the cell: protein stability varies within the cytoplasm and from organelle to organelle; folding and unfolding of proteins are modulated by the cell; protein structure is modulated; and many types of protein-protein, protein-substrate, and other protein-biomolecule interactions (e.g., IDP signaling proteins with nucleic acid-binding partner) fluctuate in time and space within the cell. "Downhill" folding and binding mechanisms are expected to be particularly susceptible to cell-induced fluctuations. By definition, "downhill" reactions have negligible free energy barriers [61, 62] (which one), and populations can be strongly shifted along the reaction coordinate even by small perturbations. It has even been suggested that downhill reactions can serve as continuously tunable "rheostats" [63], as opposed to the classic picture of on-off switches (e.g., genetic switches) proposed as the major mechanism for regulation of complex networks within the cell. If so, it may turn out that the cell is doing a substantial amount of analog computing along with digital on-off processing and that continuous fluctuations, as opposed to discrete states, can play an important role in optimizing life processes. A new generation of single-molecule experiments [64], in-cell experiments, experiments that highlight the role of "biological" water in protein-protein interactions, as well as whole-cell simulation [65, 66] will reveal how important these fluctuations really are.

Acknowledgments This work was supported by the National Science Foundation, grant MCB-1413256.

References

1. Borgia A, Wensley BG, Soranno A, Nettels D, Borgia MB, Hoffmann A, Pfeil SH, Lipman EA, Clarke J, Schuler B (2012) Localizing internal friction along the reaction coordinate of protein folding by combining ensemble and single-molecule fluorescence spectroscopy. Nat Commun 3:1195. doi:10.1038/ncomms2204
2. Ozkan SB, Dill KA, Bahar I (2002) Fast-folding protein kinetics, hidden intermediates, and the sequential stabilization model. Protein Sci Public Protein Soc 11(8):1958–1970. doi:10.1110/ps.0207102
3. Roskoski R Jr (2005) Structure and regulation of Kit protein-tyrosine kinase – the stem cell factor receptor. Biochem Biophys Res Commun 338(3):1307–1315. doi:10.1016/j.bbrc.2005.09.150
4. Radmark O, Werz O, Steinhilber D, Samuelsson B (2007) 5-Lipoxygenase: regulation of expression and enzyme activity. Trends Biochem Sci 32(7):332–341. doi:10.1016/j.tibs.2007.06.002
5. Born B, Kim SJ, Ebbinghaus S, Gruebele M, Havenith M (2009) The terahertz dance of water with the proteins: the effect of protein flexibility on the dynamical hydration shell of ubiquitin. Faraday Discuss 141:161–173, discussion 175–207
6. Halle B (2004) Protein hydration dunamics in solution: a critical survey. Philos Trans R Soc Lond Ser B Biol Sci 359(1448):1207–1223; discussion 1223–1204, 1323–1208. doi:10.1098/rstb.2004.1499
7. Knab J, Chen JY, Markelz A (2006) Hydration dependence of conformational dielectric relaxation of lysozyme. Biophys J 90(7):2576–2581. doi:10.1529/biophysj.105.069088
8. Fenimore PW, Frauenfelder H, McMahon BH, Parak FG (2002) Slaving: solvent fluctuations dominate protein dynamics and functions. Proc Natl Acad Sci U S A 99(25):16047–16051. doi:10.1073/pnas.212637899
9. Ebbinghaus S, Gruebele M (2011) Protein folding landscapes in the living cell. J Phys Chem Lett 2(4):314–319. doi:10.1021/jz101729z
10. Wirth AJ, Platkov M, Gruebele M (2013) Temporal variation of a protein folding energy landscape in the cell. J Am Chem Soc 135(51):19215–19221. doi:10.1021/ja4087165
11. Guo MH, Gelman H, Gruebele M (2014) Coupled protein diffusion and folding in the cell. Plos One 9(12). doi:10.1371/journal.pone.0113040
12. Dhar A, Girdhar K, Singh D, Gelman H, Ebbinghaus S, Gruebele M (2011) Protein stability and folding kinetics in the nucleus and endoplasmic reticulum of eucaryotic cells. Biophys J 101(2):421–430. doi:10.1016/j.bpj.2011.05.071
13. Wirth AJ, Gruebele M (2013) Quinary protein structure and the consequences of crowding in living cells: leaving the test-tube behind. Bioessays 35(11):984–993. doi:10.1002/bies.201300080
14. Ando T, Skolnick J (2010) Crowding and hydrodynamic interactions likely dominate in vivo macromolecular motion. Proc Natl Acad Sci U S A 107(43):18457–18462. doi:10.1073/pnas.1011354107
15. McConkey EH (1982) Molecular evolution, intracellular organization, and the quinary structure of proteins. Proc Natl Acad Sci U S A 79(10):3236–3240
16. Papoian GA, Ulander J, Wolynes PG (2003) Role of water mediated interactions in protein-protein recognition landscapes. J Am Chem Soc 125(30):9170–9178. doi:10.1021/ja034729u
17. De Biasi S, Villa ML (1985) Ultrastructural aspects of the primary in vitro antibody response of mononuclear cells in human blood. Acta Anat (Basel) 124(1–2):1–7
18. Koster AJ, Klumperman J (2003) Electron microscopy in cell biology: integrating structure and function. Nat Rev Mol Cell Biol Suppl:SS6–SS10
19. Dhar A, Prigozhin M, Gelman H, Gruebele M (2012) Studying IDP stability and dynamics by fast relaxation imaging in living cells. In: Dunker VNUAK (ed) Intrinsically disordered protein analysis, vol 1. Springer, Berlin, pp 101–111

20. Ghosh K, Dill KA (2009) Computing protein stabilities from their chain lengths. Proc Natl Acad Sci U S A 106(26):10649–10654. doi:10.1073/pnas.0903995106
21. Ruthardt N, Lamb DC, Brauchle C (2011) Single-particle tracking as a quantitative microscopy-based approach to unravel cell entry mechanisms of viruses and pharmaceutical nanoparticles. Mol Ther 19(7):1199–1211. doi:10.1038/mt.2011.102
22. Regner BM, Vucinic D, Domnisoru C, Bartol TM, Hetzer MW, Tartakovsky DM, Sejnowski TJ (2013) Anomalous diffusion of single particles in cytoplasm. Biophys J 104(8):1652–1660. doi:10.1016/j.bpj.2013.01.049
23. Kormos BL, Benitex Y, Baranger AM, Beveridge DL (2007) Affinity and specificity of protein U1A-RNA complex formation based on an additive component free energy model. J Mol Biol 371(5):1405–1419. doi:10.1016/j.jmb.2007.06.003
24. Wang Y, Li C, Pielak GJ (2010) Effects of proteins on protein diffusion. J Am Chem Soc 132 (27):9392–9397. doi:10.1021/ja102296k
25. Ghaemmaghami S, Oas TG (2001) Quantitative protein stability measurement in vivo. Nat Struct Biol 8(10):879–882. doi:10.1038/nsb1001-879
26. Huang B, Bates M, Zhuang X (2009) Super-resolution fluorescence microscopy. Annu Rev Biochem 78:993–1016. doi:10.1146/annurev.biochem.77.061906.092014
27. Ha T, Tinnefeld P (2012) Photophysics of fluorescent probes for single-molecule biophysics and super-resolution imaging. Annu Rev Phys Chem 63:595–617. doi:10.1146/annurev-physchem-032210-103340
28. Ebbinghaus S, Dhar A, McDonald JD, Gruebele M (2010) Protein folding stability and dynamics imaged in a living cell. Nat Methods 7(4):319–323. doi:10.1038/nmeth.1435
29. Guo M, Xu Y, Gruebele M (2012) Temperature dependence of protein folding kinetics in living cells. Proc Natl Acad Sci U S A 109(44):17863–17867. doi:10.1073/pnas.1201797109
30. Gelman H, Platkov M, Gruebele M (2012) Rapid perturbation of free-energy landscapes: from in vitro to in vivo. Chemistry 18(21):6420–6427. doi:10.1002/chem.201104047
31. Homouz D, Perham M, Samiotakis A, Cheung MS, Wittung-Stafshede P (2008) Crowded, cell-like environment induces shape changes in aspherical protein. Proc Natl Acad Sci U S A 105(33):11754–11759. doi:10.1073/pnas.0803672105
32. Zhou HX (2013) Polymer crowders and protein crowders act similarly on protein folding stability. FEBS Lett 587(5):394–397. doi:10.1016/j.febslet.2013.01.030
33. Zhou HX (2004) Loops, linkages, rings, catenanes, cages, and crowders: entropy-based strategies for stabilizing proteins. Acc Chem Res 37(2):123–130. doi:10.1021/ar0302282
34. Dhar A, Samiotakis A, Ebbinghaus S, Nienhaus L, Homouz D, Gruebele M, Cheung MS (2010) Structure, function, and folding of phosphoglycerate kinase are strongly perturbed by macromolecular crowding. Proc Natl Acad Sci U S A 107(41):17586–17591. doi:10.1073/pnas.1006760107
35. Hong J, Gierasch LM (2010) Macromolecular crowding remodels the energy landscape of a protein by favoring a more compact unfolded state. J Am Chem Soc 132(30):10445–10452. doi:10.1021/ja103166y
36. Chen E, Christiansen A, Wang Q, Cheung MS, Kliger DS, Wittung-Stafshede P (2012) Effects of macromolecular crowding on burst phase kinetics of cytochrome c folding. Biochemistry 51 (49):9836–9845. doi:10.1021/bi301324y
37. Mikaelsson T, Aden J, Johansson LB, Wittung-Stafshede P (2013) Direct observation of protein unfolded state compaction in the presence of macromolecular crowding. Biophys J 104(3):694–704. doi:10.1016/j.bpj.2012.12.020
38. Mittal J, Best RB (2008) Thermodynamics and kinetics of protein folding under confinement. Proc Natl Acad Sci U S A 105(51):20233–20238. doi:10.1073/pnas.0807742105
39. Schlesinger AP, Wang Y, Tadeo X, Millet O, Pielak GJ (2011) Macromolecular crowding fails to fold a globular protein in cells. J Am Chem Soc 133(21):8082–8085. doi:10.1021/ja201206t
40. Guzman I, Gelman H, Tai J, Gruebele M (2013) The extracellular protein VlsE is destabilized inside cells. J Mol Biol Under Rev 426:11–20

41. Denos S, Dhar A, Gruebele M (2012) Crowding effects on the small, fast-folding protein lambda6-85. Faraday Discuss 157:451–462. doi:10.1039/c2fd20009k
42. Rogovskyy AS, Bankhead T (2013) Variable VlsE is critical for host reinfection by the Lyme disease spirochete. PLoS One 8(4), e61226. doi:10.1371/journal.pone.0061226
43. Perham M, Stagg L, Wittung-Stafshede P (2007) Macromolecular crowding increases structural content of folded proteins. FEBS Lett 581(26):5065–5069. doi:10.1016/j.febslet.2007.09.049
44. Guzman I, Gelman H, Tai J, Gruebele M (2014) The extracellular protein VlsE is destabilized inside cells. J Mol Biol 426(1):11–20. doi:10.1016/j.jmb.2013.08.024
45. Cheung MS, Klimov D, Thirumalai D (2005) Molecular crowding enhances native state stability and refolding rates of globular proteins. Proc Natl Acad Sci U S A 102 (13):4753–4758. doi:10.1073/pnas.0409630102
46. Dhar A, Ebbinghaus S, Shen Z, Mishra T, Gruebele M (2010) The diffusion coefficient for PGK folding in eukaryotic cells. Biophys J 99:L69–L71. doi:10.1016/j.bpj.2010.08.066
47. Ferguson N, Pires JR, Toepert F, Johnson CM, Pan YP, Volkmer-Engert R, Schneider-Mergener J, Daggett V, Oschkinat H, Fersht A (2001) Using flexible loop mimetics to extend Phi-value analysis to secondary structure interactions. Proc Natl Acad Sci U S A 98 (23):13008–13013. doi:10.1073/pnas.221467398
48. Zhang W, Ganguly D, Chen J (2012) Residual structures, conformational fluctuations, and electrostatic interactions in the synergistic folding of two intrinsically disordered proteins. PLoS Comput Biol 8(1), e1002353. doi:10.1371/journal.pcbi.1002353
49. Monteith WB, Cohen RD, Smith AE, Guzman-Cisneros E, Pielak GJ (2015) Quinary structure modulates protein stability in cells. Proc Natl Acad Sci U S A 112(6):1739–1742. doi:10.1073/pnas.1417415112
50. Thirumalai D, Lorimer GH (2001) Chaperonin-mediated protein folding. Annu Rev Biophys Biomol Struct 30:245–269. doi:10.1146/annurev.biophys.30.1.245
51. Uversky VN, Gillespie JR, Fink AL (2000) Why are "natively unfolded" proteins unstructured under physiologic conditions? Proteins Struct Funct Genet 41:415–427
52. Csermely P, Palotai R, Nussinov R (2010) Induced fit, conformational selection and independent dynamic segments: an extended view of binding events. Trends Biochem Sci 35 (10):539–546. doi:10.1016/j.tibs.2010.04.009
53. Ganguly D, Zhang W, Chen J (2012) Synergistic folding of two intrinsically disordered proteins: searching for conformational selection. Mol Biosyst 8(1):198–209. doi:10.1039/c1mb05156c
54. Zhou HX (2010) From induced fit to conformational selection: a continuum of binding mechanism controlled by the timescale of conformational transitions. Biophys J 98(6):L15–L17. doi:10.1016/j.bpj.2009.11.029
55. Eisenmesser EZ, Bosco DA, Akke M, Kern D (2002) Enzyme dynamics during catalysis. Science 295(5559):1520–1523. doi:10.1126/science.1066176
56. Eisenmesser EZ, Millet O, Labeikovsky W, Korzhnev DM, Wolf-Watz M, Bosco DA, Skalicky JJ, Kay LE, Kern D (2005) Intrinsic dynamics of an enzyme underlies catalysis. Nature 438(7064):117–121. doi:10.1038/nature04105
57. Vopel T, Makhatadze GI (2012) Enzyme activity in the crowded milieu. PLoS One 7(6), e39418. doi:10.1371/journal.pone.0039418
58. Pozdnyakova I, Wittung-Stafshede P (2010) Non-linear effects of macromolecular crowding on enzymatic activity of multi-copper oxidase. Biochim Biophys Acta 1804(4):740–744. doi:10.1016/j.bbapap.2009.11.013
59. Moran-Zorzano MT, Viale AM, Munoz FJ, Alonso-Casajus N, Eydallin GG, Zugasti B, Baroja-Fernandez E, Pozueta-Romero J (2007) Escherichia coli AspP activity is enhanced by macromolecular crowding and by both glucose-1,6-bisphosphate and nucleotide-sugars. FEBS Lett 581(5):1035–1040. doi:10.1016/j.febslet.2007.02.004
60. Verkman AS (2002) Solute and macromolecule diffusion in cellular aqueous compartments. Trends Biochem Sci 27(1):27–33

61. Bryngelson JD, Onuchic JN, Socci ND, Wolynes PG (1995) Funnels, pathways, and the energy landscape of protein folding: a synthesis. Proteins 21(3):167–195. doi:10.1002/prot.340210302
62. Yang WY, Gruebele M (2003) Folding at the speed limit. Nature 423(6936):193–197. doi:10.1038/nature01609
63. Garcia-Mira MM, Sadqi M, Fischer N, Sanchez-Ruiz JM, Munoz V (2002) Experimental identification of downhill protein folding. Science 298(5601):2191–2195. doi:10.1126/science.1077809
64. König I, Zarrine-Afsar A, Aznauryan M, Soranno A, Wunderlich B, Dingfelder F, Stüber JC, Plückthun A, Nettels D, Schuler B (2015) Single-molecule spectroscopy of protein conformational dynamics in live eukaryotic cells. Nat Methods 12:773–779. doi:10.1038/nmeth.3475
65. McGuffee SR, Elcock AH (2010) Diffusion, crowding & protein stability in a dynamic molecular model of the bacterial cytoplasm. PLoS Comput Biol 6(3):e1000694. doi:10.1371/journal.pcbi.1000694
66. Roberts E, Stone JE, Luthey-Schulten Z (2013) Lattice microbes: high-performance stochastic simulation method for the reaction–diffusion master equation. J Comput Chem 34(3):245–255. doi:10.1002/jcc.23130

Chapter 11
Membrane-Targeted Nanotherapy with Hybrid Liposomes for Cancer Cells Leading to Apoptosis

Ryuichi Ueoka, Yoko Matsumoto, Hideaki Ichihara, and Yuji Komizu

Abstract We have produced hybrid liposomes (HL) which can be prepared by sonication of a mixture of vesicular and micellar molecules in a buffer solution. The physical properties of HL such as size, shape, and membrane fluidity can be controlled by changing the constituents and compositional ratio. We have employed HL for chemotherapy and interesting results are as follows: (A) The uniform and stable structure of HL composed of L-α-dimyristoylphosphatidylcholine (DMPC) and polyoxyethylenedodecyl ether ($C_{12}(EO)_n$) with a diameter of 80 nm was revealed. (B) The remarkable inhibitory effects of HL on the growth of various tumor cells were attained in vitro. (C) Induction of apoptosis by HL was obtained and the pathway of apoptosis induced by HL was clarified. (D) A good correlation between the membrane fluidity of HL and inhibitory effects of HL for tumor cells was obtained. (E) Significantly chemotherapeutic effects were obtained using mice model of carcinoma after the treatment with HL without any side effects in vivo. (F) In clinical applications, prolonged survival and remarkable reduction of neoplasm were attained in patients with lymphoma after the treatment with HL without any side effects after the approval of the bioethics committee.

Keywords Hybrid liposomes • Antitumor effect • Apoptosis • Chemotherapy • Membrane fluidity

11.1 Introduction

"Molecular science of fluctuations toward biological functions" should be important for understanding the origin of life. Biological membranes provide compartment of defined sizes, shapes, and microenvironments. They organize living matter in cells and create a fluid two-dimensional matrix. Such cellular functions as recognition, fusion, endocytosis, exocytosis, intercellular interaction, and transport

R. Ueoka (✉) • Y. Matsumoto • H. Ichihara • Y. Komizu
Division of Applied Life Science, Graduate School of Engineering, Sojo University, 4-22-1 Ikeda, Kumamoto-shi, Kumamoto 860-0082, Japan
e-mail: ueoka@life.sojo-u.ac.jp

M. Terazima et al. (eds.), *Molecular Science of Fluctuations Toward Biological Functions*, DOI 10.1007/978-4-431-55840-8_11

are all membrane-mediated processes. In general, membranes of plant and animal cells are typically composed of 40–50 % lipids and 50–60 % proteins. There are wide variations in the types of lipids including phospholipids and proteins as well as in their ratios. Especially, such variations in lipids should take an important role for the activation of bioactive substances. Membrane fluctuations of cancer cells are very different from those of normal cells. In general, the membranes of cancer cells are more fluid as compared with normal cells.

Liposomes are microscopic spheres made from fatty materials, predominantly phospholipids. Because of their similarity to phospholipid domains of cell membranes and an ability to incorporate and/or carry bioactive substances, liposomes can be used for the enzymological (for instance, membrane-bound enzyme models) and medical (for instance, drug delivery systems) applications.

It is well known that liposomes are used as drug carriers. The drugs include antitumor agents, hormones, and immunomodulators [5, 41]. In particular, polyoxyethyleneglycol (PEG)–phosphatidylcholine (PC) liposomes have been found to be effective for prolonging blood circulation [1, 19]. On the other hand, we have recently produced specific hybrid liposomes (HL) composed of vesicular and micellar molecules; they are free from any contamination from organic solvents and stable for a longer period. Changing the composition of HL can control the physical properties of these liposomes, such as size, membrane fluidity, phase transition temperature, and hydrophobicity [46, 50, 51, 54]. A schematic representation of HL is shown in Fig. 11.1.

In the course of our study on the liposomes, the following interesting results have been obtained. (a) Stereochemical control of the enantioselective hydrolysis of amino acid esters could be established by temperature regulation and by changing the composition of HL [7, 43, 46, 50, 51, 54]. (b) Inhibitory effects of HL including flavonoids [51], antitumor drugs [18, 26] sugar surfactants [30, 32, 56], or polyunsaturated fatty acids [8, 42] on the growth of tumor cells in vitro and in vivo have been obtained. (c) Remarkably high inhibitory effects of HL on the growth of tumor cells in vitro have been obtained without drugs [12, 27–29, 37]. (d) Induction of apoptosis by HL for tumor cells has been obtained [14, 28, 29]. (e) Significantly prolonged survival and remarkable reduction of tumor volume were obtained using mice model of carcinoma after the treatment with HL without any drug in vivo

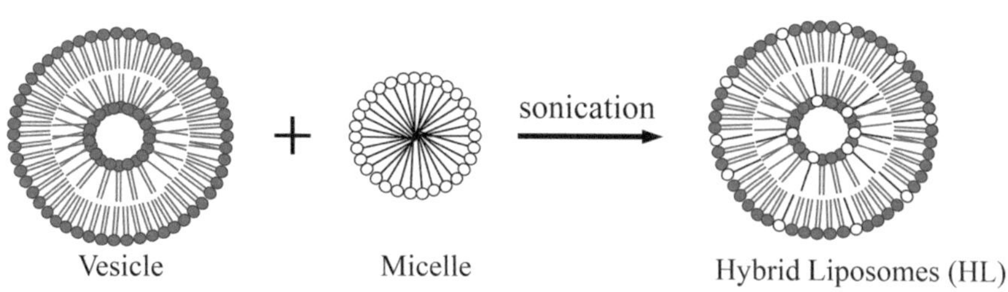

Fig. 11.1 Schematic representation of HL

[9, 15–17, 55]. (f) No toxicity of HL was observed in normal cells in vitro and in normal rats in vivo without any side effects [9, 18, 55].

In this study, we focus on the uniform and stable HL composed of L-α-dimyristoylphosphatidylcholine (DMPC) and polyoxyethylenealkyl ether ($C_{12}(EO)_n$) having inhibitory effects on the growth of tumor cells in vitro, in vivo, and in clinical applications.

11.2 History of Hybrid Liposomes: Enzymological Application

In 1985, we have produced new type "hybrid liposomes (HL)" composed of vesicular and micellar molecules for the first time [46]. The physical properties of these liposomes, such as size, shape, membrane fluidity, phase transition temperature, and hydrophobicity, can be readily controlled by changing the constituents and composition of HL. HL have attracted special interests as a tool for enzymological and medical applications. First, we have attempted steric control for the hydrolysis of amino acid esters in HL composed of synthetic surfactants as membrane-bound enzyme models [46, 50, 51].

The stereoselective hydrolytic cleavage of amino acid esters has attracted considerable attention in connection with understanding the origins of the stereoselectivity observed with proteolytic enzymes [4]. In the course of our study on the enantioselective hydrolysis of amino acid esters catalyzed by active peptides in micelles, vesicles, and HL, we emphasized that the stereochemical control could be established by changing amino acid residues covalently induced into substrates and catalysts, and composition of aggregates, and by regulating temperature and ionic strength of reaction media. In particular, the perfect enantioselectivity (enantiomer rate ratio $k^{L}{}_{a,obsd}/k^{D}{}_{a,obsd} = \infty$) was attained in the hydrolysis of the long-chain substrates (C_{12}-D(L)-Phe-PNP) by the active tripeptide (Z-PheHisLeu) in HL composed of synthetic double-chain surfactant ($2C_{14}Br$) and single-chain one (CTAB) at $\mu = 0.02$ (Fig. 11.2) [47–49]. This is the first example of successful enzyme-like catalysis observed in the enantioselective hydrolysis of amino acid esters. The complete L-enantiomer-selective catalysis has been achieved in specific HL systems [43, 47–49, 52, 53]. Importantly, excellent correlations between the enantioselective catalysis by active peptides (enzyme models) and the physical properties of HL (reaction fields) were observed. For example, (a) the enhancement of enantioselectivity was in good harmony with the apparent mean hydrodynamic diameters of HL [43, 46–51]. (b) The temperature dependence of enantioselectivity was bell-shaped with a maximum around the phase transition temperature of HL [25, 50, 51]. (c) The conformational change of peptide catalysts was observed along with the change in microenvironment of HL [43, 45, 50, 51]. These results suggest that the enantioselective catalysis can be attributed to optimization in the enzyme

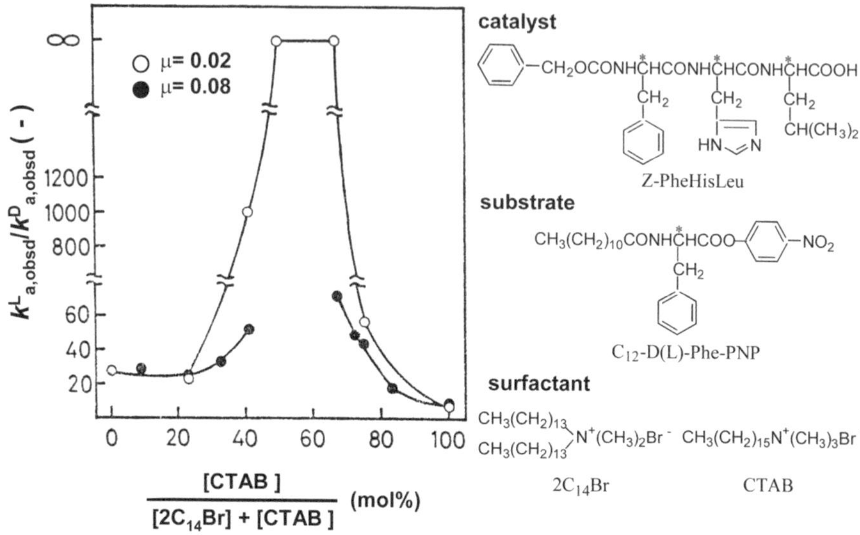

Fig. 11.2 Composition and ionic strength (μ) dependences of enantioselectivity ($k^{L}_{a,obsd}/k^{D}_{a,obsd}$) for the hydrolysis of C_{12}-D(L)-Phe-PNP catalyzed by Z-PheHisLeu in HL composed of $2C_{14}Br$ and CTAB at 25 °C and pH 7.6. [C_{12}-D(L)-Phe-PNP] = 1.0×10^{-5} M, [Z-PheHisLeu] = 5.0×10^{-5} M, [$2C_{14}Br$] = 1.0×10^{-5} M

model conformation by controlling the microenvironment of membrane matrix in HL.

We have extended the enantioselective hydrolysis of amino acid esters catalyzed by peptide catalysts into HL composed of native lipids (phosphatidylcholine) and nonionic surfactants. Significantly, it was obvious that the stereochemical control of enantioselective catalysis could be established by optimizing the temperature and/or composition of native lipid membrane systems as well as that of synthetic systems [7, 44, 52–54].

On the other hand, these HL having native lipids (constituents of biological membranes) as main components were free from any contamination from organic solvents that appear to be toxic for the central nervous system. Also, they are small in size (less than ~100 nm) which can avoid the reticular endothelial system (RES) in vivo [24]. Then, we have attempted to apply HL to anticancer chemotherapy in vitro, in vivo, and in clinical applications, as described in the following sections.

11.3 Application for Drug Delivery System

Liposomes have recently attracted considerable attention in connection with reducing the toxicity of antitumor drugs. There have been numerous studies that microcapsules containing chemotherapy agents show an excellent antitumor activity

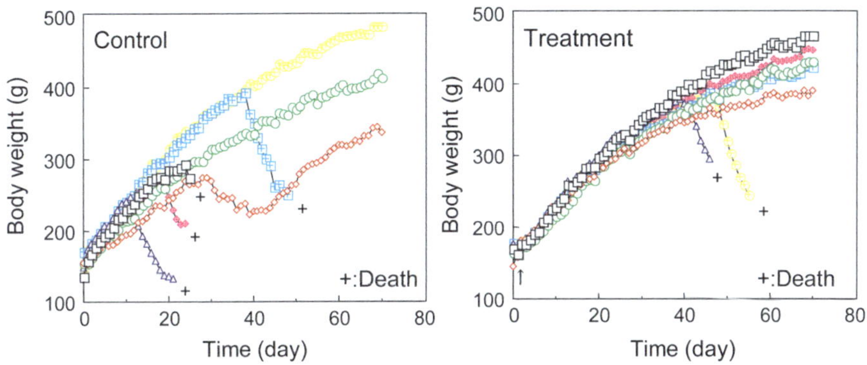

Fig. 11.3 Body weight change in rats of meningeal gliomatosis model treated with intrathecal infusion of HL composed of 90 mol% DMPC/10 mol% Tween 20 including BCNU after inoculation of glioma cells

against certain types of experimental tumors [23]. However, most of these approaches have been clinically unsuccessful, because the microcapsules were trapped by RES before reaching the targeting cells [2]. So, we have attempted to improve the pharmacological activity and/or to minimize the side effect of chemotherapy agents including 1,3-bis(2-chloroethyl)-1-nitrosourea (BCNU) by the incorporation of drugs into HL [18]. Intrathecal administration of HL composed of DMPC and polyoxyethylene (20) sorbitan monolaurate (Tween 20) including BCNU had a remarkable chemotherapeutic effect in a rat model of meningeal gliomatosis induced by intrathecal inoculation of rat C6 glioma cells (Fig. 11.3). Moreover, no side effects were observed for normal rats treated with the intrathecal infusion of HL with BCNU. The intrathecal half-life of HL including BCNU was longer than that of BCNU solubilized with 5 % dextrose/water. This is another advantage of intrathecally injected HL including BCNU.

11.4 Antitumor Effects In Vitro

We examined the inhibitory effects of HL on the growth of various tumor (mouse melanoma (B-16 F0), human breast tumor (MDA-MB-453), human B lymphoma (RAJI), and rat gastric carcinoma (CoRa 622 G6)) cells on the basis of 2-[4-iodophenyl]-3-[4-nitrophenyl]-5-[2,4-disulfophenyl]-2H-tetrazolium monosodium salt (WST-1) assay [13]. HL-23 were prepared by dissolving both DMPC and $C_{12}(EO)_{23}$ in 5 % glucose solution with sonication at 300 W and 45 °C for 5 min, followed by filtration with a 0.20 μm filter. The tumor cells (2.0–5.0 × 10^4 viable cells/ml) were cultured for 48 h in a humidified 5 % CO_2 incubator at 37 °C after adding the sample solutions. Then WST-1 solutions were added to the cells and the absorbance at a wavelength of 450 nm was measured using a spectrophotometer. The inhibitory effects were evaluated by $A_{mean}/A_{control}$, where A_{mean} and $A_{control}$ denote the absorbance of water-soluble formazan, which was useful as an indicator of cell viability, in the presence and absence of sample solutions, respectively. Fifty percent inhibitory concentration (IC_{50}) values of HL-23 were from one fourth to a half of those of the single-component DMPC liposomes, indicating that the inhibitory effects of HL-23 are large when compared with those of the single-component DMPC liposomes [10, 15, 33, 34]. These results indicate that inhibitory effects of HL-23 should be advantage as compared with single-component DMPC liposomes.

We also examined the morphology of HL composed of 95 mol% DMPC and 5 mol% $C_{12}(EO)_{23}$ on the basis of electron microscopy and dynamic light scattering measurements. The electron micrograph of HL shows the presence of spherical vesicles (Fig. 11.4). Interestingly, a clear stock solution of HL having a hydrodynamic diameter of 80 nm with single and narrow distributions could be kept over 30 days, as shown in Fig. 11.5. These results indicate that the uniform and stable structure of HL could be obtained and suggest that HL could avoid RES in vivo.

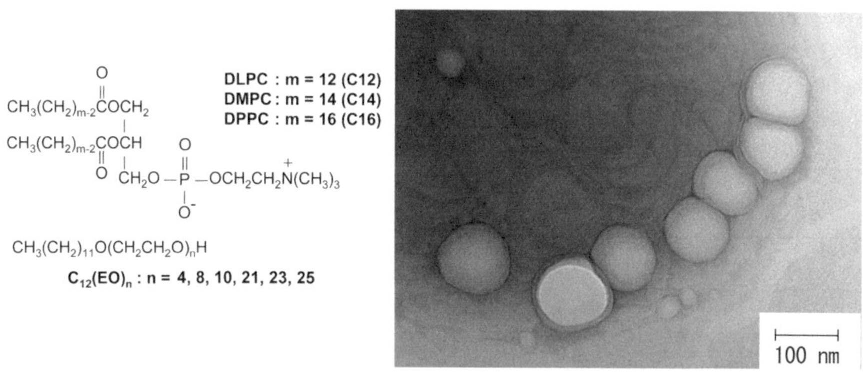

Fig. 11.4 An electron micrograph of HL composed of 95 mol% DMPC and 5 mol% $C_{12}(EO)_{23}$

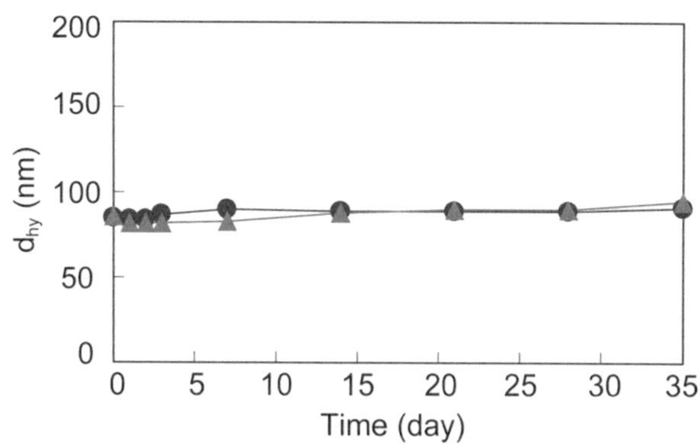

Fig. 11.5 Time course of d_{hy} change for HL composed of 95 mol% DMPC and 5 mol% $C_{12}(EO)_{23}$ (●; [DMPC] = 10 mM and ▲; [DMPC] = 50 mM)

11.5 Two Methylene Groups in Phospholipids Distinguish Between Apoptosis and Necrosis

We examined morphological change in human promyelocytic leukemia (HL-60) cells after the treatment with HL composed of phospholipids having the same hydrophilic head group (phosphatidylcholine group) and different hydrophobic alkyl chains (L-α-dilauroylphosphatidylcholine (C12: DLPC), L-α-dimyristoyl-phosphatidylcholine (C14: DMPC), L-α-dipalmitoylphosphatidylcholine (C16: DPPC)), and polyoxyethylene(23)dodecyl ether ($C_{12}(EO)_{23}$) using a time-lapse video. As shown in Fig. 11.6, the formation of bleb and corpuscle indicating characteristic feature of apoptosis were observed for HL of 90 mol% DMPC/10 mol% $C_{12}(EO)_{23}$ [35, 36, 57]. On the other hand, swelling of cells and dissolving of cell membrane, that is necrosis, were observed for HL of 90 mol% DLPC/10 mol % $C_{12}(EO)_{23}$. It is worthy to note that two methylene groups of acyl chains in phosphatidylcholines could distinguish between apoptosis and necrosis. Neither apoptosis nor necrosis was observed for HL of 90 mol% DPPC/10 mol% $C_{12}(EO)_{23}$.

Furthermore, apoptotic or necrotic rates of HL-60 cells after the treatment with HL were examined using flow cytometry. The results are shown in Fig. 11.7. Interestingly, almost the same large rates of apoptotic and viable cells were obtained using HL of DMPC/$C_{12}(EO)_{23}$. On the other hand, a large ratio of necrosis and almost no viable cells were obtained for HL-60 cells after the treatment with HL of DLPC/$C_{12}(EO)_{23}$. In the case of HL of DPPC/$C_{12}(EO)_{23}$, a small ratio of apoptotic and necrotic cells was obtained along with the large ratio of viable cells.

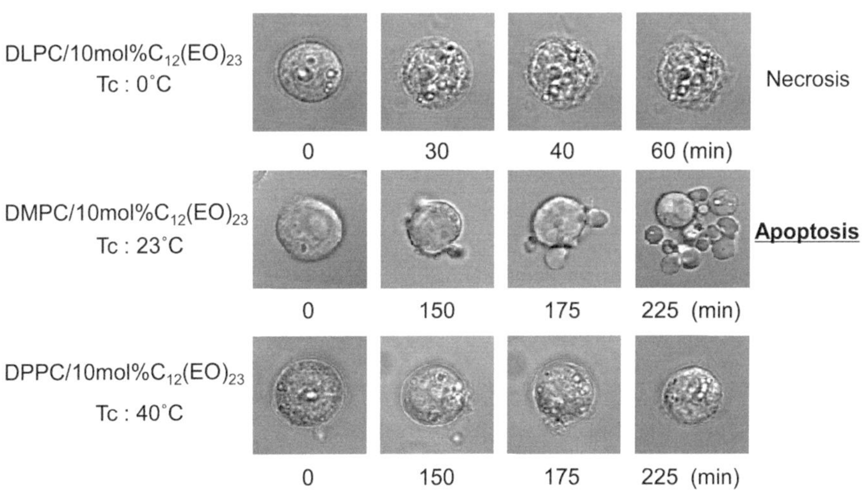

Fig. 11.6 Morphological changes in HL-60 cells treated with HL. Photographs taken 80× magnification under confocal laser microscope. [Phospholipid] = 300 μM. [$C_{12}(EO)_{23}$] = 33 μM

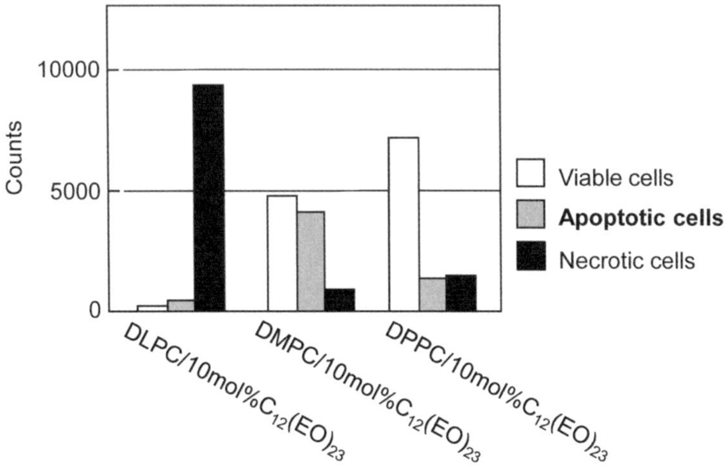

Fig. 11.7 Apoptotic and necrotic rate for HL-60 cells treated with HL. [Phospholipid] = 300 μM. [$C_{12}(EO)_{23}$] = 33 μM

Marked inhibitory effects of HL of DMPC/$C_{12}(EO)_{23}$ on the growth of HL-60 cells and induction of apoptosis were obtained without any antitumor drugs. It is noteworthy that two methylene groups in phospholipids could distinguish between apoptosis and necrosis in human leukemia cells.

11.6 Signaling Pathway for Apoptosis

The induction of apoptosis for HL-60 cells by HL composed of 90 mol% DMPC and 10 mol% $C_{12}(EO)_{10}$ was examined using flow cytometry, microphysiometer, immunoprecipitate, and western blot [31]. The time course of DNA fragmentation in HL-60 cells treated with HL was detected by flow cytometry with propidium iodide staining method. The results are shown in Fig. 11.8a. Apoptotic DNA reached a constant state 5 h after the treatment with HL. In addition, microphysiometry analysis (Fig. 11.8b) detects cellular responses by measuring the rate of acidification, indicating that induction of apoptosis had started after 1.5 h and completed for 6 h. A transient increase of acidification was detected after the treatment with HL (the first arrow from the left), which means that many protons were released from the cells by the opening of ion channels. Continuous treatment slowly increased the acidification and it reached a maximum (the second arrow from the left) followed by a gradual decrease (the third arrow from the left). These results indicate that a certain amount of HL accumulated in the cells and then induced apoptosis.

It is well known that the apoptosis signal is transduced by sequential activation of caspase family cysteine proteases. We examined caspase cascade for apoptosis

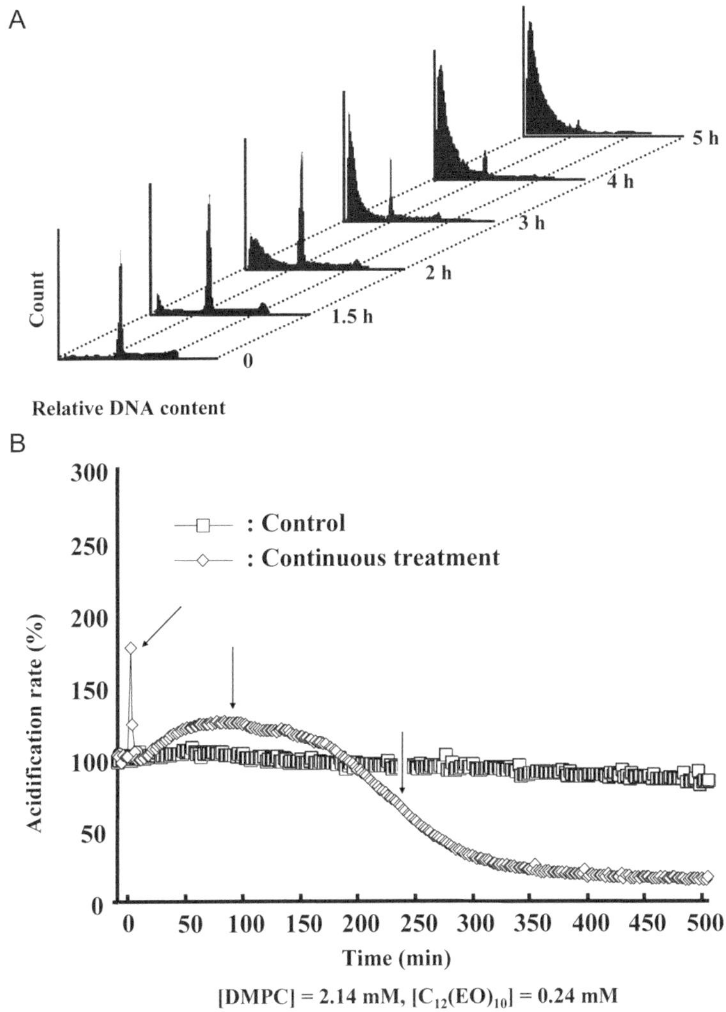

Fig. 11.8 (**a**) Detection of apoptotic cells treated with HL. (**b**) Acidification-rate response of HL-60 cells after the treatment with HL

induced by HL. Figure 11.9a shows concentration dependence of caspase inhibitors for apoptotic DNA rate of HL-60 cells. These results suggest that HL should induce apoptosis through the activation of caspase-8, 9, and 3. We also examined the time course of caspase activity for apoptosis induced by HL. The results are shown in Fig. 11.9b, suggesting that the apoptosis signal should be firstly carried out through the mitochondrial pathway and secondly through caspase-8 pathway.

Next, expression of Fas was observed, while Fas-associated death domain protein (FADD) was observed (Fig. 11.9c). These results suggest that the apoptosis signal induced by HL pass through Fas and FADD.

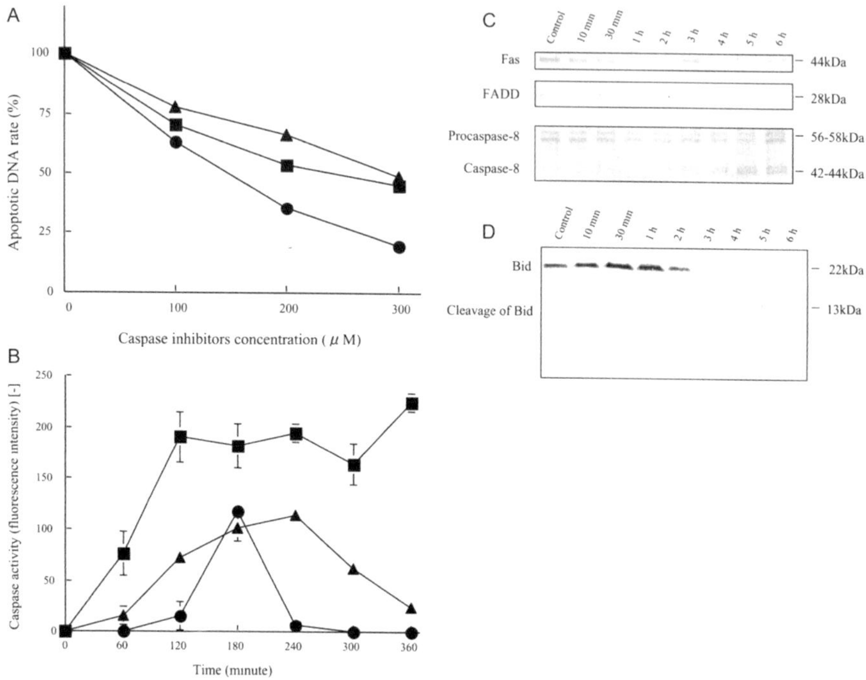

Fig. 11.9 (**a**) Concentration dependence of caspase inhibitors on apoptotic DNA rate for HL-60 cells treated with HL (●; caspase-3 inhibitor, ■; caspase-8 inhibitor, and ▲; caspase-9 inhibitor). (**b**) Caspases activity of HL-60 cells treated with HL (●; caspase-8, ■; caspase-3, and ▲; caspase-9). (**c**) Activation of Fas and FADD in HL-60 cells treated with HL. (**d**) Activation of Bid in HL-60 cells treated with HL

We also examined activation of BH3 interacting domain death agonist (Bid) in HL-60 cells by using western blotting analysis (Fig. 11.9d). Induction of Bid cleavage was observed after the treatment with hybrid liposomes for 3 h. This result suggests that the apoptosis signal could be transduced to mitochondria through Bid. However, the apoptotic signal through the mitochondrial pathway should be much faster than caspase-8 pathway.

We performed flow cytometric analysis of the mitochondrial transmembrane potential in HL-60 cells treated with HL to clarify the involvement of mitochondria pathway. The results are shown in (Fig. 11.10a). Interestingly, mitochondrial transmembrane potential decreased by adding HL, suggesting that the mitochondrial pathway is also implicated in HL-induced apoptosis. The cytochrome c released from mitochondria was detected 10 min after the treatment of HL (Fig. 11.10b). We further analyzed the effect of HL on the activation of caspase-9, which acts downstream of the mitochondrial pathway, using western blotting analysis. HL induced cleavage of procaspase-9 for 30 min after the treatment with HL.

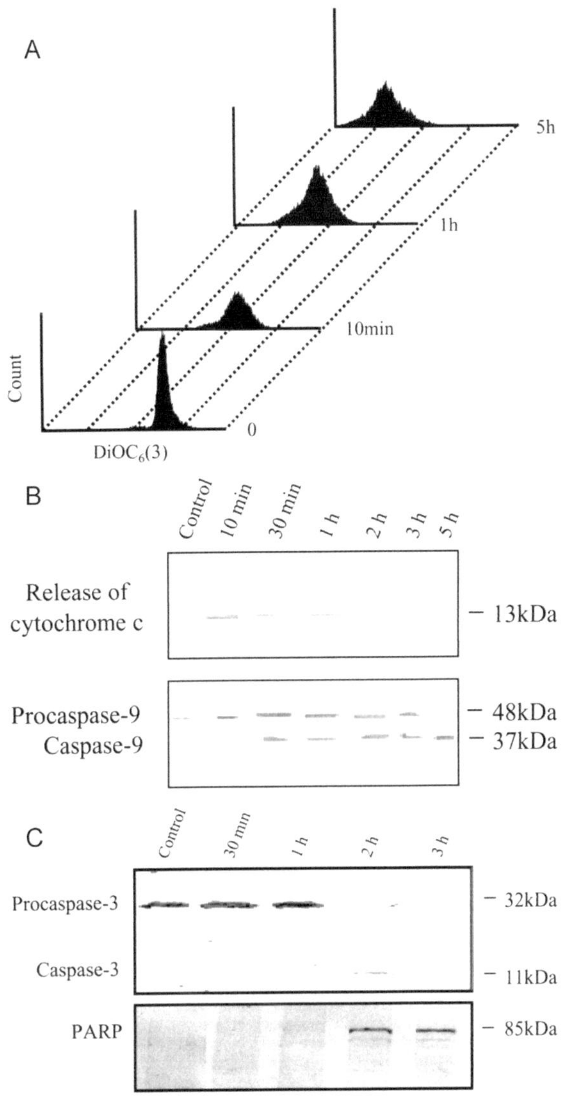

Fig. 11.10 (**a**) Decrease in mitochondrial membrane potential for HL-60 cells treated with HL. (**b**) Detection of mitochondrial cytochrome c derived from cytosol release and procaspase-9 cleavage in HL-60 cells treated with HL. (**c**) Detection of poly ADP ribose polymerase (PARP) and procaspase-3 cleavage in HL-60 cells treated with HL

We examined time-dependent caspase-3 activation by western blotting with an antibody that reacts both with pro- and activated-caspase-3 (Fig. 11.10c). Caspase-3 activation was detected 2 h after the treatment with HL. In addition, we detected 85 kDa breakdown product of poly (ADP ribose) polymerase (PARP), which is a substrate of activated caspase-3, using western blotting analysis. The p85 breakdown proteins (85 kDa) were detected 2 h after the treatment with HL. These data together implicate caspase-3 activation in apoptosis induced by HL in HL-60 cells.

We propose the mechanism of apoptosis induced by HL as shown in Fig. 11.11. That is, HL fused and accumulated into tumor cell membranes, and the apoptosis

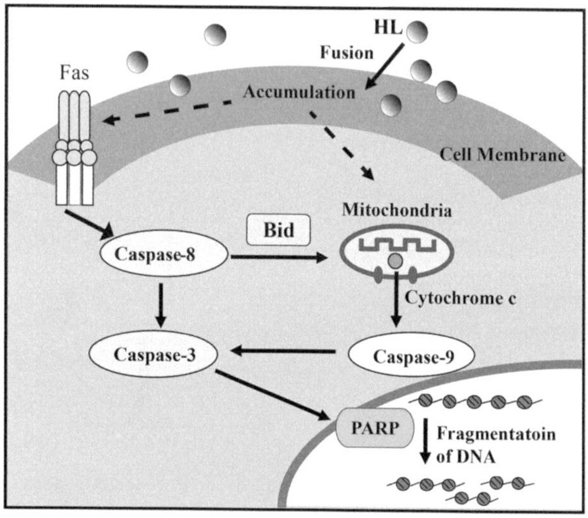

Fig. 11.11 Schematic representation of the signaling pathway for apoptosis induced by HL

signal passed firstly through mitochondria, caspase-9, and caspase-3, secondly through Fas, caspase-8, and caspase-3, and then reached the nucleus [31, 57]. We also reported that the mechanism of apoptosis induced by HL for lung cancer (RERF-LC-OK and A549) cells (adherent cells) was similar to that for HL-60 (non-adherent cell) cells [31].

11.7 Membrane Targeted Antitumor Mechanism

Recently, lipid-mediated apoptosis for tumor cells has been observed [3, 6, 39, 40]. However, as far as we know, there is no report on antitumor effects in relation to membrane fluidity of liposomes. Here, we describe the first successful experiment resulting in a good correlation between antitumor activity on the growth of human colon tumor (WiDr) cells and membrane fluidity of hybrid liposomes (HL-n) composed of 90 mol% DMPC and 10 mol% $C_{12}(EO)_n$ ($n=4$, 8, 10, 21, 23, and 25) [20].

We evaluated the fluidity of HL-n from fluorescence polarization (P) of 1,6-diphenyl-1,3,5-hexatriene (DPH). The fluorescence depolarization is caused by the molecular motion of the fluorescence probe, which reflects the microviscosity of the surrounding region. The emission at 430 nm originating from DPH was monitored upon excitation at 360 nm using a fluorescence polarization spectrophotometer. P of DPH into HL-n was calculated by $P=(I_{vv}-C_f I_{vh})/(I_{vv}+C_f I_{vh})$, where I is the fluorescence intensity and the subscripts v and h refer to the orientations, vertical and horizontal, respectively, for the excitation and analyzer polarizers in this sequence: e.g., I_{vh} indicates the fluorescence intensity measured with a vertical excitation polarizer and a horizontal analyzer polarizer.

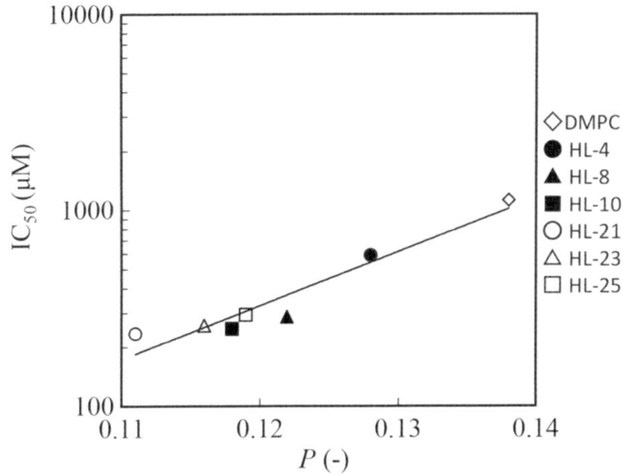

Fig. 11.12 Relationship between *P* of DPH in HL-n and IC_{50} of HL-n on the growth of WiDr cells

C_f is the grating correction factor, given by I_{hv}/I_{hh}. The results are shown in Fig. 11.12. It is worthy to note that a good correlation between the *P* values and IC_{50} was obtained. This indicates that HL-n having larger fluidity could suppress greater the growth of tumor cells.

Here, to get evidence about the fusion and accumulation of HL-n into tumor cell membranes, total internal reflection fluorescence (TIRF) microscopy was examined using 1-palmitoyl-2-[12-(7-nitro-2-1,3-benzoxadiazol-4-yl)amino]dodecanoyl]-*sn*-glycero-3-phosphocholine (NBDPC) as a fluorescence probe with a fluorescence microscope. TIRF micrographs of hepatocellular carcinoma (HCC: Hep-G2 and HuH-7) and normal liver (HC) cells after the treatment with 90 mol% DMPC/ 10 mol% $C_{12}(EO)_{23}$ (HL-23) including NBDPC are shown in Fig. 11.13a. In addition, fluorescence intensity of NBDPC incorporated into HL-23 was measured as shown in Fig. 11.13b. Interestingly, the fluorescence intensity of NBDPC incorporated into HL-23 in tumor cell membranes drastically increased after 10 min, though that in normal cell was almost constant. These results suggest that HL-23 could selectively fuse and accumulate into HCC cells but not so much into normal liver cells [22]. As shown in Fig. 11.13c, the mean *P* values of DPH-labeled both HCC cells were significantly decreased as compared to normal HC cells, which suggest that both HCC cell membranes are more fluid than those of normal liver cells. It was also found that the degree of accumulation of HL-23 into HCC cells was positively correlated well with the fluidity of cell membranes. These results suggest that HL-23 could selectively fuse and accumulate into the membranes of HCC cells due to their larger fluidity compared to those of normal liver cells (Fig. 11.13d).

Furthermore, we examined the IC_{50} of HL-23 on the growth of various tumor (cervical, colon, lung, liver, and stomach) cells and membrane fluidity of these tumor cells [21]. The results are shown in Fig. 11.14. The good linear correlation between antitumor activity (IC_{50}) and membrane fluidity of various tumor cells was

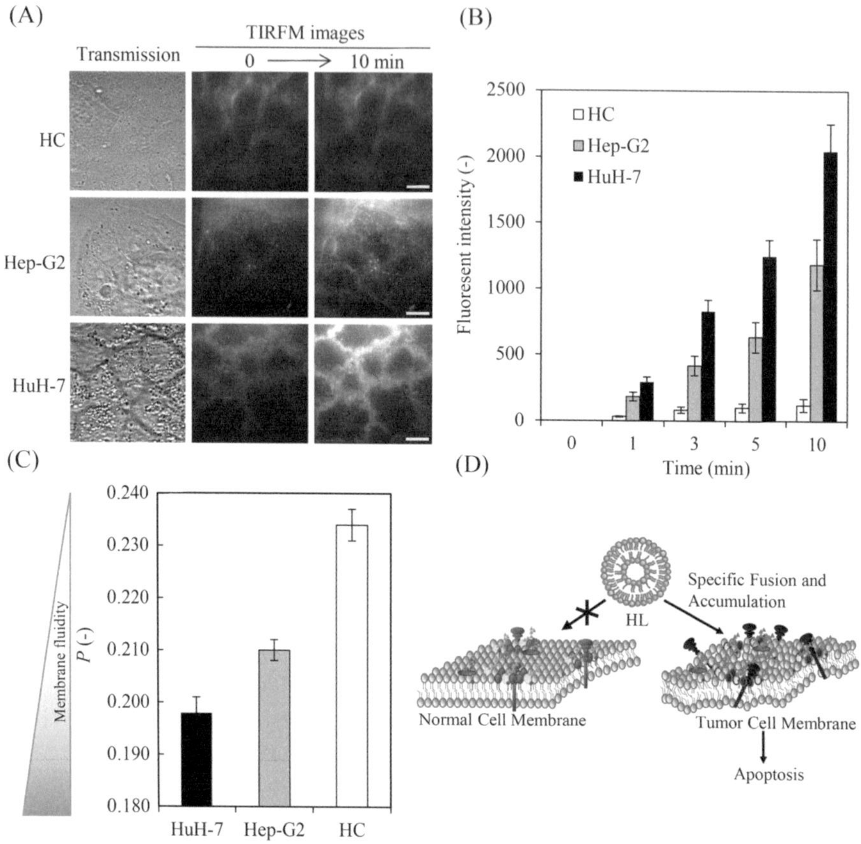

Fig. 11.13 Selective accumulation of HL-23 into tumor cells. (**a**) TIRF micrographs of HCC and HC cells after the treatment with HL-23. (**b**) Fluorescent intensity of NBDPC incorporated into HL-23 in HCC and HC cells. (**c**) *P* values of DPH in HCC and HC cells. (**d**) A schematic representation of membrane-targeted nanotherapy with HL-23 for tumor cells leading to apoptosis

observed. This result suggests that the antitumor activity should be related to membrane fluidity of various tumor cells.

A good correlation between membrane fluidity of HL-n and IC_{50} for the growth of tumor cells was observed for the first time. Furthermore, HL-23 distinguished between tumor cells and normal ones which had higher and lower membrane fluidities, respectively, then fused and accumulated preferentially into the membranes of tumor cells. It is worthy to note that a good relationship between antitumor activity and membrane fluidity was obtained for various tumor cells [57]. Thus, we have demonstrated that HL-n should target tumor cell membranes in relation to their lipid fluidity that provide the possibility of novel nanotherapy from a viewpoint of different biophysical characteristics of tumor cells and normal ones.

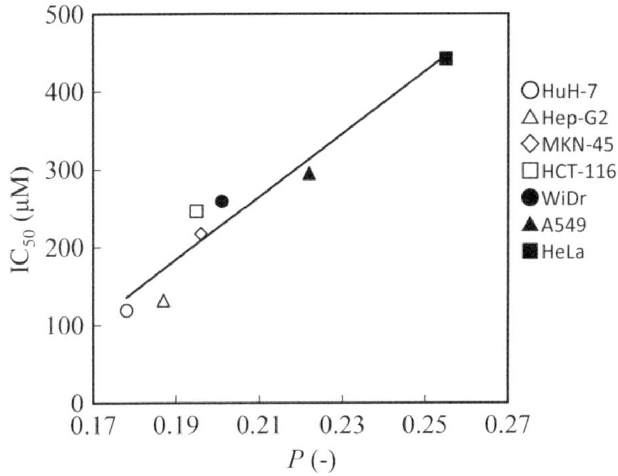

Fig. 11.14 Relationship between IC_{50} of HL-23 on the growth of various tumor cells and *P* of DPH in various tumor cells

11.8 Therapeutic Effects and Toxicity In Vivo

Screening for therapeutic effects is generally carried out in mice or rats bearing tumor cells. We have already reported the inhibitory effects of HL in mice intraperitoneally inoculated with mice B16 melanoma and B lymphoma cells in vivo [10, 15–17, 28, 29, 35–37]. Here, we examined therapeutic effects of HL using a mouse model of Lewis lung carcinoma [15], neuroblastoma [38], and human breast carcinoma (unpublished result) in vivo. The animals were handled during the study in accordance with the guidelines for animal's experiment of Japanese law.

The therapeutic effects of 90 mol% DMPC/10 mol% $C_{12}(EO)_{10}$ (HL-10) using a mouse model of carcinoma were examined in vivo. HL-10 were intraperitoneally administered once each day for 13 days after Lewis lung carcinoma (LLC) cells (1.0×10^5 cells) were inoculated intraperitoneally to C57BL/6 mice. The median lifespan was calculated using the following equation: median lifespan = (median survival days for treatment group)/(median survival days for control group) × 100. The results are shown in Fig. 11.15. Interestingly, remarkable prolonged survival (>400 %) of mice was obtained after the treatment with HL-10. The survival times in the treated groups with the HL-10 were more than 1 year [52]. This result suggests that the lung carcinoma mice model treated with HL should be recovered completely.

Next, we examined the therapeutic effects of 90 mol% DMPC/10 mol% $C_{12}(EO)_n$ (HL-n; $n = 21$, 23, and 25) using hepatic metastasis mice model [53]. HL-n were intraperitoneally administered once a day for 14 days after mice neuroblastoma (C1300N18) cells (5.0×10^4 cells) were intrasplenic inoculated into the A/J mice. A large number of hepatic metastasis was observed in the control group, whereas no hepatic metastasis was observed after the treatment with HL-25 (Fig. 11.16a). Remarkable prolonged survival (140 ~ 190 %) was obtained in hepatic metastasis mice after the treatment with HL (Fig. 11.16b). These results

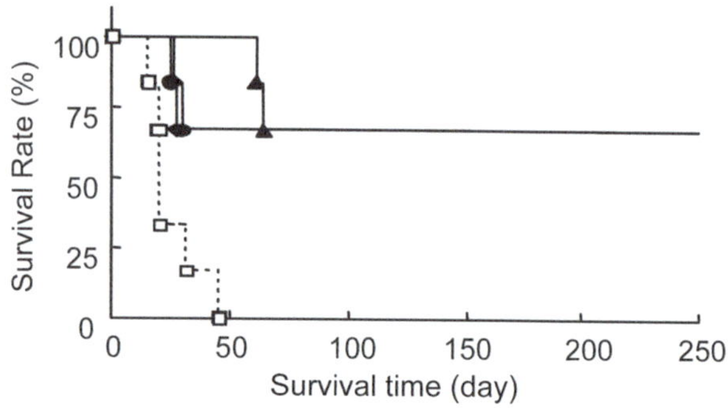

Fig. 11.15 Survival curves of mice treated with HL after the intraperitoneal inoculation of Lewis lung carcinoma (LLC) cells. Control (■), dose for DMPC was 67.8 mg/kg (♦), 339 mg/kg (●), and 678 mg/kg (▲), respectively

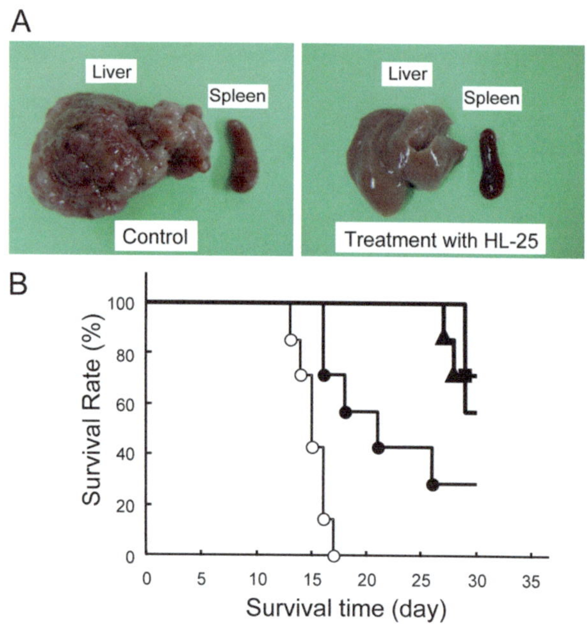

Fig. 11.16 Therapeutic effect of HL-n ($n = 21$, 23, and 25) for hepatic metastasis mice model. (**a**) Photographs of liver and spleen in hepatic metastasis mice model after the treatment with HL-25. (**b**) Survival curves of mice treated with HL-n after the intrasplenic inoculation of mice neuroblastoma cells. Control (○), HL-21 (▲), HL-23 (●), and HL-25 (■)

suggest that HL should be effective for inhibiting metastasis of mice neuroblastoma cells to liver.

Furthermore, we examined the therapeutic effects of 95 mol% DMPC/5 mol% $C_{12}(EO)_{21}$ (HL-21) using xenograft mice models after the subcutaneous inoculation of human breast tumor (MDA-MB-453) cells in vivo (unpublished result). MDA-MB-453 cells (5.0×10^6 cells) suspended into Matrigel were subcutaneously inoculated to dorsal of nude mice. HL-21 was intravenously administered once each day for 2 weeks. The tumor volume was calculated using the equation of $V = 0.5 \times a^2 \times b$, where a and b denote the smallest and longest superficial diameter, respectively. The results are shown in Fig. 11.17a. Photographs of tumor in xenograft mice model are shown in Fig. 11.17b. The tumor volume in the treatment group was almost constant, although that in the control group increased. After the

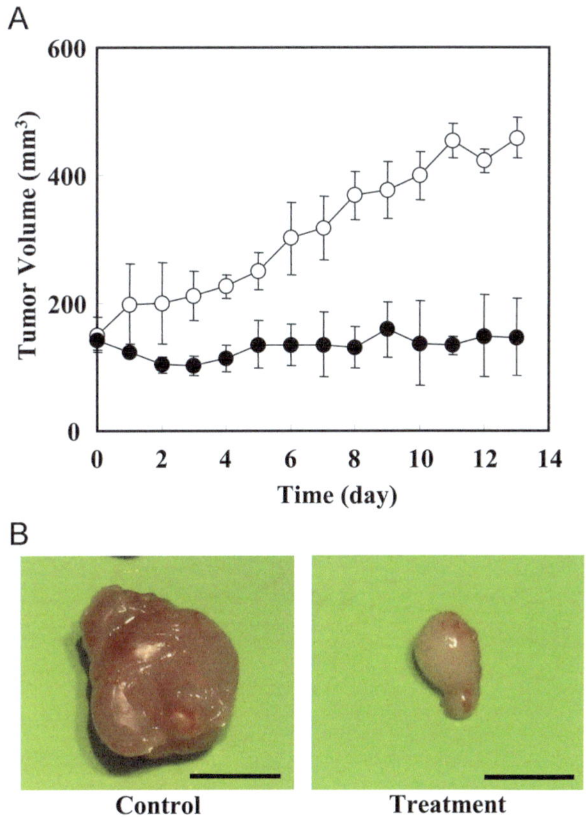

Fig. 11.17 Therapeutic effects of HL-21 for xenograft mice models after the subcutaneous inoculation of human breast rumor (MDA-MB-453) cells. (**a**) The tumor volume of mice treated with HL-21. Control (○) and HL-21 (●). (**b**) Photographs of tumor in xenograft mice model

administration of HL-21 for 2 weeks, the median of tumor volume was 458 ± 32 mm^3 and 147 ± 61 mm^3 in the control and treatment groups, respectively. It is noteworthy that the remarkable reduction of tumor volume (70 %) in mice inoculated MDA-MB-453 cells was obtained after the treatment with HL-21.

Furthermore, we examined the therapeutic effects of HL-25 using mice of acute lymphatic leukemia (ALL) with peritoneal dissemination after the intraperitoneal treatment with HL-25. The results are shown in Fig. 11.18 [11]. The photographs show that the mice of the control group had distended abdominal regions (Fig. 11.18a). Interestingly, massive ascites was obtained in the peritoneal cavity of the mice of the control group after dissection. The results are shown in Fig. 11.18b. The mice treated with DMPC (0.45 ± 0.16 g, $p < 0.01$) and HL-25 (0.62 ± 0.08 g, $p < 0.01$) had a significantly lower volume of ascites compared with the mice in the control group (2.74 ± 0.46 g). Statistical significance between DMPC liposomes and HL-25 was not obtained. Furthermore, survival rates are shown in Fig. 11.19. The median survival time for mice treated with HL-25 was higher than 420 days without death, although that for mice in the control group was 84.7 ± 2.2. On the other hand, the median survival time for mice treated with

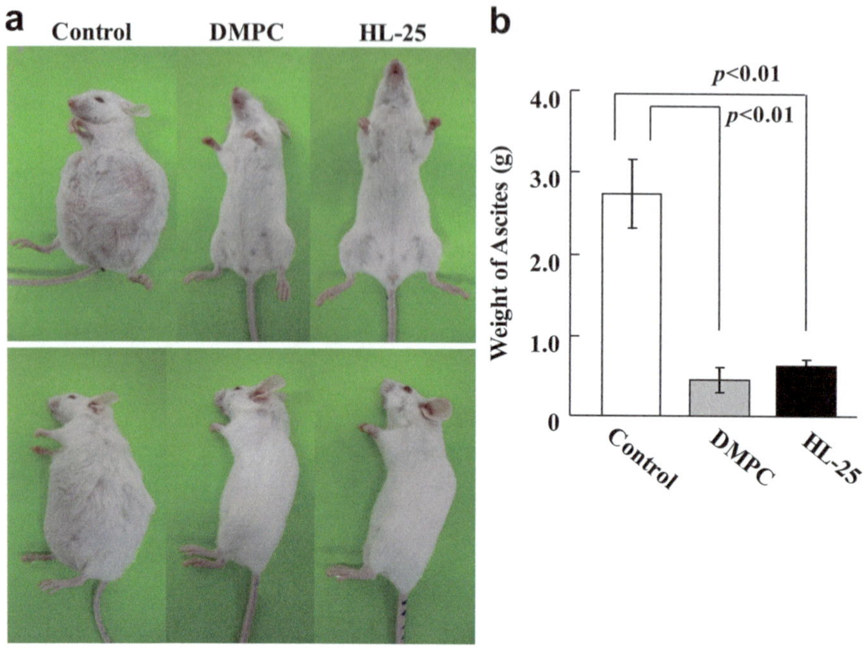

Fig. 11.18 Therapeutic effects of HL-25 for model mice of ALL with peritoneal dissemination in vivo. (**a**) Photographs of ascites-bearing mice of the control group or mice treated with HL-25 for 14 days after being inoculated with MOLT-4 cells intraperitoneally. (**b**) The weight of ascites in mice inoculated with MOLT-4 cells. Dose for DMPC: 136 mg/kg

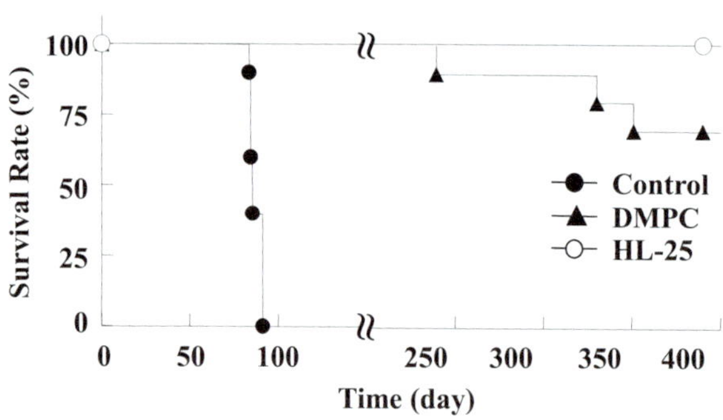

Fig. 11.19 Survival curves for model mice of ALL with peritoneal dissemination treated with HL-25 after the intraperitoneal inoculation of MOLT-4 cells. Dose for DMPC: 136 mg/kg

DMPC was 385 ± 62, and the statistical significance between DMPC liposomes and HL-25 was obtained ($p < 0.05$). It is noteworthy that a significantly prolonged survival (>400 %, $p < 0.01$) was obtained in the mice treated with HL-25. These results indicate that HL-25 could strongly inhibit the growth of MOLT-4 cells in vivo.

The safety of 95 mol% DMPC/5 mol% $C_{12}(EO)_{23}$ (HL-23) was examined using normal rats. Six rats were assigned to each group by the stratified continuous randomization method and HL-23 (dose for DMPC was 67.8, 136, and 203 mg/kg) were intravenously administered via a vein once a day for 6 months. The rats were weighed during the experimental period. The numbers of red and white blood cells of rats treated with HL-23 were within normal limits. In addition, all of the other biochemical parameters, such as ALP, GOT, and GPT activities, as well as levels of albumin, urea, nitrogen, creatinine, glucose, total protein, calcium, inorganic phosphorus, sodium, potassium, and chloride, were not significantly different from those observed in the controls. Furthermore, no weight loss was observed in the rats. These results indicate that HL should have no side effects in vivo.

11.9 Clinical Application

Clinical applications of HL without any drug for 10 patients with lymphoma (2 patients), gastric cancer (2 patients), kidney cancer (one patient), mammary cancer (one patient), pharyngeal cancer (one patient), hepatoma (one patient), gallbladder cancer (one patient), and rectal cancer (one patient) were examined after passing the committee of bioethics at the different hospitals. Informed consents were obtained in accordance with the Declaration of Helsinki. We report the pilot study using 95 mol% DMPC/5 mol% $C_{12}(EO)_{23}$ (HL-23) for one patient with advanced stage B lymphoma who did not recover by any chemotherapeutics after the approval of the Bioethics Committee at the National Kumamoto Hospital. HL-23 was administered intravenously once every day at a dose of 11.0–16.5 mg/kg for DMPC for 10 months. HL-23 was also locally administered (two times/week, dose for DMPC was 0.5 mg/kg) to the lymph node neoplasm (solid tumor) for 2 months and photographs were taken using ultra sonical echo [10, 57]. There were no abnormal findings post administration on routine blood test and hematochemistry (data not shown). A prolonged survival, more than 1 year, was attained in this patient after the intravenous injection of HL-23 without any side effects. It should be noted that a remarkable reduction of solid tumor was observed after the local administration (two times/week) of HL-23, as shown in Fig. 11.20a.

Moreover, we examined induction of apoptosis in solid tumor using the TUNEL method. Tissue sections were prepared from solid tumor and stained by a TUNEL-based method. A fluorescence micrograph is shown in Fig. 11.20b. The green color was obtained from a solid tumor after the injection of HL-23, indicating that HL-23 could induce apoptosis in solid tumor. During the period of treatment, no abnormal findings were obtained on the basis of various hematological and biochemical tests. These results demonstrate that HL should be safe and effective in clinical applications.

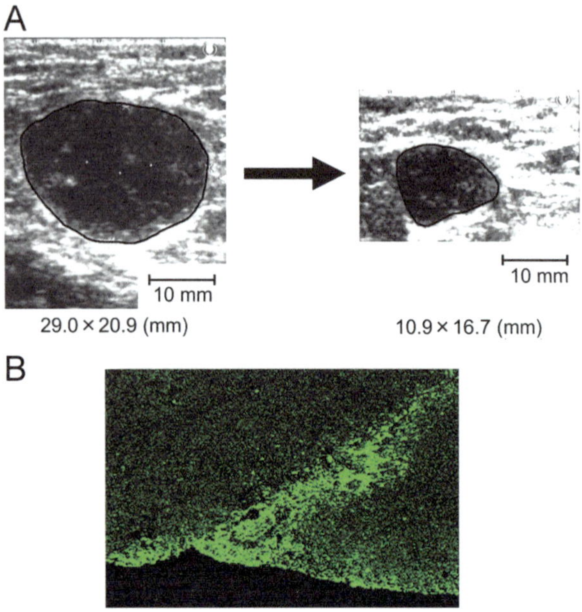

Fig. 11.20 A pilot clinical test of HL in a patient with recurrent malignant lymphoma. (**a**) Lymph node neoplasm (solid tumor) after the local administration of HL for 2 months. (**b**) A fluorescence micrograph of solid tumor using the TUNEL method after the treatment with HL

11.10 Conclusion

A summary of the noteworthy aspects of this study is as follows: (A) Two methylene groups in phospholipids distinguish between apoptosis and necrosis for tumor cells were confirmed. High inhibitory effects of hybrid liposomes (HL) themselves composed of 90 mol% DMPC/10 mol% $C_{12}(EO)_{23}$ on the growth of tumor cells were obtained without any drug. It is also of interest that the induction of apoptosis was obtained by HL. On the other hand, necrosis was observed for HL of 90 mol% DLPC/10 mol% $C_{12}(EO)_{23}$. In the case of HL composed of 90 mol% DPPC and 10 mol% $C_{12}(EO)_{23}$, neither apoptosis nor necrosis was observed. From the above-mentioned results, we have focused on chemotherapy with HL composed of DMPC and $C_{12}(EO)_n$ in vitro, in vivo, and in clinical applications. (B) Induction of apoptosis by HL was obtained using flow cytometry, DNA agarose gel electrophoresis, and fluorescence micrograph. The pathway of apoptosis induced by HL was clarified. That is, HL fused and accumulated in tumor cell membranes, and the apoptotic signal passed firstly through mitochondria, caspase-9, and caspase-3, secondly through Fas, caspase-8, and caspase-3, and then reached the nucleus. (C) A good correlation between the *P* values of 1,6-diphenyl-1,3,5-hexatriene incorporated into HL-n ($n = 4$, 8, 10, 21, 23, and 25) and 50 % inhibitory concentration (IC_{50}) was obtained. This indicates that HL-n having larger fluidity could suppress greater the growth of tumor cells. (D) A good linear correlation between IC_{50} and membrane fluidity of various tumor (cervical, rectal, colon, lung, liver, brain, and stomach) cells was also observed. This demonstrates that growth inhibition and apoptosis for tumor cells by HL-23 provides the possibility of therapy from a viewpoint of biophysical characteristics of tumor cell membranes. (E) Significantly prolonged survival and remarkable reduction of tumor volume were obtained using mice model of carcinoma after the treatment with HL-23 without any drug in vivo. There were no abnormal findings on blood test, hematochemistry, relative organ weight, and autopsy of normal rats after administering HL-23 on the basis of chronic toxicity tests. (F) In clinical applications, prolonged survival was attained in patients with lymphoma after the intravenous injection of HL-23 with a diameter of 80 nm, which could avoid RES, without any side effects after the approval of the bioethics committee. Remarkable reduction of neoplasm was obtained after the local administration of HL-23.

In conclusion, the membrane targeted chemotherapy with drug-free HL was established without any side effects for the first time.

References

1. Allen TM, Hansen CJ, Martin F, Redemann C, Yan YA (1991) Liposomes containing synthetic lipid derivatives of poly(ethylene glycol) show prolonged circulation half-lives in vivo. Biochem Biophys Acta 1066(1):29–36
2. Dinney CPN, Utsugi T, Fidler IJ, von Eschenbach AC, Killion JJ (1992) Immunotherapy of murine renal adenocarcinoma by systemic administration of liposomes containing the synthetic macrophage activator cgp 31362 or cgp 19835a in combination with interleukin 2 or – interferon. Cancer Res 52(5):1155–1161
3. Diomede L, Colotta F, Piovani B, Re F, Modest EJ, Salmona M (1993) Induction of apoptosis in human leukemic cells by the ether lipid 1-octadecyl-2-methyl-rac-glycero-3-phosphocholine. A possible basis for its selective action. Int J Cancer 53(1):124–130
4. Fendler JH (ed) (1982) Membrane mimetic chemistry. Wiley, New York
5. Fidler IJ, Lopez-Berestein G (eds) (1989) Liposomes in the therapy of infectious diseases and cancer. Alan R Liss, New York
6. Fujimoto K, Iwasaki C, Kawaguchi H, Yasugi E, Oshima M (1999) Cell membrane dynamics and the induction of apoptosis by lipid compounds. FEBS Lett 446(1):113–116
7. Goto K, Matsumoto Y, Ueoka R (1995) Hybrid liposomes coupled to steric control with high enantioselectivity. J Org Chem 60(11):3342–3346
8. Goto K, Tanaka Y, Matsumoto Y, Ueoka R (2008) Induction of apoptosis of human tumor cells by hybird liposomes including docosahexaenoic acid. Bioorg Med Chem Lett 18 (6):1880–1883
9. Ichihara H, Nagami H, Yamamoto K, Matsumoto Y, Ueoka R (2003) Chemotherapy with hybrid liposomes without any drug in vivo. Yakugaku Zasshi 123(1):25–34
10. Ichihara H, Nagami H, Kiyokawa T, Matsumoto Y, Ueoka R (2008) Chemotherapy using hybrid liposomes along with induction of apoptosis. Anticancer Res 28(2):1187–1196
11. Ichihara H, Ueno J, Umebayashi M, Matsumoto Y, Ueoka R (2011) Chemotherapy with hybrid liposomes for acute lymphatic leukemia leading to apoptosis in vivo. Int J Pharm 406 (1–2):173–178
12. Imamura C, Kemura Y, Matsumoto Y, Ueoka R (1997) Remarkable inhibitory effects of hybrid liposomes on the growth of HL-60 cells coupled to induction of apoptosis. Biol Pharm Bull 20(10):1119–1121
13. Ishiyama M, Shiga M, Sasamoto K, Mizoguchi M (1993) A new sulfonated tetrazolium salt that produces a highly water-soluble formazan dye. Chem Pharm Bull 41(6):1118–1122

14. Iwamoto Y, Matsumoto Y, Ueoka R (2005) Induction of apoptosis of human lung carcinoma cells by hybrid liposomes containing polyoxyethylenedodecyl ether. Int J Pharm 292 (1–2):231–239
15. Kanno A, Kodama R, Terada Y, Matsumoto Y, Ueoka R (1998) Chemotherapy with hybrid liposomes for cancer: an experimental study. Drug Deliv Syst 13(1):101–105
16. Kanno A, Terada Y, Tsuzaki Y, Matsumoto Y, Ueoka R (1999) Therapeutic effects of hybrid liposomes on mice model of lung carcinoma. Drug Deliv Syst 14(1):37–42
17. Kanno A, Tsuzaki Y, Miyagi M, Matsumoto Y, Ueoka R (1999) Chemotherapy with hybrid liposomes for melanomatosis. Biol Pharm Bull 22(9):1013–1014
18. Kitamura I, Kochi M, Matsumoto Y, Ueoka R, Kuratsu J, Ushio Y (1996) Intrathecal chemotherapy with 1,3-bis(2-chloroethyl)-1-nitorsourea encapsulated into hybrid liposomes for meningeal gliomatosis: an experimental study. Cancer Res 56(17):3986–3992
19. Klibanov AL, Maruyama K, Torchilin VP, Huang L (1990) Amphipathic polyethyleneglycols effectively prolong the circulation time of liposomes. FEBS Lett 268(1):235–237
20. Komizu Y, Matsumoto Y, Ueoka R (2006) Membrane targeted chemotherapy with hybrid liposomes for colon tumor cells leading to apoptosis. Bioorg Med Chem Lett 16 (23):6131–6134
21. Komizu Y, Nakata S, Goto K, Matsumoto Y, Ueoka R (2011) Membrane-targeted nanotherapy with hybrid liposomes for tumor cells leading to apoptosis. ACS Med Chem Lett 2(4):275–279
22. Komizu Y, Ueoka H, Ueoka R (2012) Selective accumulation and growth inhibition of hybrid liposomes to human hepatocellular carcinoma cells in relation to fluidity of plasma membranes. Biochem Biophys Res Commun 418(1):81–86
23. Konno H, Suzuki H, Tadakuma T, Kumai K, Yasuda T, Kubota T, Ohta S, Nagaike K, Hosokawa S, Ishibiki K et al (1987) Antitumor effect of adriamycin entrapped in liposomes conjugated with anti-human alpha-fetoprotein monoclonal antibody. Cancer Res 47 (16):4471–4477
24. Lasic DD (ed) (1993) Liposomes: from physics to applications. Elsevier Science Publishing Company, Amsterdam
25. Matsumoto Y, Ueoka R (1990) Studies on thermodynamics for hydrolysis. 3. Isokinetic temperature related to molecular location of reactants in coaggregates. J Org Chem 55 (22):5797–5799
26. Matsumoto Y, Yamada E, Hirano J, Oshige M, Iio M, Iwahara M, Ueoka R (1993) Specific inhibitory effect of hybrid liposomes on growth of human lymphoma human lymphocyte B hybridoma cells in vitro. Biol Pharm Bull 16(2):213–215
27. Matsumoto Y, Imamura C, Ito T, Taniguchi C, Ueoka R (1995) Specific hybrid liposomes composed of phosphatidylcholine and polyoxyethylenealkyl ether with markedly enhanced inhibitory effect on the growth of tumor cells in vitro. Biol Pharm Bull 18(10):1456–1458
28. Matsumoto Y, Kato T, Kemura Y, Tsuchiya M, Yamamoto M, Ueoka R (1999) Intracellular responses of hybrid liposomes against leukemia cells related to apoptosis with antitumor activity. Chem Lett 28(1):53–54
29. Matsumoto Y, Kato T, Iseki S, Suzuki H, Nakano K, Iwahara M, Ueoka R (1999) Remarkably enhanced inhibitory effects of hybrid liposomes on the growth of specific tumor cells. Bioorg Med Chem Lett 9(14):1937–1940
30. Matsumoto Y, Kato T, Suzuki H, Hirose S, Naiki Y, Hirashima M, Ueoka R (2000) Highly specific inhibitory effects of three-component hybrid liposomes including sugar surfactants on the growth of glioma cells. Bioorg Med Chem Lett 10(23):2617–2619
31. Matsumoto Y, Iwamoto Y, Matsushita T, Ueoka R (2005) Novel mechanism of hybrid liposomes-induced apoptosis in human tumor cells. Int J Cancer 115(3):377–382
32. Matsumoto Y, Tanaka Y, Goto K, Ueoka R (2008) Specific fixed aqueous layer of three-component hybrid liposomes related to inhibition of hepatoma cells growth. Chem Lett 37 (1):118–119

33. Matsuoka Y, Nagata M, Komizu Y, Ichihara H, Kawase S, Sawada Y, Matsumoto Y, Ueoka R (2008) Inhibitory effects of hybrid liposomes on the growth of gastric tumor cells established from cotton rats. Yakugaku Zasshi 128(10):1485–1492
34. Nagami H, Yamamoto K, Ichihara H, Matsumoto Y, Ueoka R (2003) Induction of apoptosis by hybrid liposomes without antitumor drugs toward human breast tumor cells. J Chem Eng Jpn 36(9):1127–1129
35. Nagami H, Nakano K, Ichihara H, Matsumoto Y, Ueoka R (2006) Two methylene groups in phospholipids distinguish between apoptosis and necrosis for tumor cells. Bioorg Med Chem Lett 16(4):782–785
36. Nagami H, Matsumoto Y, Ueoka R (2006) Chemotherapy with hybrid liposomes for lymphoma without drugs in vivo. Int J Pharm 315(1–2):167–172
37. Nakano K, Iwamoto Y, Takata W, Matsumoto Y, Ueoka R (2002) Specific accumulation and growth inhibitory effects of hybrid liposomes to hepatoma cells in vitro. Bioorg Med Chem Lett 12(22):3251–3254
38. Negishi Y, Yoshizawa J, Yamazaki Y, Matsumoto Y, Ueoka R (2005) Inhibitory effect of drug-free hybrid liposomes on metastasis of mouse neuroblasoma. J Jpn Soc Pediatr Surg 41 (7):930–937
39. Obeid LM, Knardic CM, Karolak LA, Hannun YA (1993) Programmed cell death induced by ceramide. Science 259(5102):1769–1771
40. Ohta H, Yatomi Y, Seweeney EA, Hakomori S, Igarashi Y (1994) A possible role of sphingosine in induction of apoptosis by tumor necrosis factor-alpha in human neutrophils. FEBS Lett 355(3):267–270
41. Papahadyopoulos D (ed) (1978) Liposomes and their uses in biology and medicine. New York Academy of Sciences, New York
42. Tanaka Y, Goto K, Matsumoto Y, Ueoka R (2008) Remarkably high inhibitory effects of docosahexaenoic acid incorporated into hybrid liposomes on the growth of tumor cells along with apoptosis. Int J Pharm 359(1–2):264–271
43. Tanoue O, Baba M, Tokunaga Y, Goto K, Matsumoto Y, Ueoka R (1999) Steric-control for the enantioselective hydrolysis of amino acid esters with membrane-bound enzyme models. Tetrahedron Lett 40(11):2129–2132
44. Tanoue O, Ichihara H, Goto K, Matsumoto Y, Ueoka R (2003) Steric-control for the enantioselective hydrolysis of amino acid esters in hybrid membrane systems. Chem Pharm Bull 51(2):224–226
45. Ueoka R, Matsumoto Y (1984) Isokinetic discrimination of the reaction field for the enantioselective hydrolysis. -Hydrophobic interactions in the micellar and bilayer systems-. J Synth Org Chem Jpn 42(12):1088–1094
46. Ueoka R, Moss RA, Swarup S, Matsumoto Y, Strauss G, Murakami Y (1985) Extraordinary micellar enantioselectivity coupled to altered aggregate structure. J Am Chem Soc 107 (7):2185–2186
47. Ueoka R, Matsumoto Y, Yoshino T, Hirose T, Kikuchi J, Murakami Y (1986) Enzyme-like enantioselective catalysis in the specific coaggregate system of vesicular and micellar surfactants. Chem Lett 15(1):127–130
48. Ueoka R, Matsumoto Y, Yoshino T, Watanabe N, Omura K, Murakami Y (1986) Remarkable steric control for the enantioselective cleavage of amino acid esters. Chem Lett 15 (10):1743–1746
49. Ueoka R, Matsumoto Y, Yoshino T, Hirose T (1986) Enantioselective and diastereoselective cleavage of amino acid esters in coaggregates of controlled size. Tetrahedron Lett 27 (10):1183–1186
50. Ueoka R, Matsumoto Y, Moss AR, Swarup S, Sugii A, Harada K, Kikuchi J, Murakami Y (1988) Membrane matrix for the hydrolysis of amino acid esters with marked enantioselectivity. J Am Chem Soc 110(5):1588–1595

51. Ueoka R, Matsumoto Y, Oyama H, Takekuma H, Iio M (1988) Effects of flavonoids in hybrid membranes on the activity of tripeptide as an enzyme model and their inhibition of hybridoma growth in vitro. Chem Pharm Bull 36(11):4640–4643
52. Ueoka R, Dozono H, Matsumoto Y, Cho M, Kitahara K, Kato Y (1990) Steric control for the enantioselective hydrolysis of amino acid esters in coaggregates composed of phosphatidylcholine and Triton X-100. Chem Pharm Bull 38(1):219–220
53. Ueoka R, Matsumoto Y, Dozono H, Yano Y, Hirasa H, Goto K, Kato Y (1990) Remarkable steric control for the enantioselective hydrolytic-cleavage of amino acid esters responding to ionic strength. Tetrahedron Lett 31(37):5335–5538
54. Ueoka R, Yamada E, Yamashita O, Matsumoto Y, Kato Y (1991) Composition- and temperature-sensitive membranes composed of phosphatidylcholine and Triton X-100 for the remarkably enhanced stereoselective-hydrolysis. Tetrahedron Lett 32(45):6597–6600
55. Ueoka R, Matsumoto Y, Kanno A, Tsuzaki K, Ichihara H (2000) Marked therapeutic effect of dimyristoylphosphatidylcholine liposomes on carcinoma mice model in vivo. Biol Pharm Bull 23(10):1262–1263
56. Ueoka R, Matsumoto Y, Hirose S, Goto K, Goto M, Furusaki S (2002) Remarkably enhanced inhibitory effects of three-component hybrid liposomes including sugar surfactants on the growth of lung carcinoma cells. Chem Pharm Bull 50(4):563–565
57. Ueoka R, Matsumoto Y, Goto K, Ichihara H, Komizu Y (2011) Membrane targeted chemotherapy with hybrid liposomes for tumor cells leading to apoptosis. Curr Pharm Des 17 (17):1709–1719

Chapter 12
Antitumor Complexes Formed by Oleic Acid and Molten Globule Intermediates of Proteins

Kunihiro Kuwajima and Takashi Nakamura

Abstract *H*uman α-lactalbumin *m*ade *le*thal to *t*umor cells (HAMLET), a complex formed by human α-lactalbumin and oleic acid, has a unique apoptotic activity for the selective killing of tumor cells. It has been hypothesized that HAMLET expresses its antitumor activity in the stomach of breast-fed infants, thereby protecting the infants from tumor development, and in this case the protein portion of HAMLET is in a flexible molten globule state. On the other hand, the primary biological function of α-lactalbumin in its rigid native structure is to modify the specificity of galactosyltransferase to produce lactose in mammary glands. α-Lactalbumin thus provides a unique example, in which a single globular protein has two independent biological functions in quite different locations. In this article, we summarize the historical background and recent progress of the studies on HAMLET and related protein-fatty acid complexes. It is shown that oleic acid forms an antitumor complex not only with α-lactalbumin but also with various globular proteins in the molten globule state by nonspecific hydrophobic interactions, although the strength of the activity varies somewhat depending on the protein species. Similarly, not only oleic acid but also various cytotoxic fatty

K. Kuwajima (✉)
Okazaki Institute for Integrative Bioscience and Institute for Molecular Science, 5-1 Higashiyama, Myodaiji, Okazaki, Aichi 444-8787, Japan

Department of Functional Molecular Science, School of Physical Sciences, The Graduate University for Advanced Studies (Sokendai), Myodaiji, Okazaki, Aichi 444-8787, Japan

The Center for the Promotion of Integrated Sciences, The Graduate University for Advanced Studies (Sokendai), Shonan Village, Hayama, Kanagawa 240-0193, Japan

School of Computational Sciences, Korea Institute for Advanced Study (KIAS), Dongdaemun-gu, Seoul 130-722, South Korea

Department of Physics, School of Science, University of Tokyo, 7-3-1 Hongo, Bunkyo-ku, Tokyo 113-0033, Japan
e-mail: kuwajima@ims.ac.jp; kuwajima@kk.alumni.u-tokyo.ac.jp

T. Nakamura
Okazaki Institute for Integrative Bioscience and Institute for Molecular Science, 5-1 Higashiyama, Myodaiji, Okazaki, Aichi 444-8787, Japan

Fuji Pharma Co., Ltd., 5-7 Sanban-cho, Chiyoda-ku, Tokyo 102-0075, Japan

M. Terazima et al. (eds.), *Molecular Science of Fluctuations Toward Biological Functions*, DOI 10.1007/978-4-431-55840-8_12

acids (mono- and polyunsaturated *cis* fatty acids) are bound to α-lactalbumin in the molten globule state and exhibit the antitumor activities. It is thus concluded that the protein portion of these complexes is not the origin of their cytotoxicity but plays a role as the delivery carrier of cytotoxic fatty acid molecules into tumor cells across the cell membrane.

Keywords HAMLET • α-Lactalbumin • NMR • Molten globule • Oleic acid • Antitumor complex • Apoptosis

12.1 Introduction

The classical structure-function paradigm in protein science assumes that the specific three-dimensional structure of a protein is required for expression of its biological function. Therefore, one gene is translated into a protein amino acid sequence, from which the unique three-dimensional structure of the protein is organized, and finally the biological function is expressed. This classical view of the structure-function paradigm has been the conceptual basis for protein-folding studies [1] and a number of structural-genomics projects [2–4]. However, as written in the chapter by Xue and Uversky [5] in this book, this classical view is changing now. More than a decade ago, Dunker and his colleagues proposed a protein trinity hypothesis [6], in which proteins can exist in a trinity of structures (the ordered state, the molten globule, and the random coil), and any of the three states can correspond to a biologically functional native protein structure [6].

Therefore, it is now increasingly evident that not only rigid ordered structures but also fluctuating non-ordered structures are biologically functional. A number of examples have been reported in the form of enzymatic molten globules [7–18], in which an enzyme protein exhibits significant or even full enzymatic activity in the molten globule state under a physiological condition. The enzymatic molten globule often has a broad substrate specificity, which might be important for changing the specificity during the enzyme evolution [14–18]. There are also many examples of functional molten globules other than the enzymatic molten globules. Many bacterial toxins exhibit membrane-insertion and pore-forming activities during endocytosis or translocation across biological membranes, and the biologically functional state for these activities of the toxins is often the molten globule state [19–26]. Similar functional molten globules are also found in a number of transporter proteins, which transport a nonpolar ligand across biological membranes [27–29]. Finally, a significant portion (30–50 %) of proteins in eukaryotic cells are intrinsically disordered [5, 6, 30, 31]. Many of the intrinsically disordered proteins fold into a specific three-dimensional structure when bound to their targets, but some of them exhibit biological functions in the non-folded disordered state [5, 6, 30, 31].

*H*uman α-lactalbumin *m*ade *le*thal to *t*umor cells, called HAMLET, is a protein-oleic acid complex that selectively kills tumor cells and primary immature cells without affecting healthy mature cells [32–39]. The protein portion of HAMLET is in a molten globule-like state under a physiological condition, and hence it has been considered to be a novel type of functional molten globule [32, 40, 41]. This review will summarize the historical background and recent progress of studies on HAMLET and related protein-fatty acid complexes, with a primary focus on our recent study on the structure of HAMLET and the molecular mechanisms of its cytotoxicity [42], which was performed with support from the "fluctuations and biological functions" project.

12.2 Biological Functions and Physical Properties of α-Lactalbumin

α-Lactalbumin is a major whey protein of mammalian milk, consisting of 123 amino acid residues with a molecular weight of 14,200 [43] (Fig. 12.1). There are exceptions to these numbers, however—e.g., the rat protein has 140 residues, the rabbit protein 122 residues, and the red-necked wallaby protein 121 residues [43]. α-Lactalbumin contains four disulfide bonds [43] and a specific Ca^{2+}-binding site with a binding constant of 10^7–10^8 M^{-1} [43–47]. The primary biological function of α-lactalbumin is to modify the specificity of an enzyme, galactosyltransferase, which catalyzes the transfer of galactose from UDP-galactose to an *N*-acetylglucosamine moiety of glycoproteins [37, 43, 48]. When α-lactalbumin is bound to galactosyltransferase, the substrate specificity of the enzyme is changed from *N*-acetylglucosamine to glucose, resulting in the production of lactose [37, 43, 48]. Since galactosyltransferase is an integral membrane protein located in the *trans*-Golgi with the catalytic site projected into the lumen of the Golgi apparatus

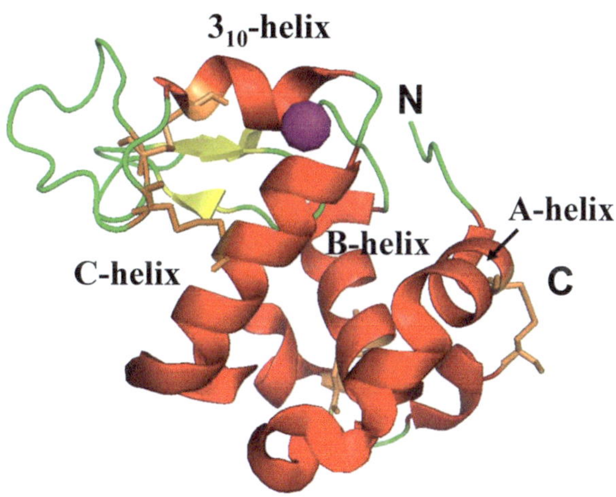

Fig. 12.1 The three-dimensional structure of holo human α-lactalbumin. The X-ray crystallographic structure of human α-lactalbumin (PDB ID: 3B0I) [44]. α-Helices, β-strands, and loops are shown by *red*, *yellow*, and *green*, respectively. The magenta sphere between the N-terminal end of the C-helix and the C-terminal end of the 3_{10}-helix represents the bound Ca^{2+} ion

of lactating mammary glands [37, 43, 48], for some time it had been unclear why α-lactalbumin is secreted into milk in such a large amount (~1 g/l); after all, there is almost no lactose production within milk. Now this question has been answered, since α-lactalbumin has been shown to perform the secondary biological function of producing an antitumor complex (HAMLET in human α-lactalbumin) in the stomach of breast-fed infants (see below). Therefore, α-lactalbumin is a unique example, in which a single globular protein has two independent biological functions in quite different locations: (1) substrate-specificity modification of the lactose synthase system in mammary glands and (2) formation of the antitumor complex in the breast-fed infant stomach.

In protein-folding studies, α-lactalbumin is a typical model protein, and it exhibits a stable folding intermediate at an early stage of kinetic refolding from the fully unfolded state in concentrated denaturant (guanidinium chloride or urea) [49, 50]. The formation of the intermediate occurs within a dead time (~1 ms) of conventional kinetic refolding experiments [50]; recently, Okabe et al. succeeded in direct observation of this early stage by slowing the refolding kinetics using a silica gel [51]. α-Lactalbumin is homologous to C-type lysozyme in terms of both the amino acid sequences and the three-dimensional structures [52, 53]; accordingly, there have been many comparative studies of α-lactalbumin and lysozyme [37, 43, 52–57]. The folding intermediate of α-lactalbumin is equivalent to an equilibrium unfolding intermediate [50, 58–60], which has characteristics typical of the molten globule state [50, 61], i.e., (1) the presence of a native-like secondary structure and native-like backbone fold, (2) the absence of the specific tertiary packing of amino acid side chains, (3) a compact molecular shape with a radius only 10–20 % larger than that in the native state, and (4) the presence of a loosely organized hydrophobic core, which is accessible to solvent and hence well characterized by ANS (8-anilino-1-naphthalenesulfonic acid) binding [50, 61, 62]. α-Lactalbumin thus provides a useful model of the molten globule, and the molten globule state of this protein has been studied extensively [46, 47, 50, 61, 63–68].

The molten globule state of α-lactalbumin is stably populated under a mildly denaturing condition, and hence the state is observed as the acid-denatured state (<pH 3) [46, 50, 64–68], the alkaline-denatured state (>pH 11) [46, 50, 69], the denatured state induced by an inorganic-salt denaturant [50], and the equilibrium unfolding intermediate accumulated at an intermediate concentration of a strong denaturant (e.g., 2 M guanidinium chloride) [50]. The removal of the bound Ca^{2+} stabilizes the molten globule state relative to the native state [50, 70], and apo-α-lactalbumin forms the molten globule state in the absence of any stabilizing ions (e.g., NaCl) at neutral pH [71]. The molten globule state of α-lactalbumin shows rapid and tight binding to phospholipid membranes, and the binding of the α-lactalbumin molten globule to membranes induces insertion of a protein portion into the membrane interior and fusion of membrane vesicles [72–82]. These properties of the α-lactalbumin molten globule must be closely related to its formation of the HAMLET complex with oleic acid and the molecular mechanisms of its cytotoxicity.

12.3 HAMLET and Its Beneficial Properties

In 1995, Håkansson et al. [83] reported that there was a component that selectively induced apoptosis of tumor cells in human milk, and rather surprisingly, this component was identified as multimeric α-lactalbumin [83]. The apoptosis-inducing antitumor complex of α-lactalbumin has a molten globule-like conformation, as revealed by UV absorbance, fluorescence, and circular dichroism spectroscopy [40]. However, the molten globule state of α-lactalbumin alone does not have any toxicity to tumor cells, and oleic acid is required as a cofactor for formation and stabilization of the antitumor complex [41]. Svensson et al. [41] thus first succeeded in production of the antitumor complex from purified human α-lactalbumin and oleic acid by using oleic acid-pre-conditioned ion-exchange column chromatography, and they named this complex HAMLET (*h*uman *α*-lactalbumin *le*thal to *t*umor cells). Essentially the same antitumor complex can also be produced from α-lactalbumin of other mammalian species, including the bovine, goat, equine, porcine, and camel proteins, when treated with oleic acid [84, 85], and the complexes made from bovine and goat α-lactalbumin are called BAMLET (*b*ovine *α*-lactalbumin *m*ade *le*thal to *t*umor cells) [86] and GAMLET (*g*oat *α*-lactalbumin *m*ade *le*thal to *t*umor cells) [42], respectively.

Although the in vivo synthesis of HAMLET during milk digestion has not yet been proven, it is strongly suggested that HAMLET is produced in the breast-fed infant stomach, protecting infants from tumor development [32, 87]. α-Lactalbumin is the most abundant whey protein in human milk [43, 88], and oleic acid is the most abundant fatty acid, occurring as triacylglycerides in human milk [84, 89]. The acidic environment of the stomach of a breast-fed infant is thus very favorable for the production of HAMLET, because α-lactalbumin forms the molten globule state at acid pH (see above), and significant amounts of milk triacylglycerides are digested by lingual and gastric lipases, releasing oleic acid [89–91]. The generally held idea that gastric lipase preferentially hydrolyzes short- and medium-chain fatty acid was recently modified, and the main fatty acids released by gastric lipase in the stomach were shown to be oleic acid and palmitic acid [90, 91]. It is thus quite reasonable to postulate that HAMLET is produced in the breast-fed infant stomach. HAMLET has also been used successfully for cancer therapy in treating glioblastomas [92], skin papillomas [93], and bladder cancer [94].

Despite the above-described beneficial properties of HAMLET, the molecular mechanisms of the formation and stabilization of HAMLET as well as the structure of this protein complex, including the oleic acid-binding site, were not well understood prior to our recent study [42]. Several issues, including (1) how many oleic acid molecules are bound to α-lactalbumin in HAMLET, (2) whether HAMLET is monomeric or oligomeric, (3) where the oleic acid-binding site is located in α-lactalbumin in the molten globule-like state, and (4) whether the oleic acid binding to the molten globule-like α-lactalbumin is specific or nonspecific, remained obscure. In this article, we address these issues, with a focus on our recent study of the structure of HAMLET and the molecular mechanisms of its cytotoxicity [42].

12.4 Preparation of HAMLET

The original method for preparing HAMLET used an oleic acid-pre-conditioned ion-exchange column [41] and was based on the purification procedure of the antitumor complex from a human milk casein fraction [40]. Apo-α-lactalbumin was added to the oleic acid-conditioned column, and the HAMLET complex was eluted after 1 M NaCl [41]. Although this original column method has been regarded as standard, it is rather cumbersome, and much simpler methods for preparation of HAMLET and related protein-oleic acid complexes have been reported. Kamijima et al. [95] thus reported a heat treatment method, in which α-lactalbumin and oleic acid are directly mixed in phosphate-buffered saline, and the mixture is heated at 50 °C or 60 °C for 10 min. A similar heat treatment method was also used for production of BAMLET [95, 96], the β-lactoglobulin-sodium oleate complex [96] and the complexes of camel α-lactalbumin with oleic and linoleic acids [85]. Zhang et al. [97] reported that oleic and linoleic acids induced the molten globule-like state of bovine α-lactalbumin at mildly acidic pH (pH 4.0–4.5) to form amorphous aggregates, and the aggregates solubilized under a physiological condition exhibited cytotoxicity to tumor cells in the same manner as HAMLET. Simple mixing of an oleic acid solution into an apo-α-lactalbumin solution in the presence of ~2 mM EDTA at neutral pH (pH 7.4 or 8.3) also produced a HAMLET- or BAMLET-like complex exhibiting toxicity to tumor cells [98, 99]. Permyakov et al. [100] reported a novel method (alkaline treatment) for the preparation of HAMLET-like complexes by titrating a bovine α-lactalbumin solution with oleic acid in the presence of 1 mM EDTA at pH 12.0 and 45 °C. Because the apparent pK_a of oleic acid is 9.85 [101], the carboxyl group of oleic acid is fully deprotonated to form oleate at pH 12.0. The solution mixture at pH 12.0 was acidified down to pH 2.0, and the resultant precipitate was dialyzed and lyophilized for use as BAMLET [100].

We prepared HAMLET and GAMLET from purified human and goat α-lactalbumin and oleic acid by a slightly modified version of the heat treatment method [95]; the mixture of α-lactalbumin and oleic acid was heated at pH 2.0 and 50–60 °C for 10 min, and this was followed by removal of excess oleic acid and lyophilization [42]. We also prepared HAMLET and GAMLET by the original column method of Svensson et al. [41], and the protein complexes prepared by the two methods were compared. The antitumor activities of the HAMLET and GAMLET complexes prepared by the two methods were identical [42]. The complexes prepared by the two methods were also essentially identical with respect to their oleic acid contents, their NMR spectra, and their hydrodynamic properties as measured by a pulsed field gradient (PFG)-NMR technique [42]. Therefore, the original column method is not the only method for preparing HAMLET and

GAMLET complexes. In this report, therefore, we use the terms HAMLET, BAMLET, and GAMLET to indicate the protein-oleic acid complexes that have essentially the same antitumor activities as the original complexes prepared by the column method.

12.5 Stoichiometry of Oleic Acid Binding to α-Lactalbumin

The stoichiometry of oleic acid binding to α-lactalbumin in HAMLET was first reported to be 1 [102], but later many groups reported various stoichiometry values for HAMLET and BAMLET, and the stoichiometries so far reported range from less than 1 to 35 (the number of bound oleic acid molecules per molecule of α-lactalbumin) [98–100, 102–109]. Different experimental methods have been used for the quantification of oleic acid, including gas chromatography/mass spectrometry after esterification of oleic acid [102–105], liquid chromatography/mass spectrometry [100, 107], spectrofluorimetric titration of α-lactalbumin by oleic acid [98, 106, 109], NMR peak area analysis [103], FTIR spectroscopy after extraction of oleic acid into an organic phase [99], and a colorimetric method using a reaction with diethyldithiocarbamate [108]. The variation in the stoichiometry values of the bound oleic acids may reflect the nonspecific nature of the oleic acid binding to α-lactalbumin in the complex, and the binding stoichiometry of oleic acid depends on solution conditions and the preparation methods of the HAMLET and BAMLET complexes.

We used a simple HPLC method to quantify the stoichiometry of oleic acid to the protein in HAMLET and GAMLET [42], and the results for HAMLET are shown in Fig. 12.2. Oleic acid and the protein in the HAMLET and GAMLET complexes were first dissociated by guanidinium chloride at 6 M. An aliquot of the protein and oleic acid mixture was then applied to a reversed-phase HPLC column (octadecyl-4PW, TOSOH) and eluted with an acetonitrile linear gradient in 0.1 % TFA. α-Lactalbumin and oleic acid were detected by absorbance at 280 and 215 nm, respectively. This method is one of the simplest, and we can detect both the protein and oleic acid using the same column and the same elution conditions. The first dissociation step by 6 M guanidinium chloride was indispensable, because the direct application of HAMLET (or GAMLET) dissolved in 0.1 % TFA resulted in the presence of a small peak of the complex together with those of the dissociated protein and oleic acid. Therefore, there was a strongly bound oleic acid molecule (or molecules), which was not dissociated under the HPLC conditions, in the HAMLET and GAMLET complexes. The stoichiometries of oleic acid to the protein thus determined were 8.5 and 6.9 oleic acid molecules bound to an α-lactalbumin molecule in HAMLET and GAMLET, respectively (Fig. 12.2) [42].

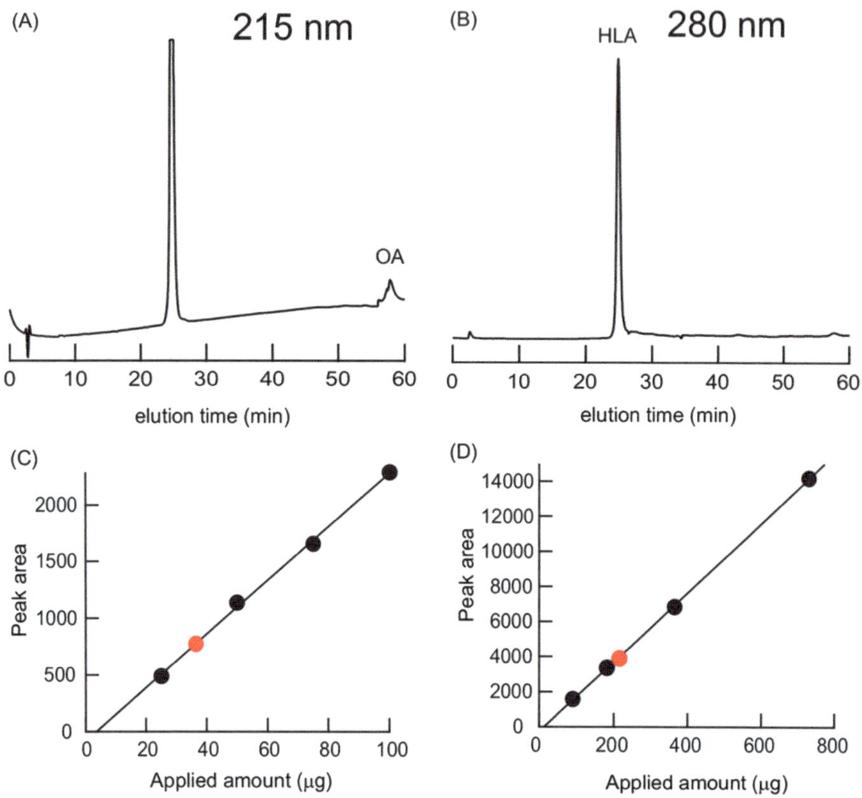

Fig. 12.2 Reversed-phase HPLC profiles of HAMLET. HAMLET dissolved in 6 M guanidinium chloride was applied to an HPLC column (octadecyl-4PW, TOSOH) and eluted with an acetonitrile linear gradient in 0.1 % TFA. (**a**) Oleic acid was eluted at 57 min as detected by absorbance at 215 nm. (**b**) α-Lactalbumin was eluted at 25 min as detected by absorbance at 280 nm. (**c**) Calibration for the oleic acid peak area measured by the HPLC analysis is shown (*black-filled circles*), and the red circle indicates the oleic acid peak of HAMLET. (**d**) Calibration for the α-lactalbumin peak area measured by the HPLC analysis is shown (*black-filled circles*), and the red circle indicates the α-lactalbumin peak of HAMLET. From the ratio of the peak areas, the stoichiometry of oleic acid to the protein was determined at 8.5

12.6 Is HAMLET Oligomeric or Monomeric?

Whether HAMLET or the other α-lactalbumin-oleic acid complexes that have similar antitumor activities are monomeric or oligomeric has been a matter of controversy. Spolaore et al. [99] reported that BAMLET is oligomeric with a hydrodynamic radius (R_h) of 37 Å at pH 7.4 and room temperature as measured by dynamic light scattering and gel filtration chromatography, and the oligomeric species displayed cytotoxicity toward tumor cells. Similar observations were also reported by Nemashkalova et al. [110] and Kohoe and Brodkor [109]. UV absorbance and transmission electron micrographs observed by Zhang et al. [97] also indicated the aggregated state of BAMLET at pH 4.3. On the other hand, Svanborg and coworkers [33, 103] reported on the basis of their PFG-NMR experiments that HAMLET existed in the monomeric form with an R_h value of 26.9 Å at pH 7 in the presence of 2 M urea. They also reported a low-resolution solution structure of HAMLET measured by small-angle X-ray scattering and showed that HAMLET (0.2–0.5 mM of the protein in phosphate-buffered saline) had a radius of gyration of 17.8 Å and a molecular mass of 15,000, indicating that HAMLET was monomeric under the conditions used [111].

To investigate the oligomeric state of HAMLET and GAMLET, we carried out PFG-NMR analyses of the HAMLET and GAMLET complexes to obtain their R_h values (pH 7.5 and 25 °C) [42]. The NMR spectrum of HAMLET was very similar to that previously reported by the Svanborg group [103], and the spectra of both HAMLET and GAMLET exhibited broad peaks at 0.745 and 1.375 ppm, which arose from the bound oleic acids, as well as the native protein peaks in the aromatic and upfield regions [42]. The analysis of the peak intensities of the individual NMR signals as a function of the field gradient strength indicated that there were three components, having different R_h values, in both HAMLET and GAMLET. These three components were native α-lactalbumin ($R_h = 20$ Å), a monomeric molten globule-like state ($R_h = 23$–25 Å), and an oligomeric species ($R_h = 57$–61 Å) (Fig. 12.3a and b), and the oligomeric species was the major species. The R_h values of the monomeric molten globule-like state were 10–20 % larger than the values (20.8–20.9 Å) known for the molten globule state of α-lactalbumin [112, 113], probably because of the presence of the bound oleic acid molecules in the monomeric HAMLET and GAMLET. The oligomeric species may consist of 13–17 protein molecules based on the volume ratios calculated from the radii of the monomeric and oligomeric HAMLET and GAMLET complexes [42].

These results were thus inconsistent with the PFG-NMR results reported by the Svanborg group, who proposed the monomeric nature of the HAMLET complex [33, 103], and this discrepancy was rather surprising, because both investigations used the same PFG-NMR techniques. The only significant difference in the experimental conditions was the presence of 2 M urea in their experiments. We therefore performed the PFG-NMR experiments in the presence of 2 M urea [42] (Fig. 12.3c and d). The results again indicated the presence of the three components, the native α-lactalbumin, the monomeric complex, and the oligomeric complex, and the R_h values of native α-lactalbumin and the monomeric complex were the same as those in the absence of urea. However, the R_h values of the oligomeric species became significantly smaller, 36.5 and 35.0 Å for HAMLET and GAMLET, respectively, at 2 M urea, indicating that significant dissociation occurred in the presence of urea. The average of the R_h values of the three components at 2 M urea was thus calculated as 26.9 Å, which was identical to the value reported by the Svanborg group [33, 103]. Therefore, it was concluded that HAMLET and GAMLET are mainly oligomeric under physiological conditions in the absence of urea [42].

Nevertheless, the discrepancy between our results and the X-ray scattering results by Ho CS et al. [111], who showed that HAMLET was monomeric in

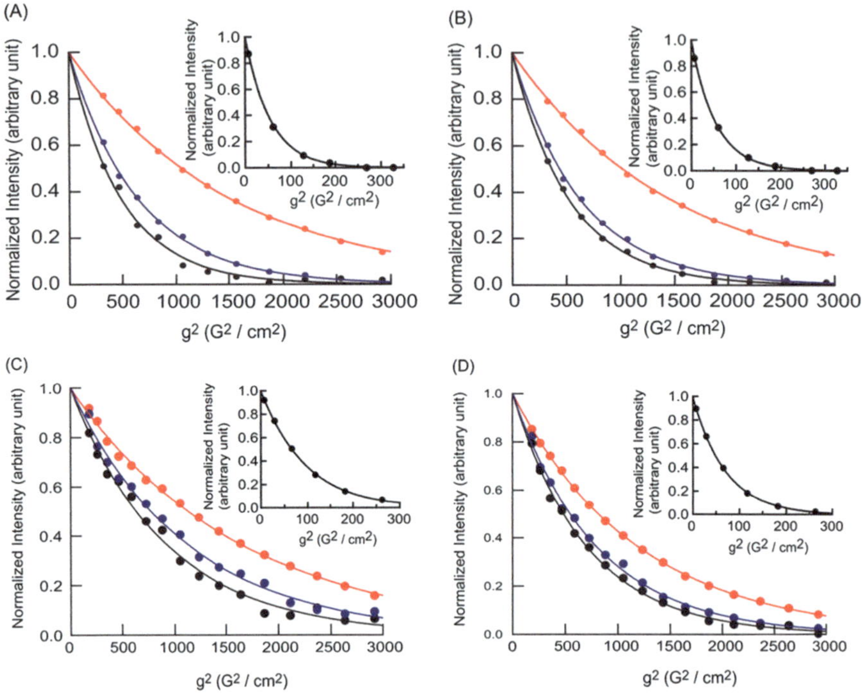

Fig. 12.3 PFG-NMR analysis of HAMLET ((**a**) and (**c**)) and GAMLET ((**b**) and (**d**)). The intensities of the NMR signals of native α-lactalbumin (*black*), the monomeric molten globule-like complex (*blue*), and the oligomeric complex (*red*) are shown as a function of the square of the field gradient strength. (**a**) and (**b**) show the results of the analysis in the absence of 2 M urea (pH 7.5 and 25 °C), and (**c**) and (**d**) show the results in the presence of urea (pH 7.5 and 25 °C). The inset in each panel shows the decay for dioxane, which was used as a reference compound, having a hydrodynamic radius of 2.12 Å (Adopted from Ref. [42])

phosphate-buffered saline, is more intractable, and we can offer no clear interpretation of this discrepancy at present. According to their report [111], however, they might have used a monomeric fraction, obtained by gel filtration, for the solution X-ray scattering experiments. If this was the case, we cannot say that HAMLET is monomeric under physiological conditions. Furthermore, their X-ray scattering patterns could have been affected by the presence of native α-lactalbumin [111], which is present in the HAMLET complex at neutral pH.

However, it should be noted that the oligomeric or monomeric status of HAMLET and the other α-lactalbumin-oleic acid complexes may depend on pH. We carried out PFG-NMR experiments at pH 2.0 and 50 °C, the conditions used for preparation of HAMLET and GAMLET by the heat treatment method [42], and under these conditions, both the complexes were in the monomeric molten globule-like state. All the NMR signals, including those of the protein and bound oleic acids, exhibited essentially identical decay curves, and the R_h values of HAMLET and GAMLET thus obtained were both 25 Å. The finding that the decay curves for the protein and oleic acid signals were coincident indicated that oleic acid was strongly bound to the protein and not in rapid exchange between the bound and free states. Recently, Rath and coworkers [105] reported small-angle X-ray scattering and dynamic light scattering studies of the BAMLET complex at pH 12. According to their data, BAMLET was monomeric at pH 12, and its molecular shape was remarkably dependent on the content of bound oleic acids, adopting an unusual coiled elongated structure at the highest oleic acid content (34 oleic acid molecules per molecule of α-lactalbumin) [105]. α-Lactalbumin forms a molten globule state under acidic (pH 2) and alkaline (pH 12) conditions (see above), and under such extreme pH conditions, the charge density of the protein molecule is highest, which may prevent the oligomerization of the protein-oleic acid complexes by electrostatic repulsion.

12.7 Identification of the Oleic Acid-Binding Site in HAMLET and GAMLET

Prior to our recent study [42], there were no direct experimental data concerning the oleic acid-binding site in the α-lactalbumin molecule of the HAMLET and the other α-lactalbumin-oleic acid complexes. To identify the oleic acid-binding site in HAMLET and GAMLET, we thus measured the ^{1}H–^{15}N heteronuclear single quantum coherence (HSQC) spectra of the monomeric molten globule-like state of HAMLET and GAMLET at pH 2.0 and 50 °C and compared these spectra with those of the free molten globule state of human and goat α-lactalbumin without the bound oleic acid at the same pH and temperature [42] (Fig. 12.4). As shown in Fig. 12.4b and d, many cross peaks appeared at the same positions between HAMLET and human α-lactalbumin and between GAMLET and goat α-lactalbumin, indicating that HAMLET and GAMLET were in the molten globule-like state under these conditions. However, several cross peaks showed clear differences, and these cross peaks are encircled in Fig. 12.4b and d. From these results, we identified the oleic acid-binding sites in HAMLET and GAMLET, as shown in Fig. 12.5a and b [42]. Rather surprisingly, the oleic acid-binding site was quite different between HAMLET and GAMLET, although both the protein-oleic acid complexes showed essentially the same antitumor activities [42]. In HAMLET, most of the residues affected by oleic acid binding were within the A- and B-helices (Fig. 12.5a), while the residues affected by oleic acid binding in GAMLET were located near the Ca^{2+}-binding site of the protein and hence at the interface between the α- and β-subdomains of α-lactalbumin [42] (Fig. 12.5b).

Although the oleic acid-binding site was markedly different between HAMLET and GAMLET, the binding site in each complex was reasonably correspondent with the most structured region in the molten globule state of each protein. From previous hydrogen/deuterium (H/D)-exchange studies, the peptide NH protons most highly protected in the molten globule state of human α-lactalbumin were

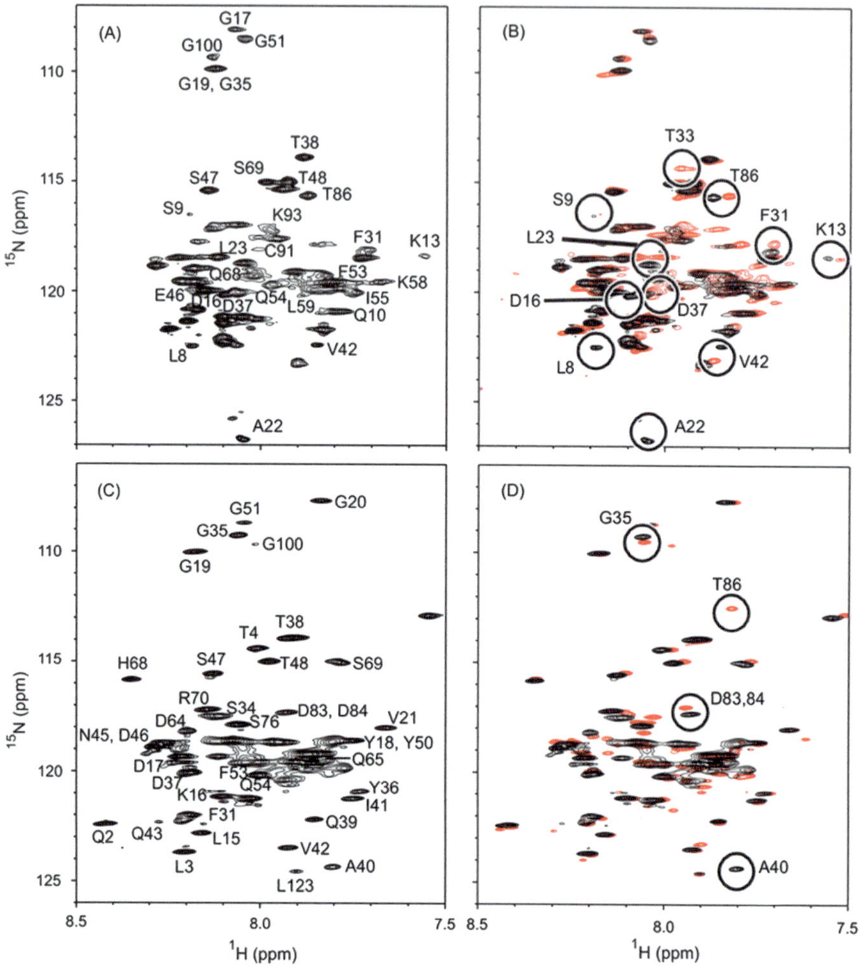

Fig. 12.4 The ^{1}H–^{15}N HSQC spectra of human (**a**) and goat (**c**) α-lactalbumin in the free molten globule state and the spectra of HAMLET (**b**) and GAMLET (**d**) (pH 2.0 and 50 °C). In (**b**) and (**d**), the spectra of HAMLET and GAMLET, shown in *red*, are superimposed on the spectra in the free molten globule state, shown in *black*. The NMR signal assignments are indicated with a one-letter amino acid code and residue number. The cross peaks that show a difference between the free molten globule state and the α-lactalbumin-oleic acid complex (HAMLET or GAMLET) are encircled in (**b**) and (**d**) (Adopted from Ref. [42])

located in the A- and B-helices [114], while they were located in the C-helix in the molten globule state of goat α-lactalbumin [57] (Fig. 12.5c and d). These regions are thus coincident with the oleic acid-binding sites in HAMLET and GAMLET (Fig. 12.5a and b). Because the structured region in the molten globule state may correspond to a loosely organized hydrophobic core accessible to solvent water, the above results demonstrate that oleic acid is not recognized by a specific structure

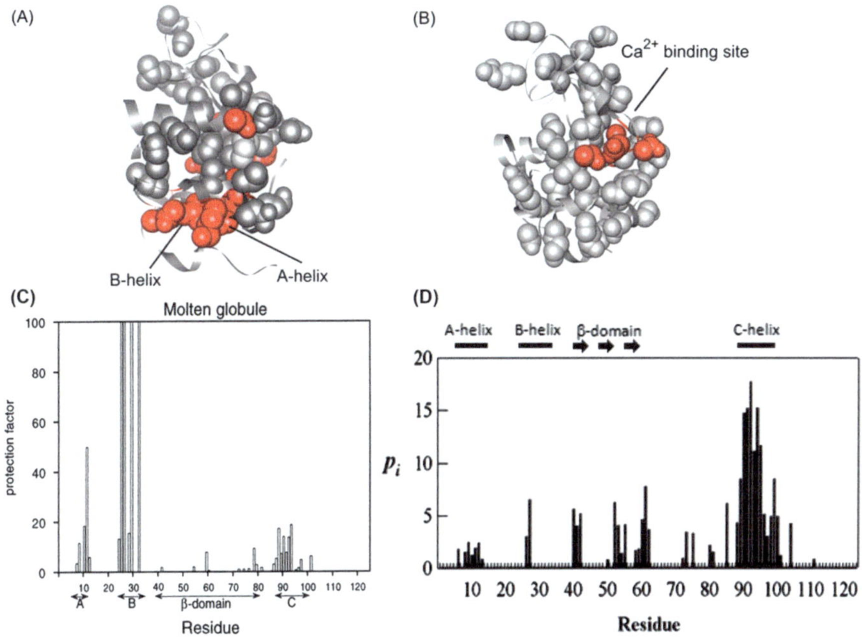

Fig. 12.5 The oleic acid-binding site of α-lactalbumin in HAMLET (**a**) and GAMLET (**b**) and the distribution of H/D-exchange protection factors for the free molten globule state. (**a**) and (**b**) The binding sites were determined by differences in cross peaks between the free molten globule state and the α-lactalbumin complex (HAMLET or GAMLET; Fig. 12.4b and d). The amino acid residues represented by the space-filling model are those whose cross peaks in the ^{1}H–^{15}N HSQC spectra are assigned (Fig. 12.4a and c) (Adopted from Ref. [42]). (**c**) and (**d**) The H/D-exchange protection factors of the peptide amide protons are shown as a function of the residue number for human α-lactalbumin ($pH_{obs} = 2$ and 5 °C) (**c**) and goat α-lactalbumin ($pH_{obs} = 1.7$ and 25 °C) (**d**) ((**c**) Adopted from Ref. [114] and (**d**) modified from Ref. [57])

but rather is recognized by nonspecific hydrophobic interactions in the loosely organized hydrophobic core in the molten globule state. The notion that the binding of oleic acid to the protein component is hydrophobic in nature is also supported by the fact that HAMLET and the other related complexes can be prepared effectively under opposite pH conditions, i.e., pH 2 [42] and pH 12 [100], because α-lactalbumin is oppositely charged at these pH values, indicating that the electrostatic interaction is not a main contributor to the production of HAMLET.

Recently, Xie et al. [115] studied the effect of charge-specific chemical modifications of lysine side chains on the formation of the BAMLET complex. They observed that upon converting the original positively charged lysine residues to negatively charged citraconyl or neutral acetyl groups, the binding of oleic acid to the protein was eliminated and concluded that electrostatic interactions play an essential role in the binding of oleic acid with α-lactalbumin in the BAMLET complex [115]. However, according to their report, the chemically modified α-lactalbumin had lost the capacity to bind to ANS, indicating that the loosely organized hydrophobic core of the protein in the molten globule state was disrupted by citraconylation and acetylation. Therefore, their results can also be interpreted in terms of the disruption of the loosely organized hydrophobic core, which is indispensable for the binding to oleic acid, rather than by the simple electrostatic effect. Fang et al. [116] studied the effect of pH on the oleic acid-binding ability of bovine α-lactalbumin by the heat treatment between pH 2 and 9 and found that the protein bound the greatest number of oleic acid molecules over the pH range of 7.0–9.0; this result cannot be interpreted by the electrostatic effect, because the protein was negatively charged under these conditions.

The location of the oleic acid-binding site in the A- and B-helices in HAMLET may not be inconsistent with the H/D-exchange and limited proteolysis studies of HAMLET reported by Casbarra et al. [117], in which they observed that the α-subdomain was resistant to proteolysis while the β-subdomain was more flexible and susceptible to proteolysis. The A- and B-helices of human α-lactalbumin are already significantly protected in the molten globule state [114], and the oleic acid binding to these helices further stabilizes the helical structures and makes the α-subdomain more resistant to proteolysis.

12.8 Antitumor Complexes Formed by Oleic Acid and Other Proteins in the Molten Globule State

Our conclusion that oleic acid molecules in HAMLET and GAMLET are recognized by nonspecific hydrophobic interactions has been supported by recent findings that the complexes between oleic acid and the other proteins that form the molten globule-like partially unfolded state express similar antitumor activities. These protein complexes include those formed by equine and canine milk lysozymes [42, 118, 119], bovine β-lactoglobulin [96, 110, 120], pike parvalbumin [110, 120], apomyoglobin [42], β_2-microglobulin [42], and bovine lactoferrin [121]. Equine and canine milk lysozymes are both Ca^{2+}-binding lysozymes and homologues of α-lactalbumin [54, 55, 57, 122–126]. Among these protein-oleic acid complexes, the equine lysozyme-oleic acid complex was prepared by the original column method used for HAMLET [118, 119], but all the other complexes were prepared by simple mixing of the protein and oleic acid followed by heat treatment [42, 121] or alkaline treatment [110, 120]. Interestingly, the antitumor activities, measured by the half lethal doses for tumor cells, of bovine lactoferrin were about 10 times higher than those of BAMLET [121]. Although the strength of the antitumor activity was thus dependent on the protein species, all the protein-oleic acid complexes expressed similar antitumor activities [42, 96, 110, 118–121]. Large fragments of α-lactalbumin, which were prepared by limited proteolysis [127] or gene manipulation [107], had a partially folded molten globule-like conformation in the presence of oleic acid and exhibited similar antitumor activities.

That similar antitumor complexes were formed by a variety of protein molecules and protein fragments in the molten globule-like state indicates that there is no specific binding site for oleic acid. The molten globule state in general has a loosely organized hydrophobic core, which is accessible to solvent and hence detected by ANS binding [50, 61, 62]. It is thus concluded that oleic acid in these antitumor complexes is bound to the loosely organized hydrophobic core of the molten globule state by nonspecific hydrophobic interactions and stabilizes the molten globule structure of the protein portion, and hence the complexes keep the molten globule structure under physiological conditions, whereas the free proteins without oleic acid fold into the rigid native structures.

The expression of similar antitumor activities by the complexes formed between oleic acid and a variety of proteins in the molten globule state also suggests that the protein portion of these complexes may not be the origin of their cytotoxicity to tumor cells. There is now ample evidence that oleic acid itself is cytotoxic and promotes apoptosis in various tumor cells [128–133], and the mitochondrion-related apoptotic behaviors, such as the reduction in mitochondrial membrane potential and cytochrome *c* release, observed after treatment with oleic acid were also observed in the HAMLET-induced apoptosis in tumor cells [134–136]. Brinkmann et al. [137] also reported that the cell-killing mechanisms of the BAMLET complex and of oleic acid alone examined by flow cytometry were very similar to each other. It is thus strongly suggested that the protein portion of HAMLET and the other HAMLET-like protein-oleic acid complexes probably works as a delivery carrier of the cytotoxic oleic acid to tumor cells [38, 39, 99, 108, 127]. As mentioned in the Introduction section above, it is becoming increasingly evident that not only the rigid native structures but also the flexible molten globule structures are biologically functional, and many of the functional molten globules play roles in the translocations of proteins or hydrophobic ligands across membrane bilayers [19–29]; in addition, the interaction of HAMLET with artificial and natural membranes has also been reported [138–141]. We thus conclude that the molten globule-like state in HAMLET and the other related complexes plays a role in transferring the hydrophobic oleic acid molecules into tumor cells across the cell membranes.

12.9 Antitumor Complexes Formed by Other Unsaturated Fatty Acids and α-Lactalbumin

It has been well established that not only oleic acid but also other monounsaturated and polyunsaturated fatty acids are cytotoxic to tumor cells [142–150], and several mechanisms have been proposed to play roles in the toxicity, including activation of the caspase cascade coupled with mitochondrial membrane depolarization [129, 132, 151], lipid peroxidation and production of reactive oxygen species [131, 152–155], downregulation of the expression of genes involved in several metabolic pathways [156], and downregulation of the protein expression of Blc-2 and the Bim mRNA level [157], as well as end points such as loss of membrane integrity and DNA fragmentation [158]. If the protein portion of HAMLET and the other protein-oleic acid complexes acts as an oleic acid-transferring carrier, α-lactalbumin in the molten globule state may also act as a carrier of the other monounsaturated and polyunsaturated fatty acids in a similar manner.

We therefore prepared complexes between human α-lactalbumin and various fatty acids (stearic acid, oleic acid, *cis*-vaccenic acid, linoleic acid, α-linolenic acid, γ-linolenic acid, arachidonic acid, and docosahexaenoic acid) by the heat treatment method at pH 2.0 [42] and investigated their levels of killing activity against L1210 leukemia cells by the trypan blue assay (Nakamura et al., unpublished data) (Fig. 12.6). The results in Fig. 12.6b show that all the α-lactalbumin-fatty acid complexes examined, except for the complex with stearic acid, exhibited significant cytotoxicity to the tumor cells, strongly suggesting that α-lactalbumin in the molten globule state formed stable complexes with these fatty acids, and acted as a transferring carrier of the cytotoxic fatty acids to the tumor cells. The absence of significant cytotoxicity of the protein-stearic acid complex is not surprising, because saturated fatty acids such as stearic acid are known to be much less cytotoxic [142, 143, 145, 152, 158].

The above results, however, were not consistent with the results reported by Svensson et al. [102], who studied the antitumor activities of the complexes formed between human α-lactalbumin and a variety of fatty acids, including palmitoleic acid, palmitelaidic acid, stearic acid, oleic acid, elaidic acid, *cis*-vaccenic acid, *trans*-vaccenic acid, petroselinic acid, linoleic acid, α-linolenic acid, γ-linolenic acid, 11-eicosenoic acid, and arachidonic acid. They prepared the complexes by the column method [41] and found that only the complexes with the C18:1:9 *cis* fatty acid (oleic acid) and the C18:1:11 *cis* fatty acid (*cis*-vaccenic acid) showed significant cytotoxicity to L1210 tumor cells. The reason for the discrepancy between their results and ours remains unclear, but differences in the methods used to prepare the α-lactalbumin-fatty acid complexes may have played a role. The simple mixing and heat treatment method [42] is effective for the preparation of the complexes between α-lactalbumin and various fatty acids, while the column method, which was originally developed for purification of the antitumor complex from human milk [41, 83], might be effective only for the complexes with oleic acid and *cis*-vaccenic acid, because the ion-exchange column may have different affinities for the complexes with different fatty acids, and hence the elution conditions used for the complex with oleic acid (HAMLET) may not always be useful for the complexes with other fatty acids.

Recently, Brinkmann et al. [159] also studied the cytotoxicity of the complexes between bovine α-lactalbumin and six different fatty acids (oleic, *cis*-vaccenic, linoleic, palmitoleic, stearic, and elaidic acids). They prepared the complexes by the heat treatment method and reported that all the unsaturated *cis* fatty acids (oleic,

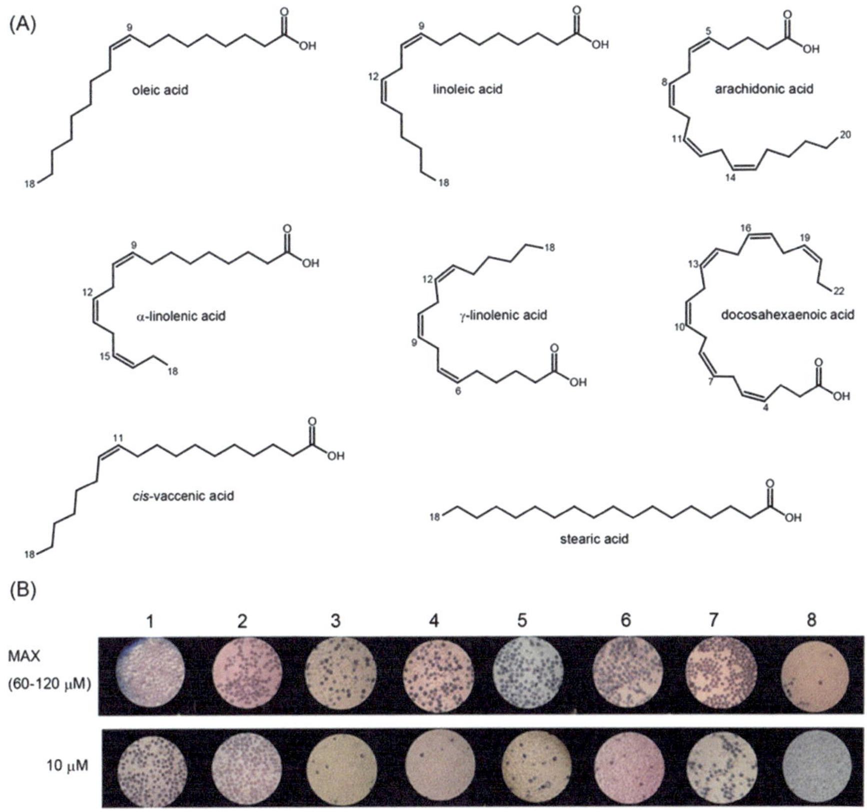

Fig. 12.6 Antitumor activities of the complexes between human α-lactalbumin and various fatty acids as measured by the trypan blue assay. (**a**) The fatty acids used for preparation of the complexes. (**b**) Images of L1210 cells obtained by the trypan blue assay with an optical microscope after 4 h of incubation with the α-lactalbumin-fatty acid complexes. The concentrations shown at left are the molar concentrations of the protein. The numbers at the top indicate the fatty acid used for the complex formation: *1* oleic acid (60 μM), *2* linoleic acid (100 μM), *3* arachidonic acid (80 μM), *4* α-linolenic acid (120 μM), *5* γ-linolenic acid (120 μM), *6* docosahexaenoic acid (110 μM), *7 cis*-vaccenic acid (110 μM), and *8* stearic acid (70 μM), where the numbers in the parentheses indicate the protein concentrations, as measured by absorbance at 280 nm, of the α-lactalbumin-fatty acid complexes (the *first row* in (**b**)). The protein concentration in the *second row* is all 10 μM. All the complexes were prepared by the heat treatment method [42]

cis-vaccenic, linoleic, and palmitoleic acids) were significantly cytotoxic to human promyelocytic leukemia-derived (HL-60) cells, which was fully consistent with our results. The antitumor cell activities of the α-lactalbumin-linoleic acid complex were also reported by Zhang et al. [97] and Atri et al. [85].

12.10 Conclusions

HAMLET, the human α-lactalbumin-oleic acid complex that selectively kills tumor cells without affecting healthy mature cells, has been considered to be a novel type of the functional molten globule states. The marked difference in the location of the oleic acid-binding site between HAMLET and GAMLET, and their locations in the loosely organized hydrophobic cores of the individual proteins in the molten globule state, led us to conclude that oleic acid molecules are recognized by nonspecific hydrophobic interactions. This conclusion was further strengthened by the following facts: (1) Similar antitumor activities were expressed by the complexes formed between oleic acid and a variety of globular proteins in the molten globule state, and (2) similar antitumor complexes were also formed by other unsaturated *cis* fatty acids and α-lactalbumin in the molten globule state. From these facts, we also conclude that the protein portion of these complexes is not the origin of their cytotoxicity to tumor cells and that the protein portion plays a role as the delivery carrier of cytotoxic fatty acid molecules into tumor cells across the cell membrane. The selectivity of HAMLET and the other protein-fatty acid complexes toward tumor cells is probably attributable to the dynamic and flexible nature of the tumor cell membrane, which may readily permit the translocation of the fatty acid across the cell membrane by the complexes. It is thus expected that a protein in the molten globule state could be used as a novel type of drug delivery carrier, which may selectively transfer a hydrophobic ligand into tumor or immature cells across the flexible cell membrane.

Acknowledgments This work was supported by a Grant-in-Aid for Scientific Research on Innovative Areas from the Ministry of Education, Culture, Sports, Science and Technology of Japan (grant number 23107725) and by a Grant-in-Aid for Scientific Research (C) from the Japan Society for the Promotion of Science (grant number 25440075).

References

1. Anfinsen CB, Scheraga HA (1975) Experimental and theoretical aspects of protein folding. Adv Protein Chem 29:205–300
2. Orengo CA, Todd AE, Thornton JM (1999) From protein structure to function. Curr Opin Struct Biol 9:374–382
3. Burley SK, Almo SC, Bonanno JB, Capel M, Chance MR, Gaasterland T, Lin D, Sali A, Studier FW, Swaminathan S (1999) Structural genomics: beyond the human genome project. Nat Genet 23:151–157
4. Yokoyama S, Matsuo Y, Hirota H, Kigawa T, Shirouzu M, Kuroda Y, Kurumizaka H, Kawaguchi S, Ito Y, Shibata T, Kainosho M, Nishimura Y, Inoue Y, Kuramitsu S (2000) Structural genomics projects in Japan. Prog Biophys Mol Biol 73:363–376
5. Xue B, Uversky VN (2016) Unfoldomes and unfoldomics: introducing intrinsically disordered proteins. In: Terazima M, Kataoka M, Ueoka R, Okamoto Y (eds) Molecular science of fluctuations toward biological functions. Chapter 6. Springer, Tokyo, pp. 125–150 (this volume)

6. Dunker AK, Lawson JD, Brown CJ, Williams RM, Romero P, Oh JS, Oldfield CJ, Campen AM, Ratliff CM, Hipps KW, Ausio J, Nissen MS, Reeves R, Kang C, Kissinger CR, Bailey RW, Griswold MD, Chiu W, Garner EC, Obradovic Z (2001) Intrinsically disordered protein. J Mol Graph Model 19:26–59
7. Cai S, Singh BR (2001) Role of the disulfide cleavage induced molten globule state of type a botulinum neurotoxin in its endopeptidase activity. Biochemistry 40:15327–15333
8. Kukreja R, Singh B (2005) Biologically active novel conformational state of botulinum, the most poisonous poison. J Biol Chem 280:39346–39352
9. Kumar R, Kukreja RV, Li L, Zhmurov A, Kononova O, Cai S, Ahmed SA, Barsegov V, Singh BR (2014) Botulinum neurotoxin: unique folding of enzyme domain of the most-poisonous poison. J Biomol Struct Dyn 32:804–815
10. Pervushin K, Vamvaca K, Vogeli B, Hilvert D (2007) Structure and dynamics of a molten globular enzyme. Nat Struct Mol Biol 14:1202–1206
11. Roca M, Messer B, Hilvert D, Warshel A (2008) On the relationship between folding and chemical landscapes in enzyme catalysis. Proc Natl Acad Sci U S A 105:13877–13882
12. Prasad M, Thomas JL, Whittal RM, Bose HS (2012) Mitochondrial 3β-hydroxysteroid dehydrogenase enzyme activity requires reversible pH-dependent conformational change at the intermembrane space. J Biol Chem 287:9534–9546
13. Rahaman H, Zhou S, Dodia C, Feinstein SI, Huang S, Speicher D, Fisher AB (2012) Increased phospholipase A2 activity with phosphorylation of peroxiredoxin 6 requires a conformational change in the protein. Biochemistry 51:5521–5530
14. Vamvaca K, Vögeli B, Kast P, Pervushin K, Hilvert D (2004) An enzymatic molten globule: efficient coupling of folding and catalysis. Proc Natl Acad Sci U S A 101:12860–12864
15. Woycechowsky KJ, Choutko A, Vamvaca K, Hilvert D (2008) Relative tolerance of an enzymatic molten globule and its thermostable counterpart to point mutation. Biochemistry 47:13489–13496
16. Bemporad F, Gsponer J, Hopearuoho HI, Plakoutsi G, Stati G, Stefani M, Taddei N, Vendruscolo M, Chiti F (2008) Biological function in a non-native partially folded state of a protein. EMBO J 27:1525–1535
17. Saha R, Dasgupta S, Banerjee R, Mitra-Bhattacharyya A, Soll D, Basu G, Roy S (2012) A functional loop spanning distant domains of glutaminyl-tRNA synthetase also stabilizes a molten globule state. Biochemistry 51:4429–4437
18. Honaker MT, Acchione M, Zhang W, Mannervik B, Atkins WM (2013) Enzymatic detoxication, conformational selection, and the role of molten globule active sites. J Biol Chem 288:18599–18611
19. Bychkova VE, Pain RH, Ptitsyn OB (1988) The 'molten globule' state is involved in the translocation of proteins across membranes? FEBS Lett 238:231–234
20. van der Goot FG, Gonzalez-Manas JM, Lakey JH, Pattus F (1991) A 'molten-globule' membrane-insertion intermediate of the pore-forming domain of colicin A. Nature 354:408–410
21. Ren J, Kachel K, Kim H, Malenbaum SE, Collier RJ, London E (1999) Interaction of diphtheria toxin T domain with molten globule-like proteins and its implications for translocation. Science 284:955–957
22. Man P, Montagner C, Vitrac H, Kavan D, Pichard S, Gillet D, Forest E, Forge V (2010) Accessibility changes within diphtheria toxin T domain when in the functional molten globule state, as determined using hydrogen/deuterium exchange measurements. FEBS J 277:653–662
23. Vecsey-Semjen B, Mollby R, van der Goot FG (1996) Partial C-terminal unfolding is required for channel formation by staphylococcal alpha-toxin. J Biol Chem 271:8655–8660
24. Hamada D, Kato T, Ikegami T, Suzuki KN, Hayashi M, Murooka Y, Honda T, Yanagihara I (2005) EspB from enterohaemorrhagic *Escherichia coli* is a natively partially folded protein. FEBS J 272:756–768

25. Faudry E, Job V, Dessen A, Attree I, Forge V (2007) Type III secretion system translocator has a molten globule conformation both in its free and chaperone-bound forms. FEBS J 274:3601–3610
26. Krantz BA, Trivedi AD, Cunningham K, Christensen KA, Collier RJ (2004) Acid-induced unfolding of the amino-terminal domains of the lethal and edema factors of anthrax toxin. J Mol Biol 344:739–756
27. Ptitsyn OB, Zanotti G, Denesyuk AL, Bychkova VE (1993) Mechanism of pH-induced release of retinol from retinol-binding protein. FEBS Lett 317:181–184
28. Gasymov OK, Abduragimov AR, Yusifov TN, Glasgow BJ (1998) Structural changes in human tear lipocalins associated with lipid binding. Biochim Biophys Acta 1386:145–156
29. Bose HS, Whittal RM, Baldwin MA, Miller WL (1999) The active form of the steroidogenic acute regulatory protein, StAR, appears to be a molten globule. Proc Natl Acad Sci U S A 96:7250–7255
30. Dyson HJ, Wright PE (2005) Intrinsically unstructured proteins and their functions. Nat Rev Mol Cell Biol 6:197–208
31. Uversky VN, Dunker AK (2010) Understanding protein non-folding. Biochim Biophys Acta 1804:1231–1264
32. Mok KH, Pettersson J, Orrenius S, Svanborg C (2007) HAMLET, protein folding, and tumor cell death. Biochem Biophys Res Commun 354:1–7
33. Mossberg AK, Mok KH, Morozova-Roche LA, Svanborg C (2010) Structure and function of human α-lactalbumin made lethal to tumor cells (HAMLET)-type complexes. FEBS J 277:4614–4625
34. Min S, Meehan J, Sullivan LM, Harte NP, Xie Y, Davey GP, Svanborg C, Brodkorb A, Mok KH (2012) Alternatively folded proteins with unexpected beneficial functions. Biochem Soc Trans 40:746–751
35. Ho CSJ, Rydstrom A, Trulsson M, Balfors J, Storm P, Puthia M, Nadeem A, Svanborg C (2012) HAMLET: functional properties and therapeutic potential. Future Oncol 8:1301–1313
36. Barbana C, Sanchez L, Perez MD (2011) Bioactivity of α-lactalbumin related to its interaction with fatty acids: a review. Crit Rev Food Sci Nutr 51:783–794
37. Brew K (2013) α-Lactalbumin. In: McSweeney PLH, Fox PF (eds) Advanced dairy chemistry: proteins: basic aspects, vol 1A, 4th edn. Springer, New York, pp 261–271
38. Brinkmann CR, Thiel S, Otzen DE (2013) Protein-fatty acid complexes: biochemistry, biophysics and function. FEBS J 280:1733–1749
39. Fontana A, Spolaore B, Polverino de Laureto P (2013) The biological activities of protein/oleic acid complexes reside in the fatty acid. Biochim Biophys Acta 1834:1125–1143
40. Svensson M, Sabharwal H, Håkansson A, Mossberg AK, Lipniunas P, Leffler H, Svanborg C, Linse S (1999) Molecular characterization of α-lactalbumin folding variants that induce apoptosis in tumor cells. J Biol Chem 274:6388–6396
41. Svensson M, Håkansson A, Mossberg AK, Linse S, Svanborg C (2000) Conversion of α-lactalbumin to a protein inducing apoptosis. Proc Natl Acad Sci U S A 97:4221–4226
42. Nakamura T, Aizawa T, Kariya R, Okada S, Demura M, Kawano K, Makabe K, Kuwajima K (2013) Molecular mechanisms of the cytotoxicity of human α-lactalbumin made lethal to tumor cells (HAMLET) and other protein-oleic acid complexes. J Biol Chem 288:14408–14416
43. McKenzie HA, White FHJ (1991) Lysozyme and α-lactalbumin: structure, function, and interrelationships. Adv Protein Chem 41:173–315
44. Makabe K, Nakamura T, Kuwajima K (2013) Structural insights into the stability perturbations induced by N-terminal variation in human and goat α-lactalbumin. Protein Eng Des Sel 26:165–170
45. Hiraoka Y, Segawa T, Kuwajima K, Sugai S, Murai N (1980) α-lactalbumin—a calcium metalloprotein. Biochem Biophys Res Commun 95:1098–1104
46. Kronman MJ (1989) Metal-ion binding and the molecular conformational properties of α-lactalbumin. Crit Rev Biochem Mol Biol 24:565–667

47. Permyakov EA, Berliner LJ (2000) α-Lactalbumin: structure and function. FEBS Lett 473:269–274
48. Takase K, Ebner KE (1984) Interaction of galactosyltransferase with α-lactalbumin and substrates. Curr Top Cell Regul 24:51–62
49. Kuwajima K, Hiraoka Y, Ikeguchi M, Sugai S (1985) Comparison of the transient folding intermediates in lysozyme and α-lactalbumin. Biochemistry 24:874–881
50. Arai M, Kuwajima K (2000) Role of the molten globule state in protein folding. Adv Protein Chem 53:209–282
51. Okabe T, Tsukamoto S, Fujiwara K, Shibayama N, Ikeguchi M (2014) Delineation of solution burst-phase protein folding events by encapsulating the proteins in silica gels. Biochemistry 53:3858–3866
52. Qasba PK, Kumar S (1997) Molecular divergence of lysozymes and α-lactalbumin. Crit Rev Biochem Mol Biol 32:255–306
53. Pike ACW, Brew K, Acharya KR (1996) Crystal structures of guinea-pig, goat and bovine α-lactalbumin highlight the enhanced conformational flexibility of regions that are significant for its action in lactose synthase. Structure (Camb) 4:691–703
54. Nitta K, Sugai S (1989) The evolution of lysozyme and α-lactalbumin. Eur J Biochem 182:111–118
55. Sugai S, Ikeguchi M (1994) Conformational comparison between α-lactalbumin and lysozyme. Adv Biophys 30:37–84
56. Radford SE, Dobson CM (1995) Insights into protein folding using physical techniques: studies of lysozyme and α-lactalbumin. Philos Trans R Soc Lond B Biol Sci 348:17–25
57. Nakamura T, Makabe K, Tomoyori K, Maki K, Mukaiyama A, Kuwajima K (2010) Different folding pathways taken by highly homologous proteins, goat α-lactalbumin and canine milk lysozyme. J Mol Biol 396:1361–1378
58. Ikeguchi M, Kuwajima K, Mitani M, Sugai S (1986) Evidence for identity between the equilibrium unfolding intermediate and a transient folding intermediate - a comparative-study of the folding reactions of α-lactalbumin and lysozyme. Biochemistry 25:6965–6972
59. Balbach J, Forge V, van Nuland NA, Winder SL, Hore PJ, Dobson CM (1995) Following protein folding in real time using NMR spectroscopy. Nat Struct Biol 2:865–870
60. Arai M, Ito K, Inobe T, Nakao M, Maki K, Kamagata K, Kihara H, Amemiya Y, Kuwajima K (2002) Fast compaction of α-lactalbumin during folding studied by stopped-flow X-ray scattering. J Mol Biol 321:121–132
61. Ptitsyn OB (1995) Molten globule and protein folding. Adv Protein Chem 47:83–229
62. Semisotnov GV, Rodionova NA, Razgulyaev OI, Uversky VN, Gripas AF, Gilmanshin RI (1991) Study of the "molten globule" intermediate state in protein folding by a hydrophobic fluorescent probe. Biopolymers 31:119–128
63. Redfield C (2004) Using nuclear magnetic resonance spectroscopy to study molten globule states of proteins. Methods 34:121–132
64. Mok KH, Nagashima T, Day IJ, Hore PJ, Dobson CM (2005) Multiple subsets of side-chain packing in partially folded states of α-lactalbumins. Proc Natl Acad Sci U S A 102:8899–8904
65. Rosner HI, Redfield C (2009) The human α-lactalbumin molten globule: comparison of structural preferences at pH 2 and pH 7. J Mol Biol 394:351–362
66. Craig PO, Gomez GE, Ureta DB, Caramelo JJ, Delfino JM (2009) Experimentally approaching the solvent-accessible surface area of a protein: insights into the acid molten globule of bovine α-lactalbumin. J Mol Biol 394:982–993
67. Kelkar DA, Chaudhuri A, Haldar S, Chattopadhyay A (2010) Exploring tryptophan dynamics in acid-induced molten globule state of bovine α-lactalbumin: a wavelength-selective fluorescence approach. Eur Biophys J 39:1453–1463
68. Litwinczuk A, Ryu SR, Nafie LA, Lee JW, Kim HI, Jung YM, Czarnik-Matusewicz B (2014) The transition from the native to the acid-state characterized by multi-spectroscopy approach: study for the holo-form of bovine α-lactalbumin. Biochim Biophys Acta 1844:593–606

69. Kuwajima K, Ogawa Y, Sugai S (1981) Role of the interaction between ionizable groups in the folding of bovine α-lactalbumin. J Biochem (Tokyo) 89:759–770
70. Ikeguchi M, Kuwajima K, Sugai S (1986) Ca^{2+}-induced alteration in the unfolding behavior of α-lactalbumin. J Biochem (Tokyo) 99:1191–1201
71. Hiraoka Y, Sugai S (1985) Equilibrium and kinetic study of sodium- and potassium-induced conformational changes of apo-α-lactalbumin. Int J Pept Protein Res 26:252–261
72. Hanssens I, Houthuys C, Herreman W, van Cauwelaert FH (1980) Interaction of α-lactalbumin with dimyristoyl phosphatidylcholine vesicles. I. A microcalorimetric and fluorescence study. Biochim Biophys Acta 602:539–557
73. Berliner LJ, Koga K (1987) α-Lactalbumin binding to membranes: evidence for a partially buried protein. Biochemistry 26:3006–3009
74. Van Dael H, Van Cauwelaert F (1988) The effect of α-lactalbumin on the thermotropic phase behaviour of phosphatidylcholine bilayers, studied by fluorescence polarization, differential scanning calorimetry and Raman spectroscopy. Biochim Biophys Acta 943:126–136
75. Kim J, Kim H (1986) Fusion of phospholipid vesicles induced by α-lactalbumin at acidic pH. Biochemistry 25:7867–7874
76. Park BS, Kim J, Kim UH, Kim H (1989) Effect of phosphatidylcholine on the α-lactalbumin-induced fusion of vesicles. Lipids 24:854–858
77. Montich GG, Marsh D (1995) Interaction of α-lactalbumin with phosphatidylglycerol. Influence of protein binding on the lipid phase transition and lipid acyl chain mobility. Biochemistry 34:13139–13145
78. Lala AK, Kaul P, Ratnam PB (1995) Membrane-protein interaction and the molten globule state: interaction of α-lactalbumin with membranes. J Protein Chem 14:601–609
79. Agasoster AV, Halskau O, Fuglebakk E, Froystein NA, Muga A, Holmsen H, Martinez A (2003) The interaction of peripheral proteins and membranes studied with α-lactalbumin and phospholipid bilayers of various compositions. J Biol Chem 278:21790–21797
80. Chenal A, Vernier G, Savarin P, Bushmarina NA, Gèze A, Guillain F, Gillet D, Forge V (2005) Conformational states and thermodynamics of α-lactalbumin bound to membranes: a case study of the effects of pH, calcium, lipid membrane curvature and charge. J Mol Biol 349:890–905
81. Halskau O, Muga A, Martinez A (2009) Linking new paradigms in protein chemistry to reversible membrane-protein interactions. Curr Protein Pept Sci 10:339–359
82. Chaudhuri A, Chattopadhyay A (2014) Lipid binding specificity of bovine α-lactalbumin: a multidimensional approach. Biochim Biophys Acta 1838:2078–2086
83. Håkansson A, Zhivotovsky B, Orrenius S, Sabharwal H, Svanborg C (1995) Apoptosis induced by a human milk protein. Proc Natl Acad Sci U S A 92:8064–8068
84. Pettersson J, Mossberg AK, Svanborg C (2006) α-Lactalbumin species variation, HAMLET formation, and tumor cell death. Biochem Biophys Res Commun 345:260–270
85. Atri MS, Saboury AA, Moosavi-Movahedi AA, Goliaei B, Sefidbakht Y, Alijanvand HH, Sharifzadeh A, Niasari-Naslaji A (2011) Structure and stability analysis of cytotoxic complex of camel α-lactalbumin and unsaturated fatty acids produced at high temperature. J Biomol Struct Dyn 28:919–928
86. Svensson M, Fast J, Mossberg AK, Duringer C, Gustafsson L, Hallgren O, Brooks CL, Berliner L, Linse S, Svanborg C (2003) α-Lactalbumin unfolding is not sufficient to cause apoptosis, but is required for the conversion to HAMLET (human α-lactalbumin made lethal to tumor cells). Protein Sci 12:2794–2804
87. Svanborg C, Agerstam H, Aronson A, Bjerkvig R, Duringer C, Fischer W, Gustafsson L, Hallgren O, Leijonhuvud I, Linse S, Mossberg AK, Nilsson H, Pettersson J, Svensson M (2003) HAMLET kills tumor cells by an apoptosis-like mechanism–Cellular, molecular, and therapeutic aspects. Adv Cancer Res 88:1–29
88. Artym J, Zimecki M (2013) Milk-derived proteins and peptides in clinical trials. Postepy Hig Med Dosw (Online) 67:800–816
89. Jensen RG (1999) Lipids in human milk. Lipids 34:1243–1271

90. Roman C, Carriere F, Villeneuve P, Pina M, Millet V, Simeoni U, Sarles J (2007) Quantitative and qualitative study of gastric lipolysis in premature infants: do MCT-enriched infant formulas improve fat digestion? Pediatr Res 61:83–88
91. Lindquist S, Hernell O (2010) Lipid digestion and absorption in early life: an update. Curr Opin Clin Nutr Metab Care 13:314–320
92. Fischer W, Gustafsson L, Mossberg AK, Gronli J, Mork S, Bjerkvig R, Svanborg C (2004) Human α-lactalbumin made lethal to tumor cells (HAMLET) kills human glioblastoma cells in brain xenografts by an apoptosis-like mechanism and prolongs survival. Cancer Res 64:2105–2112
93. Gustafsson L, Leijonhufvud I, Aronsson A, Mossberg AK, Svanborg C (2004) Treatment of skin papillomas with topical α-lactalbumin-oleic acid. N Engl J Med 350:2663–2672
94. Mossberg AK, Hou Y, Svensson M, Holmqvist B, Svanborg C (2010) HAMLET treatment delays bladder cancer development. J Urol 183:1590–1597
95. Kamijima T, Ohmura A, Sato T, Akimoto K, Itabashi M, Mizuguchi M, Kamiya M, Kikukawa T, Aizawa T, Takahashi M, Kawano K, Demura M (2008) Heat-treatment method for producing fatty acid-bound α-lactalbumin that induces tumor cell death. Biochem Biophys Res Commun 376:211–214
96. Liskova K, Auty MAE, Chaurin V, Min S, Mok KH, O'Brien N, Kelly AL, Brodkorb A (2011) Cytotoxic complexes of sodium oleate with β-lactoglobulin. Eur J Lipid Sci Technol 113:1207–1218
97. Zhang M, Yang F, Yang F, Chen J, Zheng CY, Liang Y (2009) Cytotoxic aggregates of α-lactalbumin induced by unsaturated fatty acid induce apoptosis in tumor cells. Chem Biol Interact 180:131–142
98. Knyazeva EL, Grishchenko VM, Fadeev RS, Akatov VS, Permyakov SE, Permyakov EA (2008) Who is Mr. HAMLET? Interaction of human α-lactalbumin with monomeric oleic acid. Biochemistry 47:13127–13137
99. Spolaore B, Pinato O, Canton M, Zambonin M, Polverino de Laureto P, Fontana A (2010) α-Lactalbumin forms with oleic acid a high molecular weight complex displaying cytotoxic activity. Biochemistry 49:8658–8667
100. Permyakov SE, Knyazeva EL, Leonteva MV, Fadeev RS, Chekanov AV, Zhadan AP, Hakansson AP, Akatov VS, Permyakov EA (2011) A novel method for preparation of HAMLET-like protein complexes. Biochimie 93:1495–1501
101. Kanicky JR, Shah DO (2002) Effect of degree, type, and position of unsaturation on the pK_a of long-chain fatty acids. J Colloid Interface Sci 256:201–207
102. Svensson M, Mossberg AK, Pettersson J, Linse S, Svanborg C (2003) Lipids as cofactors in protein folding: stereo-specific lipid-protein interactions are required to form HAMLET (human α-lactalbumin made lethal to tumor cells). Protein Sci 12:2805–2814
103. Pettersson-Kastberg J, Mossberg AK, Trulsson M, Yong YJ, Min S, Lim Y, O'Brien JE, Svanborg C, Mok KH (2009) α-Lactalbumin, engineered to be nonnative and inactive, kills tumor cells when in complex with oleic acid: a new biological function resulting from partial unfolding. J Mol Biol 394:994–1010
104. Brinkmann CR, Thiel S, Larsen MK, Petersen TE, Jensenius JC, Heegaard CW (2011) Preparation and comparison of cytotoxic complexes formed between oleic acid and either bovine or human α-lactalbumin. J Dairy Sci 94:2159–2170
105. Rath EM, Duff AP, Hakansson AP, Knott RB, Church WB (2014) Small-angle X-ray scattering of BAMLET at pH 12: a complex of α-lactalbumin and oleic acid. Proteins 82:1400–1408
106. Barbana C, Perez MD, Sanchez L, Dalgalarrondo M, Chobert JM, Haertle T (2006) Interaction of bovine α-lactalbumin with fatty acids as determined by partition equilibrium and fluorescence spectroscopy. Int Dairy J 16:18–25
107. Mercer N, Ramakrishnan B, Boeggeman E, Qasba PK (2011) Applications of site-specific labeling to study HAMLET, a tumoricidal complex of α-lactalbumin and oleic acid. PLoS One 6, e26093

108. Hoque M, Dave S, Gupta P, Saleemuddin M (2013) Oleic acid may be the key contributor in the BAMLET-induced erythrocyte hemolysis and tumoricidal action. PLoS One 8, e68390
109. Kehoe JJ, Brodkorb A (2014) Interactions between sodium oleate and α-lactalbumin: the effect of temperature and concentration on complex formation. Food Hydrocoll 34:217–226
110. Nemashkalova EL, Kazakov AS, Khasanova LM, Permyakov EA, Permyakov SE (2013) Structural characterization of more potent alternatives to HAMLET, a tumoricidal complex of α-lactalbumin and oleic acid. Biochemistry 52:6286–6299
111. Ho CSJ, Rydstrom A, Manimekalai MS, Svanborg C, Gruber G (2012) Low resolution solution structure of HAMLET and the importance of its α-domains in tumoricidal activity. PLoS One 7, e53051
112. Gast K, Zirwer D, Muller-Frohne M, Damaschun G (1998) Compactness of the kinetic molten globule of bovine α-lactalbumin: a dynamic light scattering study. Protein Sci 7:2004–2011
113. Redfield C, Schulman BA, Milhollen MA, Kim PS, Dobson CM (1999) α-lactalbumin forms a compact molten globule in the absence of disulfide bonds. Nat Struct Biol 6:948–952
114. Schulman BA, Redfield C, Peng ZY, Dobson CM, Kim PS (1995) Different subdomains are most protected from hydrogen exchange in the molten globule and native states of human α-lactalbumin. J Mol Biol 253:651–657
115. Xie Y, Min S, Harte NP, Kirk H, O'Brien JE, Voorheis HP, Svanborg C, Hun Mok K (2013) Electrostatic interactions play an essential role in the binding of oleic acid with α-lactalbumin in the HAMLET-like complex: a study using charge-specific chemical modifications. Proteins 81:1–17
116. Fang B, Zhang M, Jiang L, Jing H, Ren FZ (2012) Influence of pH on the structure and oleic acid binding ability of bovine α-lactalbumin. Protein J 31:564–572
117. Casbarra A, Birolo L, Infusini G, Dal Piaz F, Svensson M, Pucci P, Svanborg C, Marino G (2004) Conformational analysis of HAMLET, the folding variant of human α-lactalbumin associated with apoptosis. Protein Sci 13:1322–1330
118. Wilhelm K, Darinskas A, Noppe W, Duchardt E, Mok KH, Vukojević V, Schleucher J, Morozova-Roche LA (2009) Protein oligomerization induced by oleic acid at the solid–liquid interface—equine lysozyme cytotoxic complexes. FEBS J 276:3975–3989
119. Nielsen SB, Wilhelm K, Vad B, Schleucher J, Morozova-Roche LA, Otzen D (2010) The interaction of equine lysozyme: oleic acid complexes with lipid membranes suggests a cargo off-loading mechanism. J Mol Biol 398:351–361
120. Permyakov SE, Knyazeva EL, Khasanova LM, Fadeev RS, Zhadan AP, Roche-Hakansson H, Hakansson AP, Akatov VS, Permyakov EA (2012) Oleic acid is a key cytotoxic component of HAMLET-like complexes. Biol Chem 393:85–92
121. Fang B, Zhang M, Tian M, Jiang L, Guo HY, Ren FZ (2014) Bovine lactoferrin binds oleic acid to form an anti-tumor complex similar to HAMLET. Biochim Biophys Acta 1841:535–543
122. Nitta K, Tsuge H, Sugai S, Shimazaki K (1987) The calcium-binding property of equine lysozyme. FEBS Lett 223:405–408
123. Morozova-Roche LA (2007) Equine lysozyme: the molecular basis of folding, self-assembly and innate amyloid toxicity. FEBS Lett 581:2587–2592
124. Grobler JA, Rao KR, Pervaiz S, Brew K (1994) Sequences of two highly divergent canine type c lysozymes: implications for the evolutionary origins of the lysozyme/α-lactalbumin superfamily. Arch Biochem Biophys 313:360–366
125. Kikuchi M, Kawano K, Nitta K (1998) Calcium-binding and structural stability of echidna and canine milk lysozymes. Protein Sci 7:2150–2155
126. Koshiba T, Hayashi T, Miwako I, Kumagai I, Ikura T, Kawano K, Nitta K, Kuwajima K (1999) Expression of a synthetic gene encoding canine milk lysozyme in *Escherichia coli* and characterization of the expressed protein. Protein Eng 12:429–435

127. Tolin S, De Franceschi G, Spolaore B, Frare E, Canton M, Polverino de Laureto P, Fontana A (2010) The oleic acid complexes of proteolytic fragments of α-lactalbumin display apoptotic activity. FEBS J 277:163–173
128. Mizotani K, Inoue I (2002) The mechanism of apoptosis by the oleic acid in HeLa cells: caspase-independent pathway by induction of IkBb. J Saitama Med School 29:117–123
129. Dymkowska D, Szczepanowska J, Wojtczak L (2004) Fatty-acid-induced apoptosis in Ehrlich ascites tumor cells. Toxicol Mech Methods 14:73–77
130. Zhu Y, Schwarz S, Ahlemeyer B, Grzeschik S, Klumpp S, Krieglstein J (2005) Oleic acid causes apoptosis and dephosphorylates Bad. Neurochem Int 46:127–135
131. Dymkowska D, Szczepanowska J, Wieckowski MR, Wojtczak L (2006) Short-term and long-term effects of fatty acids in rat hepatoma AS-30D cells: the way to apoptosis. Biochim Biophys Acta 1763:152–163
132. Fernanda Cury-Boaventura M, Cristine Kanunfre C, Gorjao R, Martins de Lima T, Curi R (2006) Mechanisms involved in Jurkat cell death induced by oleic and linoleic acids. Clin Nutr 25:1004–1014
133. Yu F, Lu S, Yu F, Shi J, McGuire PM, Wang R (2008) Cytotoxic activity of an octadecenoic acid extract from *Euphorbia kansui* (Euphorbiaceae) on human tumour cell strains. J Pharm Pharmacol 60:253–259
134. Köhler C, Håkansson A, Svanborg C, Orrenius S, Zhivotovsky B (1999) Protease activation in apoptosis induced by MAL. Exp Cell Res 249:260–268
135. Köhler C, Gogvadze V, Håkansson A, Svanborg C, Orrenius S, Zhivotovsky B (2001) A folding variant of human α-lactalbumin induces mitochondrial permeability transition in isolated mitochondria. Eur J Biochem 268:186–191
136. Hallgren O, Aits S, Brest P, Gustafsson L, Mossberg AK, Wullt B, Svanborg C (2008) Apoptosis and tumor cell death in response to HAMLET (human α-lactalbumin made lethal to tumor cells). Adv Exp Med Biol 606:217–240
137. Brinkmann CR, Heegaard CW, Petersen TE, Jensenius JC, Thiel S (2011) The toxicity of bovine α-lactalbumin made lethal to tumor cells is highly dependent on oleic acid and induces killing in cancer cell lines and noncancer-derived primary cells. FEBS J 278:1955–1967
138. Zherelova OM, Kataev AA, Grishchenko VM, Knyazeva EL, Permyakov SE, Permyakov EA (2009) Interaction of antitumor α-lactalbumin-oleic acid complexes with artificial and natural membranes. J Bioenerg Biomembr 41:229–237
139. Mossberg AK, Puchades M, Halskau O, Baumann A, Lanekoff I, Chao Y, Martinez A, Svanborg C, Karlsson R (2010) HAMLET interacts with lipid membranes and perturbs their structure and integrity. PLoS One 5, e9384
140. Baumann A, Gjerde AU, Ying M, Svanborg C, Holmsen H, Glomm WR, Martinez A, Halskau O (2012) HAMLET forms annular oligomers when deposited with phospholipid monolayers. J Mol Biol 418:90–102
141. Wen H, Glomm WR, Halskau O (2013) Cytotoxicity of bovine α-lactalbumin: oleic acid complexes correlates with the disruption of lipid membranes. Biochim Biophys Acta 1828:2691–2699
142. Finstad HS, Myhrstad MC, Heimli H, Lomo J, Blomhoff HK, Kolset SO, Drevon CA (1998) Multiplication and death-type of leukemia cell lines exposed to very long-chain polyunsaturated fatty acids. Leukemia 12:921–929
143. Colquhoun A, Curi R (1998) Effects of saturated and polyunsaturated fatty acids on human tumor-cell proliferation. Gen Pharmacol 30:191–194
144. Verlengia R, Gorjao R, Kanunfre CC, Bordin S, de Lima TM, Curi R (2003) Effect of arachidonic acid on proliferation, cytokines production and pleiotropic genes expression in Jurkat cells--a comparison with oleic acid. Life Sci 73:2939–2951
145. Healy DA, Watson RW, Newsholme P (2003) Polyunsaturated and monounsaturated fatty acids increase neutral lipid accumulation, caspase activation and apoptosis in a neutrophil-like, differentiated HL-60 cell line. Clin Sci (Lond) 104:171–179

146. Serini S, Piccioni E, Merendino N, Calviello G (2009) Dietary polyunsaturated fatty acids as inducers of apoptosis: implications for cancer. Apoptosis 14:135–152
147. Colquhoun A (2010) Lipids, mitochondria and cell death: implications in neuro-oncology. Mol Neurobiol 42:76–88
148. Rovito D, Giordano C, Vizza D, Plastina P, Barone I, Casaburi I, Lanzino M, De Amicis F, Sisci D, Mauro L, Aquila S, Catalano S, Bonofiglio D, Ando S (2013) Omega-3 PUFA ethanolamides DHEA and EPEA induce autophagy through PPARγ activation in MCF-7 breast cancer cells. J Cell Physiol 228:1314–1322
149. Meng H, Shen Y, Shen J, Zhou F, Shen S, Das UN (2013) Effect of n-3 and n-6 unsaturated fatty acids on prostate cancer (PC-3) and prostate epithelial (RWPE-1) cells in vitro. Lipids Health Dis 12:160
150. Yao QH, Zhang XC, Fu T, Gu JZ, Wang L, Wang Y, Lai YB, Wang YQ, Guo Y (2014) ω-3 polyunsaturated fatty acids inhibit the proliferation of the lung adenocarcinoma cell line A549 in vitro. Mol Med Rep 9:401–406
151. Arita K, Kobuchi H, Utsumi T, Takehara Y, Akiyama J, Horton AA, Utsumi K (2001) Mechanism of apoptosis in HL-60 cells induced by n-3 and n-6 polyunsaturated fatty acids. Biochem Pharmacol 62:821–828
152. Puertollano MA, de Pablo MA, Alvarez de Cienfuegos G (2003) Polyunsaturated fatty acids induce cell death in YAC-1 lymphoma by a caspase-3-independent mechanism. Anticancer Res 23:3905–3910
153. Lindskog M, Gleissman H, Ponthan F, Castro J, Kogner P, Johnsen JI (2006) Neuroblastoma cell death in response to docosahexaenoic acid: sensitization to chemotherapy and arsenic-induced oxidative stress. Int J Cancer 118:2584–2593
154. Shinohara N, Tsuduki T, Ito J, Honma T, Kijima R, Sugawara S, Arai T, Yamasaki M, Ikezaki A, Yokoyama M, Nishiyama K, Nakagawa K, Miyazawa T, Ikeda I (2012) Jacaric acid, a linolenic acid isomer with a conjugated triene system, has a strong antitumor effect in vitro and in vivo. Biochim Biophys Acta 1821:980–988
155. Zajdel A, Wilczok A, Chodurek E, Gruchlik A, Dzierzewicz Z (2013) Polyunsaturated fatty acids inhibit melanoma cell growth in vitro. Acta Pol Pharm 70:365–369
156. Notarnicola M, Messa C, Refolo MG, Tutino V, Miccolis A, Caruso MG (2011) Polyunsaturated fatty acids reduce fatty acid synthase and hydroxy-methyl-glutaryl CoA-reductase gene expression and promote apoptosis in HepG2 cell line. Lipids Health Dis 10:10
157. Sun SN, Jia WD, Chen H, Ma JL, Ge YS, Yu JH, Li JS (2013) Docosahexaenoic acid (DHA) induces apoptosis in human hepatocellular carcinoma cells. Int J Clin Exp Pathol 6:281–289
158. Lima TM, Kanunfre CC, Pompeia C, Verlengia R, Curi R (2002) Ranking the toxicity of fatty acids on Jurkat and Raji cells by flow cytometric analysis. Toxicol In Vitro 16:741–747
159. Brinkmann CR, Brodkorb A, Thiel S, Kehoe JJ (2013) The cytotoxicity of fatty acid/-lactalbumin complexes depends on the amount and type of fatty acid. Eur J Lipid Sci Technol 115:591–600

MIX
Papier aus verantwortungsvollen Quellen
Paper from responsible sources
FSC® C105338
FSC
www.fsc.org

If you have any concerns about our products,
you can contact us on
ProductSafety@springernature.com

In case Publisher is established outside the EU,
the EU authorized representative is:
Springer Nature Customer Service Center GmbH
Europaplatz 3, 69115 Heidelberg, Germany

Printed by Libri Plureos GmbH
in Hamburg, Germany